军事英语听说教程
（新版）学习指南

（上册）

总 主 编：宋德伟
主　　审：董　伟　林　易
主　　编：任园园　李晓芳
副 主 编：陈英涛　胡宗敏　李　菲　李晓芬　林楚甜
（以汉语拼音排序）
　　　　　彭　强　史珂珮　王　青　徐若飞　杨璐夷
　　　　　张利芬　赵秋霞
编　　委：陈彦如　何明生　侯新飞　姬昱昕　刘鸿鹄
　　　　　张小艳

国防工业出版社

·北京·

内 容 简 介

《军事英语听说教程（新版）学习指南（上、下册）》（以下简称指南）紧扣主干教材《军事英语听说教程（新版）》的重点、难点，帮助学生精准理解教材内容，拓展学生的军事知识，指导学生进行听说练习，提高学生军事英语综合应用能力。本书（即上册）共12个单元，每单元设计6个独立板块，即单元导学清单、听说指导、阅读指导、词汇表、拓展学习和单元练习。单元导学清单以列表形式呈现本单元各小节的主题、学习目标、听力任务、军事知识和核心词汇。听说指导针对各小节进行相关军事知识的拓展、词汇点拨、听力文本长难句解读及口语输出任务设置。阅读指导针对主干教材课后文章进行导读，包括补充军事知识，讲解重点词汇，提供参考译文。词汇表按照单元任务的顺序编制，给出中文解释。拓展学习部分为学生提供相关的补充阅读材料。单元练习通过设置多样化的习题巩固本单元的重点词汇和军事知识，提高翻译及口语输出技能。

本套指南既可作为军队院校大学英语课程主干教材《军事英语听说教程（新版）》的配套资源供学生使用，也可作为军队院校大学英语教师的备课资料。同时，该书也可作为广大军事英语爱好者的良师益友。

图书在版编目（CIP）数据

军事英语听说教程（新版）学习指南. 上册 / 宋德伟总主编. -- 北京：国防工业出版社，2025.3.
ISBN 978-7-118-13535-0

Ⅰ. E

中国国家版本馆 CIP 数据核字第 2025H2G178 号

※

国防工业出版社出版发行
（北京市海淀区紫竹院南路23号　邮政编码100048）
河北文盛印刷有限公司印刷
新华书店经售

＊

开本 787×1092　1/16　印张 23¼　字数 440 千字
2025 年 3 月第 1 版第 1 次印刷　印数 1—1500 册　定价 88.00 元

（本书如有印装错误，我社负责调换）

国防书店：（010）88540777　　书店传真：（010）88540776
发行业务：（010）88540717　　发行传真：（010）88540762

前　言

《军事英语听说教程（新版）学习指南》是根据军队院校大学英语教学大纲要求，结合各军队院校通用军事英语教学实际，为主干教材《军事英语听说教程（新版）》编写的配套辅导用书。

本套指南聚焦军事英语基础知识积累、军事英语应用能力提升以及军事知识拓展，依据主干教材内容编排，分为上、下两册。上册对应主干教材1~12单元，下册对应主干教材13~24单元，每个单元由以下6个特色模块组成。

单元导学清单：呈现本单元各小节的主题、学习目标、听力任务、军事知识和核心词汇。

听说指导：拓展相关军事知识，讲解重难点词汇，精选听力文本中的长难句进行解读，设置口语话题并提供相关参考词汇。

阅读指导：针对主干教材课后文章进行导读，补充相关军事知识，讲解重难点词汇，提供参考译文。

词汇表：按照文中出现的先后顺序汇编整理重难点词汇。

拓展学习：补充与单元主题相关的军事语篇。

单元练习：巩固单元重难点词汇和军事知识，提高军事语篇的阅读翻译能力以及军事话题的口语表达能力。每单元的附录给出了单元练习答案。

本套指南由宋德伟任总主编，既可作为军队院校大学英语课程主干教材《军事英语听说教程（新版）》的配套资源供学生使用，也可作为军队院校大学英语教师的备课资料。同时，该书也可作为广大军事英语爱好者的良师益友。

由于编者水平及经验有限，时间仓促，编写过程中错误疏漏之处在所难免，万望广大读者批评指正，不吝赐教。

编者

2024 年 7 月

目 录

Unit 1　Boot Camp ·· 1

  Part 1　单元导学清单 ······································ 1
  Part 2　听说指导 ·· 3
  Part 3　阅读指导 ·· 18
  Part 4　词汇表 ·· 22
  Part 5　拓展学习 ·· 23
  单元练习 ·· 26
  附录：Unit 1 练习答案 ································ 28

Unit 2　Military Organisation ·················· 30

  Part 1　单元导学清单 ···································· 30
  Part 2　听说指导 ·· 31
  Part 3　阅读指导 ·· 44
  Part 4　词汇表 ·· 47
  Part 5　拓展学习 ·· 48
  单元练习 ·· 51
  附录：Unit 2 练习答案 ································ 53

Unit 3　Military Technology　55

Part 1　单元导学清单	55
Part 2　听说指导	56
Part 3　阅读指导	74
Part 4　词汇表	79
Part 5　拓展学习	81
单元练习	84
附录：Unit 3 练习答案	86

Unit 4　War Games　88

Part 1　单元导学清单	88
Part 2　听说指导	89
Part 3　阅读指导	100
Part 4　词汇表	102
Part 5　拓展学习	104
单元练习	106
附录：Unit 4 练习答案	108

Unit 5　Peacekeeping　110

Part 1　单元导学清单	110
Part 2　听说指导	111
Part 3　阅读指导	131
Part 4　词汇表	135
Part 5　拓展学习	137
单元练习	140
附录：Unit 5 练习答案	142

Unit 6　Convoy 146

  Part 1　单元导学清单 146
  Part 2　听说指导 147
  Part 3　阅读指导 166
  Part 4　词汇表 171
  Part 5　拓展学习 173
  单元练习 176
  附录：Unit 6 练习答案 178

Unit 7　Patrol 180

  Part 1　单元导学清单 180
  Part 2　听说指导 181
  Part 3　阅读指导 195
  Part 4　词汇表 199
  Part 5　拓展学习 201
  单元练习 203
  附录：Unit 7 练习答案 205

Unit 8　The Battalion 208

  Part 1　单元导学清单 208
  Part 2　听说指导 209
  Part 3　阅读指导 225
  Part 4　词汇表 230
  Part 5　拓展学习 231
  单元练习 233
  附录：Unit 8 练习答案 235

Unit 9　Parachute Regiment ……………… 237

  Part 1　单元导学清单 ……………………… 237
  Part 2　听说指导 …………………………… 238
  Part 3　阅读指导 …………………………… 250
  Part 4　词汇表 ……………………………… 254
  Part 5　拓展学习 …………………………… 256
  单元练习 …………………………………… 261
  附录：Unit 9 练习答案 …………………… 263

Unit 10　The 3rd Armoured Cavalry Regiment ……………………………… 265

  Part 1　单元导学清单 ……………………… 265
  Part 2　听说指导 …………………………… 266
  Part 3　阅读指导 …………………………… 284
  Part 4　词汇表 ……………………………… 289
  Part 5　拓展学习 …………………………… 291
  单元练习 …………………………………… 294
  附录：Unit 10 练习答案 ………………… 296

Unit 11　International HQ ……………… 299

  Part 1　单元导学清单 ……………………… 299
  Part 2　听说指导 …………………………… 300
  Part 3　阅读指导 …………………………… 314
  Part 4　词汇表 ……………………………… 318
  Part 5　拓展学习 …………………………… 319
  单元练习 …………………………………… 322
  附录：Unit 11 练习答案 ………………… 324

Unit 12　Carrier ·· 327

- Part 1　单元导学清单 ····································· 327
- Part 2　听说指导 ·· 328
- Part 3　阅读指导 ·· 343
- Part 4　词汇表 ·· 350
- Part 5　拓展学习 ·· 352
- 单元练习 ··· 355
- 附录：Unit 12 练习答案 ································ 357

参考文献 ··· 360

Unit 1 Boot Camp

 单元导学清单

模块	主题	学习目标	任务	军事知识	核心词汇
Alpha	美军新兵训练介绍	1）了解美国新兵训练的基本情况；2）掌握新兵训练常用词汇和表达；3）运用所学知识尝试介绍中国新兵训练情况。	Task 1 Task 2 Task 3 Task 5	■ boot camp ■ US Army BCT ■ Fort Benning ■ Fort Jackson (South Carolina)	basic combat training, recruit barracks, instructor, drill, weapons training, non-commissioned officer, military uniform
	枪支介绍	掌握训练常用枪支名称及其基本性能。	Task 4	■ M60 ■ AK-47 ■ M16 ■ G17T	machine gun, rifle, pistol
Bravo	军事字母表	1）识记军事字母表；2）运用军事字母表完成相关练习。	Task 1 Task 2 Task 3	■ military alphabet ■ Fort Knox ■ Fort Leonard Wood ■ Fort McClellan ■ Fort Sill	Alpha, Bravo, Charlie, Delta, Echo, Foxtrot, Golf, Hotel, India, Juliet, Kilo, Lima, Mike, November, Oscar, Papa, Quebec, Romeo, Sierra, Tango, Uniform, Victor, Whisky, X-ray, Yankee, Zulu
Charlie	训练科目与日程安排	1）了解新兵训练具体的训练科目与日程安排；2）熟练掌握与训练科目、日程安排相关的词汇与句型；3）简要介绍每周训练日程。	Task 1 Task 2 Task 3	■ field training exercise ■ obstacle course	first aid training, field training exercise, foot march, map reading, NBC training, obstacle course, weapons training, combat uniform

续表

模块	主题	学习目标	任务	军事知识	核心词汇
Charlie	军营生活	1）了解军营生活概况；2）熟练掌握军容风纪、军营日常生活等相关的词汇与句型；3）简要描述军营生活与普通日常生活作息的异同。	Task 4 Task 5 Task 6 Task 7	■ Ramstein Air Base ■ Travis Air Base ■ UNPROFOR	salute, superior, civilian, beret, cap badge, badge of rank, name tag, epaulette, camouflage, cap, polish, iron, parade
	简述军旅生涯	1）了解描述军旅生涯的构成要素；2）熟练掌握与军旅生涯相关的词汇和表达；3）运用所学知识简要描述个人的军旅经历。	Task 8 Task 9		duty station, promote, transfer, sergeant, second lieutenant, infantry, deploy, battalion, captain
Delta	12小时制与24小时制	1）了解两种计时方法的特点；2）熟练掌握并应用两种计时方法相关的表达与使用规则。	Task 1 Task 2 Task 3 Task 4 Task 5 Task 6 Task 7	■ 12-hour clock ■ 24-hour clock ■ PX	lights out, reveille, hygiene, barrack and uniform inspection
Echo	营院作息	1）了解营院的日常作息规律；2）熟练掌握并应用与营院作息相关的词汇与表达。	Task 1 Task 2 Task 3 Task 4 Task 5	■ Fitness center	sports pub, do weights, bowling
Foxtrot (Review)	新兵训练	1）了解英军与美军新训的相关知识；2）熟练掌握相关的词汇和表达；3）运用所学知识介绍外军与我军新训的异同。	Task 1 Task 2	■ armed forces ■ military organization	initiation, ethos, warfare, marksmanship, tactical, maneuver, push-up, sit-up, canoe, incorporate, discipline, leave, phase, specialty, operator, crewman

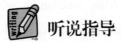 听说指导

Alpha US Army basic combat training

【军事知识】

1. boot camp 新兵营

Recruit training, more commonly known as Basic Training and colloquially called boot camp, is the initial indoctrination and instruction given to new military personnel, enlisted and officer. After completion of Basic Training, new recruits undergo Advanced Individual Training (AIT), where they learn the skills needed for their military jobs. Officer trainees undergo more detailed programs that may either precede or follow the common recruit training in an officer training academy (which may also offer a civilian degree program simultaneously) or in special classes at a civilian university.

2. US Army basic combat training 美军基础战斗训练

US Army Basic Combat Training is the program of physical and mental training required in order for an individual to become a soldier in the United States Army, United States Army Reserve, or Army National Guard. It is carried out at several different Army posts around the United States. Basic Training is designed to be highly intense and challenging. The challenge comes as much from the difficulty of physical training as it does from the required quick psychological adjustment to an unfamiliar way of life.

3. Fort Benning 本宁堡

Fort Benning is a United States Army post outside Columbus, Georgia. Fort Benning supports more than 120,000 active-duty military, family members, reserve component soldiers, retirees, and civilian employees on a daily basis. It is a power projection platform, and possesses the capability to deploy combat-ready forces by air, rail, and highway. Fort Benning is the home of the United States Army Maneuver Center of Excellence, the United States Army Armor School, United States Army Infantry School, the Western Hemisphere Institute for Security Cooperation (formerly known as the School of the Americas), elements of the 75th Ranger Regiment, 3rd Brigade – 3rd Infantry Division, and many other additional tenant units.

4. Fort Jackson (South Carolina) 杰克逊堡（南卡罗来纳州）

Fort Jackson is a United States Army installation, which TRADOC（美国陆军训练和条令司令部）operates on for Basic Combat Training (BCT), and is located in Columbia, South Carolina. This installation is named for Andrew Jackson, a United States Army General and President of the United States of America who was born in the border region of North and South Carolina.

5. M60 M60 机枪

The M60 machine gun is a family of American machine guns, firing the 7.62mm×51mm NATO cartridge. Since its production it has been in the service of the U.S. and other countries with support weapon to a squad or mounted on tanks, helicopters and other vehicles. In the U.S. military, it has largely been replaced by various versions of the M240 machine gun. However, it remains in use in every branch, as well as some other countries; it continues to be manufactured into the 21st century. The M60 can be used in both offensive and defensive configurations.

6. AK-47 AK-47 型冲锋枪

The AK-47 (Avtomat Kalashnikova-47, Kalashnikov automatic rifle, model of 1947) is the most prolific small arm of the 2nd half of the twentieth century. It had been and still is (in more or less modified form) manufactured in dozens of countries, and used in hundreds of countries and conflicts since its introduction. The total number of the AK-type rifles made worldwide during the last 60 years is estimated at over 90 million. This is a true legendary weapon, known for its extreme ruggedness, simplicity of operation and maintenance, and unsurpassed reliability even in worst conditions possible. It is used not only as a military weapon, but also as a platform for numerous sporting civilian rifles and shotguns.

7. M16 M16 美式步枪

The M16 rifle, officially designated Rifle, Caliber 5.56 mm, M16, is a family of military rifles adapted from the ArmaLite AR-15 rifle for the United States military. It has also been widely adopted by other armed forces around the world. The original M16 rifle was a 5.56 mm automatic rifle with a 20-round magazine. Adopted in July 1997, the M16A4 is the fourth generation of the M16 series. Total worldwide production of M16s is approximately 8 million, making it the most-produced firearm of its 5.56 mm caliber.

8. G17T 格洛克 17 型手枪

Glock pistol is a general term for a series of automatic pistols developed and manufactured by Glock GmbH, Austria. The original Glock 17 was developed by Austria's

Glock GMBH in 1983 at the request of the Austrian Army, and since then, the Glock pistol has become famous. Today, the Glock pistol has developed into a variety of large families, 34 types in total. This series of pistols covers the world's military, police and civilian markets, among which the armed forces from more than 40 countries and the law enforcement units in many countries are loyal users. In the civil market, it is also highly respected.

【词汇点拨】Words and Expressions

1. recruit /rɪˈkruːt/ *n.* a person who has recently joined the armed forces or the police 新兵；新警员

He spoke of us scornfully as raw recruits. 他轻蔑地称我们是新兵娃娃。

2. barracks /ˈbærəks/ *n.* [pl.] a large building or group of buildings for soldiers to live in 营房；兵营

The troops were ordered back to barracks.

3. instructor /ɪnˈstrʌktə(r)/ *n.* a person whose job is to teach sb. a practical skill or sport 教练；教官；导师

a driving instructor 驾驶教练

4. drill /drɪl/ *n.* military training in marching, the use of weapons, etc. 军事训练；操练

rifle drill 步枪操练

5. fort /fɔːt/ *n.* (NAmE) a place where soldiers live and have their training 堡垒，要塞；兵营；军营；营地

Fort Drum 德拉姆堡

6. uniform /ˈjuːnɪfɔːm/ *n.* the special set of clothes worn by all members of an organization or a group at work, or by children at school 制服；校服

a military/police/nurse's uniform 军装；警服；护士制服

soldiers in uniform 穿制服的军人

7. non-commissioned officer (*abbr.* NCO) a soldier in the army, etc. who has a rank such as sergeant or corporal, but not a high rank 士官

Non-commissioned officer salary and benefits will be increased 40 percent.

士官工资和津贴将会提高 40%。

8. company /ˈkʌmpəni/ *n.* a group of soldiers that is part of a battalion or regiment, and that is divided into two or more platoons 连（队）

The division will consist of two tank companies and one infantry company.

该师将由两个坦克连和一个步兵连组成。

9. machine gun a gun that automatically fires many bullets one after the other very quickly 机关枪；机枪

a burst/hail of machine gun fire 一阵猛烈的机关枪扫射

10. rifle /ˈraɪfl/ *n.* a gun with a long barrel which you hold to your shoulder to fire 步枪；来复枪

When I'd missed a few times, he suggested I rest the rifle on a rock to steady it.
由于我几次都没打中，他就建议我把步枪放到石头上来保持稳定。

11. pistol /ˈpɪstl/ *n.* a small gun that you can hold and fire with one hand 手枪

A starter's pistol fires only blanks.
初学者的手枪发射的只是空弹。

【词汇点拨】Proper Names

1. Fort Benning 本宁堡（美军基地，北卡罗来纳州）

2. Fort Jackson 杰克逊堡（美军基地，南卡罗来纳州）

3. North Carolina 北卡罗来纳州

4. South Carolina 南卡罗来纳州

【长难句解读】

1. The Army trains recruits at basic training units. The main basic training unit is Fort Jackson in South Carolina. (1-1)

➢ 句子分析：该句中 Army 为大写，结合本听力训练的标题 BCT in the US Army，可知此处的 Army 为缩略形式，完整形式应为 the US Army。unit 本意为"单元"，此处应为"单位、机构"。

➢ 翻译：美军的新兵训练均在各个基础训练单位进行。主要的基础训练单位是位于南卡罗来纳州的杰克逊堡。

2. The instructors are non-commissioned officers. They teach basic military skills, including drill and weapons training. (1-1)

➢ 句子分析：该句中 instructor 的常见含义为"导师、教练"。在新兵训练中，instructor 为"教官"或"教员"，负责训练与教授军事知识。

➢ 翻译：新训中的教官为军士。他们负责教授基本的军事技能，包括队列训练与武器训练。

3. Recruits are organised into companies-A (Alpha) company, B (Bravo) company, and C (Charlie) company. (1-1)

> 句子分析：该句中 be organized into 意为"被分配、编入……"。
> 翻译：新兵被编入连，连队编号为 Alpha、Bravo 和 Charlie。

4. A: They train with the G17T.

B: Pardon?

C: You know, the Glock-the G17T pistol. (1-2)

> 句子分析：train with 意为"使用……进行训练"。
> 翻译：

A：他们训练使用的枪支是 G17T。

B：什么枪？

C：格洛克手枪，格洛克 17 型手枪。

【口语输出】

1. Can you briefly introduce BCT in the US Army?

【参考词汇】volunteer, basic training unit, instructor, military skills, barracks, uniform, graduation, company

2. What was your basic training like?

【参考词汇】recruit, week, haircut, drill, weapons training, goose step, parade, political and theoretical subject

Bravo The military alphabet

【军事知识】

1. military alphabet 军事字母表

The military alphabet and NATO phonetic alphabet are the same alphabet. It is a system of letters and numbers used by the armed forces of the United States, North Atlantic Treaty Organization (NATO), and International Civil Aviation Organization, and even by civilians to spell out words and phrases or communicate in code. It is a phonetic alphabet that uses 26 code words. These words are used to ensure oral communication is clearly understood, thus preventing miscommunication.

2. Fort Knox 诺克斯堡（美军基地，肯塔基州）

Fort Knox, Kentucky is an army base with more than one hundred thousand acres stretching across three counties. It is approximately 35 miles south of Louisville, KY, and is the sixth largest community in the entire commonwealth of Kentucky. With a diverse collection of missions and units, the base has a large military community including a sizable retiree population. It is home to the Army Human Resource Center of Excellence, which is responsible for career management for all soldiers.

3. Fort Leonard Wood 伦纳德伍德堡（美军基地，密苏里州）

Fort Leonard Wood, a military community located in the beautiful south-central Missouri Ozarks, covers more than 61,000 acres. Fort Leonard Wood Garrison Command is proud to have the 1st Engineer Brigade, 3rd Chemical Brigade, 14th Military Police Brigade and many more on the installation to make Fort Leonard Wood number 1 in training. The installation was designated as Fort Leonard Wood, in honor of Major General Leonard Wood, a distinguished American soldier who served for 40 years.

4. Fort McClellan 麦克莱伦堡（美军基地，阿拉巴马州）

Fort McClellan is an army base with a proud, fascinating history dating back to the war between the Spaniards and the Americans. The army base proved advantageous during World War I and sowed even more fruit when World War II came. The War Department set to formally establish Fort McClellan in 1917, named after Major General George McClellan, the General-in-Chief of the United States Army from the years 1861 to 1862 and also the New Jersey Governor from 1878 to 1881. While it is not typical for a fort in the south to be named after a general from the north, there were a lot of strong points for using McClellan's name for the fort. For starters, McClellan was credited for the swift training and mobilization of army troops of the Potomac when the Civil War was going on. In keeping with the actions of the major general, Fort McClellan was used as a mobilization camp for quick training of soldiers to fight in World War II.

5. Fort Sill 锡尔堡（美军基地，俄克拉何马州）

Fort Sill is home to the oldest continuously operating airfield in U.S. Army history. Located in Oklahoma 85 miles away from Oklahoma City, the fort has a rich legacy of Army history stretching back to the 1800s. There are roughly 50 thousand at Fort Sill with some 20 thousand of those serving in uniform and as civilians; the remainder are associated family members. Its mission is to train artillery soldiers and train them well. On any given day, roughly nine thousand trainees attend basic training, advanced individual training, and other types of military education and training.

【词汇点拨】Words and Expressions

1. **alpha** /ˈælfə/ *n.* the first letter of the Greek alphabet (A, α) 希腊字母表的第 1 个字母

2. delta /ˈdeltə/ *n.*

① the fourth letter of the Greek alphabet (Δ, δ) 希腊字母表的第 4 个字母

② [C] an area of land, shaped like a triangle, where a river has split into several smaller rivers before entering the sea 三角洲

the Nile Delta 尼罗河三角洲

3. echo /ˈekəʊ/

① *n.* [pl. –es] the reflecting of sound off a wall or inside a confined space so that a noise appears to be repeated; a sound that is reflected back in this way 回响；回声；回音

② *v.* if a sound echoes, it is reflected off a wall, the side of a mountain, etc. so that you can hear it again 回响；回荡

The gunshot echoed through the forest. 枪炮声在林中回荡。

4. foxtrot /ˈfɒkstrɒt/ *n.* a formal dance for two people together, with both small fast steps and longer slow ones; a piece of music for this dance 狐步舞；狐步舞曲

5. Lima /ˈliːmə/ *n.* the capital of Peru; founded in 1535 by Francisco Pizarro; the capital of the Spanish colonies in South America until the 19th century 利马（秘鲁首都，由弗朗西斯科·皮萨罗于 1535 年建立，19 世纪前一直为西班牙在南美的殖民地首都）

6. Quebec /kwɪˈbek/ *n.* a heavily forested province in eastern Canada. It was settled by the French in 1608, ceded to the British in 1763, and became one of the original four provinces in the Dominion of Canada in 1867. The majority of its residents are French-speaking and it is a focal point of the French-Canadian nationalist movement, which advocated independence for Quebec. 魁北克省（加拿大东部森林密集的省份；1608 年法国人来此定居，1763 年割让给英国人，1867 年成了加拿大联邦最早的四个省之一；居民大多讲法语，是鼓吹魁北克独立的法加民族主义运动中心）

7. sierra /siˈerə/ *n.* a long range of steep mountains with sharp points, especially in Spain and America（尤指西班牙和美洲的）锯齿状山脉

the Sierra Nevada 内华达山脉

8. tango /ˈtæŋgəʊ/

① *n.* a fast South American dance with a strong beat, in which two people hold each other closely; a piece of music for this dance 探戈舞；探戈舞曲

② *v.* to dance the tango 跳探戈舞

They can rock and roll, they can tango, but they can't bop.
他们会跳摇滚，会跳探戈，就是不会跳博普舞。

9. Yankee /ˈjæŋki/ *n.*

① (*NAmE*) a person who comes from or lives in any of the northern states of the US, especially New England 美国北方人；（尤指）新英格兰人

② a soldier who fought for the Union (= the northern states) in the American Civil War（美国南北战争时的）北军士兵

10. Zulu /ˈzuːluː/ *n.*

① [C] a member of a race of black people who live in South Africa 祖鲁人（南非黑人种族）

② [U] the language spoken by Zulus and many other black South Africans 祖鲁语

【词汇点拨】Proper Names

1. Fort Knox 诺克斯堡

2. Kentucky /kenˈtʌki/ 肯塔基州（美国）

3. Fort Leonard Wood 伦纳德伍德堡

4. Missouri /mɪˈzʊəri/ 密苏里州（美国）

5. Fort McClellan 麦克莱伦堡

6. Alabama /ˌæləˈbæmə/ 阿拉巴马州（美国）

7. Oklahoma /ˌəʊkləˈhəʊmə/ 俄克拉何马州（美国）

【长难句解读】

1. The US Army has several locations for basic training, including Fort Jackson in South Carolina, Fort Knox in Kentucky, Fort Leonard Wood in Missouri, Fort McClellan in Alabama and Fort Sill in Oklahoma. (1-4)

➢ 句子分析：该句中 locations 意为 places，指美军新训的基地；including 为介词，具体介绍这些新训基地的名称及所在地。

➢ 翻译：美军设有如下五处新训基地：南卡罗来纳州的杰克逊堡，肯塔基州的诺克斯堡，密苏里州的伦纳德伍德堡，阿拉巴马州的麦克莱伦堡和俄克拉何马州的锡尔堡。

2. Fort Jackson is spelt juliet-alpha-charlie-kilo-sierra-oscar-november. (1-4)

➢ 句子分析：该句主要讲述如何用军事字母表拼读单词。首先将 Jackson 拆分为 j-a-c-k-s-o-n，然后与军事字母表中对应首字母的单词进行匹配，最后拼读出来即可。

➢ 翻译：杰克逊堡英文表达中的"Jackson"拼读为 juliet-alpha-charlie-kilo-sierra-oscar-november。

【口语输出】

1. Can you recite the whole military alphabet?

【参考词汇】Alpha, Bravo, Charlie, Delta, Echo, Foxtrot, Golf, Hotel, India, Juliet, Kilo, Lima, Mike, November, Oscar, Papa, Quebec, Romeo, Sierra, Tango, Uniform, Victor, Whisky, X-ray, Yankee, Zulu

2. How should you spell your name and your classmates' names using military alphabet?

【参考词汇】to use the military alphabet to spell the names

Charlie   To be a soldier

【军事知识】

1. field training exercise 野战训练演习

A field training exercise, generally shortened to the acronym "FTX", describes a coordinated exercise conducted by military units for training purposes.

In active duty, field training exercises are usually practice "mini-battles" which provide fairly realistic scenarios and situations based on actual situations a unit might face if deployed. While squad and platoon sized units can conduct an FTX, most of these exercises involve units ranging from a company up to a regiment or brigade. Nearly every possibility is considered during planning, and often the scenarios can be more difficult or more far-fetched than actual battles, thus sharpening the skills of those participating to a level which will surpass that of the enemy.

In basic training. Most branches of the U.S. Armed Forces implement field training exercises into their basic military training courses for enlistees and officers. In the Army and often in other branches, the last few days of basic training are used to conduct a field training exercise where recruits can practice the skills they have learned over the past several weeks of training. The purpose of this is to give soon-to-be soldiers a taste of battle before they leave basic training. This also allows instructors to look for mistakes and correct them before their recruits become active duty and potentially go to battle.

2. obstacle course 障碍训练

The military/Army obstacle course is used (mostly in recruit training) as a way to familiarize recruits with the kind of tactical movement they will use in combat, as well as for physical training, building teamwork, and evaluating problem solving skills. Typical courses involve obstacles the participants must climb over, crawl under, balance, hang, jump, etc. Often, specialized courses are made to focus on specific needs, such as night movement, assault, and bayonet training. Military courses can also contain climbing walls and rappelling walls.

3. Ramstein Air Base 拉姆斯泰因空军基地

Ramstein Air Base is a United States Air Force base in Rhineland-Palatinate, a state of Germany. It serves as headquarters for the US Air Forces in Europe (USAFE) and is also a North Atlantic Treaty Organization installation. Ramstein is located near town of Ramstein-Miesenbach, in the rural district of Kaiserslautern.

4. Travis Air Base 崔维斯空军基地

Travis Air Force Base is a United States Air Force air base under the operational control of the Air Mobility Command (AMC), located three miles (5 km) east of the central business district of Fairfield, in Solano County, California, United States. The base is named for Brigadier General Robert F. Travis, who died in the crash of a B-29 Superfortress while transporting a nuclear weapon.

5. UNPROFOR 联合国保护部队

The United Nations Protection Force (UNPROFOR) was the first United Nations peacekeeping force in Croatia, Bosnia and Herzegovina during the Yugoslav wars. It existed between the beginning of UN involvement in February 1992, and its restructuring into other forces (United Nations Preventive Deployment Force—UNPREDEP and United Nations Confidence Restoration Operation—UNCRO) in March 1995.

【词汇点拨】Words and Expressions

1. salute /səˈluːt/ *v.* to touch the side of your head with the fingers of your right hand to show respect, especially in the armed forces（尤指军队中）敬礼

The sergeant stood to attention and saluted. 中士立正敬礼。

2. superior /suːˈpɪəriə(r)/

① *adj.* ~ (to sb.) higher in rank, importance or position（在级别、重要性或职位上）更高的

my superior officer 我的上级军官

② *n.* a person of higher rank, status or position 级别（或地位、职位）更高的人；上级；上司

He's my immediate superior. 他是我的顶头上司。

3. operation /ˌɒpəˈreɪʃn/ *n.* [C, usually pl.] military activity 军事行动

He was the officer in charge of operations. 他是负责指挥作战行动的军官。

4. PT the abbreviation for "physical training" (sport and physical exercise that is taught in schools, in the army, etc.) 体格锻炼；体能训练

5. civilian /səˈvɪliən/ *n.* a person who is not a member of the armed forces or the police 平民；老百姓；庶民（常与 military 相对应）

6. badge /bædʒ/ *n.* a piece of material that you sew onto clothes as part of a uniform （制服上的）标记，标识

cap badge 帽徽

7. epaulette /ˈepəlet/ *n.* a decoration on the shoulder of a coat, jacket, etc., especially as part of a military uniform （尤指军服上的）肩章，肩饰

a large gold epaulette 一条很宽的金肩章

8. camouflage /ˈkæməflɑːʒ/ *n.* [U] a way of hiding soldiers and military equipment, using paint, leaves or nets, so that they look like part of their surroundings （军事上的）伪装，隐蔽

a camouflage jacket 迷彩夹克衫

troops dressed in camouflage 穿迷彩服的军队

9. polish /ˈpɒlɪʃ/

① *n.* a substance used when rubbing a surface to make it smooth and shiny 擦光剂；上光剂；亮光剂

furniture/floor/shoe/silver polish 家具上光漆 / 地板蜡 / 鞋油 / 银抛光剂

② *v.* ~ sth. (up) (with sth.) to make sth. smooth and shiny by rubbing it with a cloth, often with polish on it 擦光；磨光

Polish shoes regularly to protect the leather. 要经常擦鞋，以保护皮革。

10. iron /ˈaɪən/ *v.* to make clothes, etc. smooth by using an electrical device with a flat metal base that you heat it until the base is hot and then rub it over clothes （用熨斗）熨，烫平

I'll need to iron that uniform before I can wear it. 我得先把那件制服烫平再穿。

11. parade /pəˈreɪd/

① *n.* a formal occasion when soldiers march or stand in lines so that they can be examined by their officers or other important people 检阅；阅兵

a military parade 军事检阅

② *v.* to come together, or to bring soldiers together, in order to march in front of other people （使）列队行进，接受检阅

The crowds applauded as the guards paraded past. 卫队列队走过时，人群鼓掌欢迎。

12. promote /prəˈməʊt/ *v.* ~ sb. (from sth.) (to sth.) [often passive] to move sb. to a higher rank or more senior job 提升；晋升

He has been promoted to sergeant. 他已被提升为中士。

13. transfer

① /ˈtrænsfɜː(r)/ *n.* the act of moving sb./sth. from one place, group or job to another; an occasion when this happens 搬迁；转移；调动；变换

He has asked for a transfer to the frontline. 他已要求调到前线去。

② /trænsˈfɜː(r)/ *v.* ~ (sb.) (from...) (to...) to move from one job, school, situation, etc. to another; to arrange for sb. to move （使）调动；转职；转学；改变（环境）

The sergeant was transferred to Bosnia. 这个中士调到了波斯尼亚。

14. deploy /dɪˈplɔɪ/ *v.* to move soldiers or weapons into a position where they are ready for military action 部署，调度（军队或武器）

2,000 troops were deployed in the area. 那个地区部署了 2000 人的部队。

At least 5,000 missiles were deployed along the border. 边境沿线至少部署了 5000 枚导弹。

15. field training exercise a training exercise in which military skills are practiced in field conditions 野外训练演习

16. foot march 徒步行军

17. NBC training (nuclear, biological and chemical training) 核生化训练

18. combat uniform 作战服

【词汇点拨】Proper Names

Bosnia /ˈbɒzniə/ 波斯尼亚（位于原南斯拉夫中西部，现为波斯尼亚和黑塞哥维那北部的一个地区）

【长难句解读】

1. I joined the US Air Force at the age of eighteen, and went straight to Boot Camp for six weeks. (1-8)

➤ 句子分析：该句中 join 作动词，常与军队中相关的机构、部门连用，意为"入伍"或"加入陆/海/空军"。

➤ 翻译：我十八岁入伍美国空军，而后直接进入军营进行了为期六周的新训。

2. Then I was promoted to the rank of airman and deployed to Ramstein Air Base in Germany. (1-8)

> **句子分析**：该句包含两个被动语态，一为 be promoted to the rank of... 意为被提干或晋升为……；二为 be deployed to...（地点），意为被派遣至……。

> **翻译**：然后，我被晋升为二等飞行员，并被派往驻扎在德国的拉姆斯泰因空军基地。

【口语输出】

1. What is your weekly training schedule?

【参考词汇】drill, PT, first aid training, FTX, NBC training, foot march, map reading

2. What's the difference between your civilian life and your military life?

【参考词汇】get up, do exercise, follow/give orders, wear a uniform, salute superiors, weapon

3. Can you introduce your military uniform?

【参考词汇】cap badge, boots, badge of rank, name tag, epaulette, combat uniform, camouflage

4. Would you please describe your military career so far?

【参考词汇】join the army, enter, recruit training, challenging, rewarding, arduous

Delta A day in the life: the 12-hour and 24-hour clock

【军事知识】

1. 12-hour clock 12 小时制

In American and Canadian dialects, the time of day is customarily given almost exclusively using the 12-hour clock notation, which counts the hours of the day as 12, 1, ..., 11 with suffixes a.m. and p.m. distinguishing the two diurnal repetitions of this sequence.

2. 24-hour clock 24 小时制

In American and Canadian English, the term military time is a synonym for the 24-hour clock. The 24-hour clock is commonly used there only in some specialist areas (military, aviation, navigation, tourism, meteorology, astronomy, computing, logistics, emergency services, hospitals), where the ambiguities of the 12-hour notation are deemed too inconvenient, cumbersome, or dangerous.

3. PX（美国军营中的）军人服务社

The PX, or Post Exchange, is a shop on American military base with special prices. British military don't have PXs. They have NAAFIs, or Navy, Army, and Air Force Institutes. In Europe, some bases have a NATEX, or NATO Exchange.

【词汇点拨】Words and Expressions

1. reveille /rɪˈvæli/ *n.* a tune that is played to wake soldiers in the morning; the time when it is played（军队的）起床号，起床时间

2. hygiene /ˈhaɪdʒiːn/ *n.* the practice of keeping yourself and your living and working areas clean in order to prevent illness and disease 卫生

personal hygiene 个人卫生

3. lights out the time when those residents at an institution, such as soldiers in barracks or children at a boarding school, are expected to retire to bed 就寝时间

4. barrack and uniform inspection 内务与军容风纪检查

【词汇点拨】Proper Names

Wendy Phillip 温迪·菲利普（人名）

【长难句解读】

Then we clean the barrack room—make our beds, clean the toilet areas, put our things away—until 0700 hours. (1-13)

➤ 句子分析：该句讲述打扫宿舍、整理内务的一系列动作。其中 make、clean、put away 是对 clean the barrack room 做出进一步的解释说明。

➤ 翻译：然后，在7点之前，我们要完成铺床，打扫卫生间与洗漱区，归置物品等内务整理的工作。

【口语输出】

1. What are the hours of operation for the library in your campus? Please use the 12-hour clock.

【参考词汇】open, close, from Monday to Friday, Sunday, Saturday

2. What do you do every day in the camp? Please use the 24-hour clock.

【参考词汇】wake up, breakfast, class, lunch, dinner, training, personal time

Echo Hours of operation

【军事知识】

Fitness center 健身中心；健身俱乐部

Fitness center is a place with specially-designed facilities and equipment for people to maintain or improve their physical fitness. In the army, it is a commonly-seen installation used to train the soldiers and maintain their excellent physical condition.

【词汇点拨】Words and Expressions

1. **sports pub** 体育酒吧；运动酒馆

2. **do weights** 举重；负重训练

3. **bowling** /ˈbəʊlɪŋ/ *n.* a game in which players roll heavy balls (called bowls) along a special track towards a group of pins (= bottle-shaped objects) and try to knock over as many of them as possible 保龄球运动

    go bowling 去打保龄球

4. **NCO's club** 士官俱乐部

【词汇点拨】Proper Names

1. **Andy** 安迪（健身中心）

2. **Magruder** 马格鲁德（运动俱乐部）

【长难句解读】

Our hours of operation are: Monday through Friday 0500 to 2100; Saturdays 0800 until 1800; Sundays and holidays from 1000 to 1600. (1-15)

➢ 句子分析：该句中的 hours of operation 为常见搭配，意为"营业时间、运营时间"；表达从几点到几点有两种方式：...to... 或 ...until...。

➢ 翻译：以下为我们的营业时间：周一至周五从 5 点营业至 21 点，周六为 8 点至 18 点，周日和节假日为 10 点至 16 点。

【口语输出】

1. Think about the facilities in your base and pick two of them to tell their respective hours of operation.

    【参考词汇】Monday through Friday, weekend, holidays

2. What do you do on Sunday and Saturday?

    【参考词汇】fitness center, do sports, do weights, personal time, go bowling

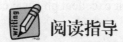 阅读指导

【文章导读】

This passage mainly introduces the basic training in the US and the UK army. After explaining the definition of basic training and its duration, it lists some common subjects of skills training and physical training. Then the possible difficulties in the whole process are also mentioned. Finally, the next phase after the graduation from the boot camp is addressed.

【军事知识】

1. armed forces 武装部队

In most countries the armed forces are divided into three or four Armed Services (also called branches): an army, a navy, and an air force. Many countries have a variation on the standard model of three or four basic Armed Services. Some nations also organize their marines, special forces or strategic missile forces as independent armed services. A nation's coast guard may also be an independent armed service of its military, although in many nations the coast guard is a law enforcement or civil agency. A number of countries have no navy, for geographical reasons, and some other variations include:

Brazil: Army, Navy, Air Force, Police, Firefighters

Canada: Army, Navy, Air Force

Chile: Army, Navy, Air Force, National Police

Egypt: Army, Navy, Air Force, Air Defense

Germany: Army, Navy, Air Force, Joint Support Service, Joint Medical Services

Hungary: Army, Air Force

India: Army, Navy, Air Force, Strategic Missile Force, Coast Guard, Paramilitary Forces

Indonesia: Army, Navy, Air Force, Coast Guard, Police

Italy: Army, Navy, Air Force, Military Police

Poland: Army, Navy, Air Force, Special Forces

People's Republic of China: Army, Navy, Air, Rocket Force, Reserve Force

Russian Federation: Ground Forces, Navy, Air Force plus three independent arms of service (Strategic Missile Troops, Aerospace Defence Troops and Airborne Troops)

South Africa: Army, Navy, Air Force, Military Health Service

The Netherlands: Army, Navy, Air Force, Military Police

United States: Army, Navy, Air Force, Marines, Coast Guard

Venezuela: Army, Navy, Air Force, National Guard, National Militia

Vietnam: Army, Navy, Air Force, Border Guard, Coast Guard

In larger armed forces the culture between the different Armed Services of the armed forces can be quite different.

Most smaller countries have a single organization that encompasses all armed forces employed by the country in question. Third-world armies tend to consist primarily of infantry, while first-world armies tend to have larger units manning expensive equipment and only a fraction of personnel in infantry units.

It is worthwhile to make mention of the term joint. In western militaries, a joint force is defined as a unit or formation comprising representation of combat power from two or more branches of the military.

2. military organisation 军事组织

Military organisation is the structuring of the armed forces of a state so as to offer military capability required by the national defence policy. In some countries paramilitary forces are included in a nation's armed forces. Armed forces that are not a part of military or paramilitary organisations, such as insurgent forces, often mimic military organizations, or use ad hoc structures.

Military organisation is hierarchical. The use of formalised ranks in a hierarchical structure came into widespread use with the Roman Army. In modern times, executive control, management and administration of military organisations is typically undertaken by the government through a government department within the structure of public administration, often known as a Department of Defense, Department of War, or Ministry of Defence. These in turn manage Armed Services that themselves command combat, combat support and service support formations and units.

【词汇点拨】Words and Expressions

1. initiation /ɪˌnɪʃiˈeɪʃn/ *n.* the act of starting sth. 开始；创始；发起

the initiation of criminal proceedings 提起刑事诉讼

2. ethos /ˈiːθɒs/ *n.* [sing.] the moral ideas and attitudes that belong to a particular group or society（某团体或社会的）道德思想，道德观

an ethos of public service 公益服务的意识

3. warfare /ˈwɔːfeə(r)/ *n.* the activity of fighting a war, especially using particular weapons or methods 战；作战；战争

air/naval/guerrilla warfare 空战 / 海战 / 游击战

countries engaged in warfare 参战国

4. marksmanship /ˈmɑːksmənʃɪp/ *n.* skill in shooting 射击术

5. tactical /ˈtæktɪkl/ *adj.* relating to tactics; relating to the battlefield 战术上的；策略上的

tactical planning 对策谋划

to have a tactical advantage 拥有战术上的优势

6. maneuver /məˈnuːvə(r)/ *n.* an armed forces training exercise; especially an extended and large-scale training exercise involving military and naval units separately or in combination 演习；演练

All the fighters landed safely on the airport after the military maneuver.

7. push-up exercises to strengthen your arms and chest muscles. They are done by lying with your face toward the floor and pushing with your hands to raise your body until your arms are straight. 俯卧撑

8. sit-up an exercise for making your stomach muscles strong, in which you lie on your back on the floor and raise the top part of your body to a sitting position 仰卧起坐

9. canoe /kəˈnuː/

① *n.* a light narrow boat which you move along in the water with a paddle 皮划艇；独木舟

② *v.* to travel in a canoe 划（或乘）独木舟

10. incorporate /ɪnˈkɔːpəreɪt/ *v.* ~ sth. (in/into/within sth.) to include sth. so that it forms a part of sth. 将……包括在内；包含；吸收；使并入

We have incorporated all the latest safety features into the design.

11. discipline /ˈdɪsəplɪn/ *n.* [U] the practice of training people to obey rules and orders and punishing them if they do not; the controlled behaviour or situation that results from this training 训练；训导；纪律；风纪

Strict discipline is imposed on army recruits.

12. leave /liːv/ *n.* a period of time when you are allowed to be away from work for a holiday/vacation or for a special reason 假期；休假

The soldiers will go home on leave once every two years.

13. phase /feɪz/ *n.* a stage in a process of change or development 阶段；时期

during the first/next/last phase 在第一 / 下一 / 最后阶段

the initial/final phase of the project 工程的初始 / 最后阶段

a critical/decisive phase 关键性 / 决定性阶段

the design phase 设计阶段

14. specialty /ˈspeʃəlti/ *n.* the special line of work you have adopted as your career 专业；专长

15. operator /ˈɒpəreɪtə(r)/ *n.* (often in compounds 常构成复合词) a person who operates equipment or a machine 操作人员；技工

a computer/machine operator 计算机 / 机器操作员

16. crewman /ˈkruːmən/ *n.* [pl.–men] a member of a crew, usually a man 乘务员，船员，驾驶员（通常为男性）

【词汇点拨】Proper Names

1. boot camp 新兵营
2. US Army 美国陆军

【参考译文】

军队中的新兵训练

基础训练或新兵训练是士兵入伍后的必修科目。在这里，新兵可以学习军事生活的常识，以及了解如何成为合格的士兵。美军的新训时长为 9 周，英军的新训持续 12 周，两军的新训科目如下：

- 军事组织，军人价值观或军人气质
- 队列与典礼（队列行进与阅兵式训练）
- 急救训练
- 识图与指南针使用
- 核生化战争
- 基础步枪射击术（300 米射击训练）
- 野外战术演习

除技能训练之外，士兵还需面临艰苦的体能训练，包括每日的跑步训练、俯卧撑、仰卧起坐和每周累计 10 千米的队列行进。障碍训练，有时也被称为军人越障训练课或信心提升课（美式英语），是新训中最大的挑战之一。基础训练过后，攀岩、独木舟之类的团队探险活动也会被纳入拓展训练。

对于新兵而言，最困难的莫过于要适应部队中严格的纪律要求。经常性的军容风纪、内务和武器装备检查，让士兵们懂得要"精益求精"。训练开始的前几周，新兵几乎没有自由时间，也不允许看书或听广播。英军士兵可在新训第六周请假外出，美军士兵则需完成全部新训内容之后方可请假离开基地。

新训结束时会举办一个结业典礼，家人也可以参加。此后战士们即将进入军旅生涯的下一个阶段：高级训练或行业训练阶段，在这期间他们将学习各自行业所需的专业知识，比如雷达操作员、坦克手或步兵。

 词汇表

Alpha US Army basic combat training
recruit 新兵；新警员
barracks 营房；兵营
instructor 教练；教官；导师
drill 军事训练；操练
fort 堡垒，要塞；兵营，军营；营地
uniform 制服；校服
non-commissioned officer (*abbr*. NCO) 军士
company 连（队）
machine gun 机关枪；机枪
rifle 步枪；来复枪
pistol 手枪

Bravo The military alphabet
alpha 希腊字母表的第1个字母
delta 三角洲
echo 回响；回声；回音
foxtrot 狐步舞；狐步舞曲
Lima 利马
Quebec 魁北克省
Sierra（尤指西班牙和美洲的）锯齿状山脉
tango 探戈舞；探戈舞曲；跳探戈舞

Yankee 美国北方人；（尤指）新英格兰人；（美国南北战争时的）北军士兵
Zulu 祖鲁人（南非黑人种族）；祖鲁语

Charlie To be a soldier
salute 敬礼
superior（在级别、重要性或职位上）更高的；上级；上司
operation 军事行动
PT 体格锻炼；体能训练
civilian 平民；老百姓；庶民
badge（制服上的）标记，标识
epaulette（尤指军服上的）肩章，肩饰
camouflage（军事上的）伪装，隐蔽
polish 上光剂；亮光剂；擦光；磨光
iron（用熨斗）熨，烫平
parade 检阅；阅兵；（使）列队行进，接受检阅
promote 提升；晋升
transfer 搬迁；转移；调动；变换；（使）调动；转职；转学；改变（环境）
deploy 部署，调度（军队或武器）

field training exercise (FTX) 野外训练演习
foot march 徒步行军
NBC training 核生化训练
combat uniform 作战服

Delta A day in the life: the 12-hour and 24-hour clock

reveille（军队的）起床号，起床时间
hygiene 卫生
lights out 就寝时间
barrack and uniform inspection 内务与军容风纪检查

Echo Hours of operation

sports pub 体育酒吧；运动酒馆
do weights 举重；负重训练；撸铁
bowling 保龄球运动
NCO's club 士官俱乐部

Foxtrot Recruit training in the armed forces

initiation 开始；创始；发起
ethos（某团体或社会的）道德思想，道德观
warfare 战；作战；战争
marksmanship 射击术
tactical 战术上的；策略上的
maneuver 演习；演练
push-up 俯卧撑
sit-up 仰卧起坐
canoe 皮划艇；独木舟；划（或乘）独木舟
incorporate 将……包括在内；包含；吸收；使并入
discipline 训练；训导；纪律；风纪
leave 假期；休假
phase 阶段；时期
specialty 专业；专长
operator 操作人员；技工
crewman 乘务员，船员，驾驶员

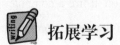 拓展学习

Military Weapons Used in Basic Combat Training

　　It wouldn't be the military if it didn't involve firing weapons. Members will get their first crack at firing actual military weapons during the last few weeks of basic combat training. Weapons training differs greatly among the different branches' basic training programs. Without a doubt, Marine Corp recruits fire the most rounds during the basic training programs. They're followed by the Army, the Air Force, the Navy, and finally the Coast Guard. Regardless of the branch, a recruit can't graduate from military basic combat training or boot camp without proving that they can handle a military weapon without shooting themselves, their classmates, or the instructors.

　　There are many types of weapons used in the United States Military, but in military basic combat training, a recruit is only required to learn about a few. If the military job requires one to know about additional weapons, and how to use them, more training will be given additionally during military job school.

M-16A2 Assault Rifle

The M-16A2 rifle is the standard military rifle used for combat. It's carried by pretty much every military member in a combat zone. Most people simply call it the "M-16". The M-16 has been around in one form or another since the Vietnam war (the first version, the M16A1, entered Army service in 1964). Its longevity is creditable to its usefulness as a general assault weapon. It's considered by many to be one of the finest military rifles ever made, although advocates of the M-4 Carbine may argue with that assessment. The rifle is lightweight, simple to operate, and puts out a lot of lead.

The M16A2 5.56mm rifle is a lightweight, air-cooled, gas-operated, magazine fed, shoulder or hip-fired weapon designed for either automatic fire (3-round bursts) or semiautomatic fire (single shot) through the use of a selector lever. The weapon has a fully adjustable rear sight. The bottom of the trigger guard opens to provide access to the trigger while wearing winter mittens or chemical protective gear. The upper receiver or barrel assembly has a fully adjustable rear sight and a compensator which helps keep the muzzle down during firing. The steel bolt group and barrel extension are designed with locking lugs which lock the bolt group to the barrel extension, allowing the rifle to have a lightweight aluminum receiver.

In basic combat training, recruits of the Army, Air Force, and Marine Corp will fire this weapon. In Navy Recruit Training, you'll fire a computerized simulator of the M-16 rifle. This simulator is almost like firing the real thing (the computerized rifle even kicks and makes a loud noise). The Coast Guard is the only branch that does not fire the M-16 rifle during basic training.

Recruits who receive classroom training, though, are given instruction on how to fire the weapon, as well as practical training for disassembly, cleaning, and reassembly. If a member of the Coast Guard gets a job that requires him or her to carry an M-16, that member will go through additional training, including actually firing the weapon.

M-4 Carbine

The M-4 combat assault rifle first entered Army service in 1997. The rifle is the standard weapon used by some Army units such as the 82nd Airborne Division and special operations units, such as Army Rangers. With a shortened barrel and collapsible stock, the M-4 is ideal for close quarter marksmanship where lightweight and quick action is required. Firing a standard 5.56 millimeter round (the same as the M-16), the weapon weighs a mere 5.6 lbs.

when empty. A revised rear sight allows for better control of the weapon out to the maximum range of the ammunition used. With the PAQ-4 (Infrared Sight) mounted on the forward rail system, the M-4 can be fitted for increased firepower.

The M-4 Carbine can also be fitted with the M-203 40mm grenade launcher. The M-203 is a lightweight, compact, breech loading, pump action, single shot launcher. The launcher consists of a hand guard and sight assembly with an adjustable metallic folding, short-range blade sight assembly, and an aluminum receiver assembly which houses the barrel latch, barrel stop, and firing mechanism. The launcher is capable of firing a variety of low velocity 40mm ammunition. The launcher also has a quadrant sight that may be attached to the M-4 carrying handle and is used when precision is required out to the maximum effective range of the weapon.

Some Army recruits (usually those in infantry training) will get a chance to carry and qualify with the M-4, instead of the M-16. Many infantry Marines will be trained on the M-4 during Marine Corp infantry training, following basic training.

M-9 Pistol

Did you know that, in combat, it's mostly officers who carry handguns? Most enlisted don't. Notable exceptions are military police and special operations forces. The M-9 pistol is the primary sidearm for all of the military services, except the Coast Guard. It entered the services in 1985 (1990 for the Army). The adoption of the M-9 pistol was the result of a congressional mandate to equip all U.S. services with a standard handgun. The M-9 meets the strict requirements for functional reliability, the speed of the first shot, rapidity of fire, the speed of reloading, range, penetration, and accuracy to 50 yards.

The pistol's components are interchangeable, allowing this weapon to be pieced together from the parts of others. Those attending Army basic combat training will fire the M-9 before graduation. The Air Force previously had those attending fire the M-9 pistol during basic training; they've since removed this requirement, as few Air Force enlisted members are required to carry a pistol in combat. The other branches do not fire this weapon during initial training.

Sig Sauer P229 DAK Pistol

While the other branches use the M-9 as their standard issue pistol, the Coast Guard belongs to the Department of Homeland Security, not the Department of Defense, and therefore uses the standard weapons used by the Department of Homeland Security. The P229 DAK pistol is the standard sidearm for the Department of Homeland Security and Coast

Guard and is a compact, double-action pistol. The pistol weighs only 6.5 pounds and fires double action only, meaning it is a safe and reliable weapon. A key feature of this pistol is quick and easy disassembly for cleaning. All one has to do is lock the slide back and remove the magazine. The DAK model also includes a double strike ability. (1076 words)

(Selected from https://www.thebalancecareers.com/military-weapons-3357161)

【Notes】

1. first crack 第一次

2. Marine Corp 海军陆战队

3. round（弹药的）（一）发；一发子弹

4. Coast Guard 海岸警卫队

5. Assault Rifle 突击步枪

6. Vietnam war 越南战争

7. longevity 持久；耐用

8. gas-operated 气动式的

9. selector lever 变速杆

10. rear sight 后瞄准器

11. bolt group 螺栓组

12. simulator 模拟器

13. carbine 卡宾枪

14. grenade launcher 枪榴弹发射器

15. pump action 压动式枪机

16. sidearm 随身携带的武器（通常指小手枪）

17. Department of Homeland Security 国土安全部

18. double-action 双动式的

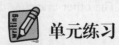 单元练习

I. Complete the following military alphabet

Alpha, __1__, Charlie, __2__, Echo, __3__, Golf, Hotel, India, __4__, Kilo, __5__, Mike, November, __6__, Papa, __7__, Romeo, __8__, Tango, Uniform, Victor, __9__, X-ray, __10__, Zulu

II. Multiple choice

1. In the US Army, there are many basic training units except _____.

   A. Fort Jackson   B. Fort Knox   C. Fort Sill   D. Travis Air Base

2. _____ refer to a large building or group of buildings for soldiers to live in.

   A. Living sectors   B. Barracks
   C. Dormitories   D. Accommodation quarters

3. In terms of length, the shortest of the following weapons is _____.

   A. machine gun   B. rifle   C. pistol   D. AK-47

4. _____ is a training exercise in which military skills are practiced in field conditions.

   A. First aid training   B. Tactical maneuvers
   C. NBC training   D. FTX

5. After joining the army, soldiers _____ their superiors.

   A. salute   B. say hello to   C. kiss   D. shake hands with

6. Which of the following can best cover you on the battlefield located in the woods?

   A. Uniform.   B. Track suit.   C. Camouflage.   D. Jacket.

7. The military time "2015" should be read _____.

   A. twenty-one-five   B. two-zero-one-five
   C. two thousand and fifteen D. twenty-fifteen

8. A soldier from the US Army can buy daily necessities in the _____.

   A. fitness center   B. sports pub   C. barracks   D. PX

9. The following phrases share the same meaning with basic combat training except _____.

   A. recruit training   B. boot camp   C. basic training   D. military training

10. A training area where soldiers have to get past various obstacles such as ditches or high walls as quickly as possible is called _____.

    A. obstacle course   B. map reading   C. drill   D. foot march

III. Translation: English to Chinese

1. Recruits in basic training live and sleep in barracks. They wear a military uniform.

2. There are other differences. As a civilian I never wore a uniform, I didn't salute my superiors and I didn't carry a weapon. I do all these things now.

3. I didn't join the army as an officer. I joined as a soldier when I was 19.

4. In 1992 I served in UNPROFOR, that's the United Nations Protection Force in

Bosnia. I worked in the UN headquarters with French, Italian and Spanish soldiers.

5. It's twenty minutes for barrack inspection and then we start training at 0720 hours. We train from 0720 hours to 1210 hours and then it's time for lunch at 1220 hours. We have one hour for lunch before we start training again.

IV. Translation: Chinese to English

除技能训练之外，士兵还需面临艰苦的体能训练，每日的跑步训练、俯卧撑、仰卧起坐和每周累计10千米的队列行进。障碍训练，有时也被称为军人越障训练课或信心提升课（美式英语），是新训中最大的挑战之一。基础训练过后，攀岩、独木舟之类的团队探险活动也会被纳入拓展训练。

V. Oral practice

1. What is Basic Combat Training? How important is it to the recruits?
2. What's your typical day in the camp?

附录：Unit 1 练习答案

I. Complete the following military alphabet

1. Bravo   2. Delta   3. Foxtrot   4. Juliet   5. Lima
6. Oscar   7. Quebec   8. Sierra   9. Whisky   10. Yankee

II. Multiple choice

1-5 DBCDA   6-10 CDDDA

III. Translation: English to Chinese

1. 新训时，新兵们都在营房内居住休息，而且需身着军装。

2. 平民生活和军人生活还有其他方面的区别。以前是平民老百姓的时候，我从不穿军装，不用给上级敬礼，也不用随身携带武器。但是，入伍以后这些都变成了必须要做的事。

3. 我19岁去当兵，入伍的时候不是军官，而是士兵。

4. 1992年，我前往波斯尼亚的联合国保护部队去执行任务。在当地设立的联合国总部办事处，我与来自法国、意大利和西班牙的士兵一起并肩战斗。

5. 二十分钟的内务检查过后，7点20分至12点10分我们进行训练。午餐时间为12点20分，用餐时长是1个小时。午餐过后，继续训练。

IV. Translation: Chinese to English

In addition to skills training, soldiers face tough physical conditioning, with daily running, pushups and sit-ups, and weekly marches of up to ten kilometers. One of the biggest challenges is the obstacle course, sometimes called the assault course or confidence course (US English). After basic training, team adventure sports like climbing and canoeing are often encouraged and incorporated into further training.

V. Oral practice

1. Basic Combat Training is the initial indoctrination and instruction given to new military personnel, enlisted and officer. After completion of Basic Training, new recruits undergo Advanced Individual Training (AIT), where they learn the skills needed for their military jobs. Various courses are arranged in order to provide the troops with the most up-to-date and comprehensive training available. In addition, soldiers are trained discipline and teamwork; it's critical for each soldier to learn how the overall structure of the military operates so they can perform the functions vital at their level and assure the success of the Platoon and higher elements. Soldiers are taught both leadership and followership as the course is focused on ensuring that soldiers have a good sense of military bearing. They should know how to follow their superiors even as they make their own decisions at the tactical level. The training is committed to providing its new recruits with the skills they need to become effective military members.

2. Here is my typical day in the camp. I wake up at 0600 hours and then we go downstairs to do the roll call for 5 minutes. When that is finished, we go back to the bathroom immediately. That's 0610 hours until 0630 hours—we have twenty minutes for personal hygiene—and then at 0640 hours we go for breakfast—from 0640 hours to 0700 hours. Then we march to the teaching building and do some reading. At 0750 hours we begin to make preparations for the class. The class starts at 0800 hours and finishes at 1150 hours. Then comes the lunch time! We have an hour for the lunch. From 1300 hours to 1400 hours, we take a little nap to ensure good performance in the afternoon. We have the afternoon classes 1430 hours until 1705 hours. When classes are finished, we go to the training ground and do the PT like running, sit-ups, push-ups and so on. Dinner time is at 1815 hours. After dinner, we will do the homework and preview new lessons. When all is done, some personal time is permitted. Lights out is at 2230 hours. That's my day in the camp. How about yours?

Unit 2　Military Organisation

单元导学清单

模块	主题	学习目标	任务	军事知识	核心词汇
Alpha	英军建制	1）了解英军建制基本知识；2）掌握英军编制单位的常用词汇和表达。	Task 1 Task 2 Task 3 Task 4	■ Military symbols	section, platoon, company, battalion, officer commanding (OC), second in command (2IC), commanding officer (CO)
	英美军衔	1）了解英军及美军军衔的基本知识；2）掌握英军军衔的表达；3）对比中国人民解放军军队编制，学习相应知识、词汇及表达。	Task 5 Task 6 Task 7 Task 8	■ Ranks in the US military	staff sergeant, sergeant first class, master sergeant, command sergeant major, sergeant major
Bravo	装甲兵、炮兵、工程兵编制	1）了解相关兵种符号及表达；2）掌握装甲兵、炮兵和工程兵编制与步兵的异同，掌握相关的词汇和表达。	Task 1 Task 2 Task 3 Task 4	■ United Nations military unit size symbols ■ Different names for same units of infantry, armour, artillery and engineer	branch, equivalent, regiment, troop, battery, squadron, battlegroup, task force
Charlie	英军军兵种	1）了解英军皇家部队构成、装备力量等；2）掌握与英军皇家部队编制和力量相关的词汇与表达。	Task 1 Task 2 Task 3 Task 4 Task 5	■ Army Air Corps ■ Royal Army Medical Corps ■ Royal Artillery	aviation, combat, armour, corps

Unit 2 Military Organisation

续表

模块	主题	学习目标	任务	军事知识	核心词汇
Charlie	美军兵种	1）了解美军构成；2）熟练掌握与美军军队构成相关的词汇与术语。	Task 3	■ The US Armed Forces	quartermaster, ordnance, signal corps, intelligence, civil affairs , logistic
Delta	威尔士王妃皇家军团	1）了解威尔士王妃皇家军团构成、装备力量等；2）熟练掌握相关的词汇和表达。	Task 1 Task 2 Task 3	■ The Princess of Wales's Royal Regiment (PWRR) ■ Airborne infantry and airmobile infantry ■ Warrior Infantry Fighting Vehicle ■ Saxon Armoured Personnel Carrier	princess, airborne, airmobile , mechanised, brigade, assign , post, attached, station, equip
Echo (Review)	英军军队编制：单位规模及职能	1）了解英军军队编制的相关知识；2）熟练掌握相关的词汇和表达；3）运用所学知识介绍中国军队编制与英军编制的异同。	Task 1 Task 2	■ British Armed Forces ■ Service branches of the U.K. armed forces	administrative, operational, division, militia, permanent, brigade, standing

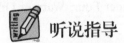

听说指导

Alpha British Army organization and ranks

【军事知识】

1. Military symbols 军用标识

All members of the military need to understand the military symbols used for map marking as they enable tactical and logistic information to be displayed on maps and in pamphlets. A common NATO "language" is used so that the information can be transferred between units of all types.

1) Military Symbol. A military symbol is a symbol or a combination of symbols completed as necessary, by letters, numbers, abbreviations, words or colour and is used to identify and distinguish a particular formation, unit, installation, organization or activity.

2) Unit. A unit is any military body whose structure is prescribed by a competent authority in a table of organization and equipment. In other words it is a body of men organized under one commander, in accordance with a separate establishment and which is not part of a larger unit.

3) Sub-Unit. A Sub-Unit is necessarily therefore a component of a larger unit and may or may not have an independent table of organization or establishment.

4) Formation. A Formation consists of two or more units grouped under the command of an Officer not below the rank of Colonel.

Military symbols are used for a number of reasons such as to save time, secrecy, space, etc.. Persons competent in the use are an asset to Commander as their ability to translate an idea into an overlay order capable of conveying clear instructions to junior commanders without any ambiguity, enhances battle procedures.

2. Ranks in the US military 美军军衔编制

Currently, the ranks of the US military fall into four categories: Army, Air Force, Navy and Marine Corps. Army, Air Force and Navy ranks are divided into six grades and fourteen levels, which are shown as follows.

Commissioned Officer: General, Lieutenant General, Major General, Brigadier General; Colonel, Lieutenant Colonel, Major; Captain, Lieutenant, second lieutenant.

Warrant Officer: Warrant Officer Four, Warrant Officer Three, Warrant Officer Two, Warrant Officer One.

Non-commissioned Officer: Master Sergeant, Sergeant Major, Command Sergeant Major, Sergeant, Corporal.

Enlisted: Private First Class, Private Second Class, new recruits (sailors: Private Third Class).

The Marine Corps ranks are of six grades and twenty-three levels, with no five-star admiral.

【词汇点拨】Words and Expressions

1. section /ˈsekʃn/ *n.* a basic military unit usually having a special function 美军（分排）；（英军）班

Corporal Smith is the commander of his section.

2. platoon /pləˈtəʊn/ *n.* a small group of soldiers which is part of a company and is led by a lieutenant 排

To the left Charlie could see another platoon ahead of him.

3. company /ˈkʌmpəni/ *n.* a group of about 120 soldiers who are usually part of a larger group 连

The division will consist of two tank companies and one infantry company.

4. battalion /bəˈtæljən/ *n.* a large group of soldiers consisting of several companies 营

Our unit was a general support artillery battalion.

5. officer commanding (OC) the commander of a sub-unit or minor unit (smaller than battalion size), principally used in the United Kingdom and Commonwealth 连长

Suddenly, however, their officer commanding decided it would be wrong to retreat.

6. second in command (2IC) a person who has the second highest rank in a group and takes charge when the leader is not there 副司令员，副指挥官，副中队长

Within 1 year he was named the second in command.

7. commanding officer (CO) an officer of any rank who controls a particular military group or operation; the commander of a battalion 指挥官；营长

You have to go to the commanding officer for special permission to leave the camp.

【词汇点拨】Proper Names

1. **Private** /ˈpraɪvət/ 列兵

2. **Lance corporal** 一等兵

3. **Corporal** /ˈkɔːpərəl/ 下士

4. **Sergeant** /ˈsɑːdʒənt/ 中士

5. **Staff sergeant** 三级军士长

6. **Company sergeant major** 连军士长

7. **Regimental sergeant major** 一级军士长

8. **Second lieutenant** 少尉

9. **Lieutenant** 中尉

10. **Captain** /ˈkæptɪn/ 上尉

11. **Major** /ˈmeɪdʒə/ 少校

12. **Lieutenant colonel** 中校

13. **Colonel** /ˈkɜːnl/ 上校

14. **Brigadier** /ˌbrɪɡəˈdɪə/ 准将（大校）

15. **Major general** 少将

16. **Lieutenant general** 中将

17. **General** 上将

【长句解读】

1. The section is the smallest element in the army. (2-1)
> 句子分析：该句中 element 原意为"元素"，在这里指"单位"。
> 翻译："班"是军队中最小的单位。

2. An infantry section has between eight and ten men.(2-1)
> 句子分析：该句中 infantry 指"步兵"。
> 翻译：一个步兵班包含八到十人。

3. The senior NCO is the regimental sergeant major (RSM). (2-1)
> 句子分析：NCO 是 non-commissioned officer 的缩写，指军士；senior NCO 指的是高级士官，包含上士、三级军士长、二级军士长及一级军士长。
> 翻译：（营的）高级士官是一级军士长。

4. This is Captain White. (2-2)
> 句子分析：在英语中，称呼军人时需要在姓前加上其军衔，如 Captain White, Corporal Parks, Sergeant Minter 等。
> 翻译：我是怀特上校。

5. Lance corporal Ducan reporting, ma'am. (2-3)
> 句子分析：sb. reporting 指的是"某某人（向您）汇报"；称呼女性上司（军官）用 madam，缩写为 ma'am。
> 翻译：三等兵邓肯向您汇报，长官（女士）。

【口语输出】

1. What do you know about ranks in the US military? Can you describe them?

【参考词汇】Army, Air Force, Navy and Marine Corps, Commissioned Officer, Warrant Officer, Non-commissioned Officer, Enlisted

2. What are the differences of military ranks and organizations among UK, US and China?

【参考词汇】names of the ranks (private first class, sergeant first class, master sergeant, command sergeant major/sergeant major...), categories of ranks, etc.

Bravo Armour, artillery and engineer formations

【军事知识】

1. United Nations military unit size symbols 联合国军用单位标识

United Nations military symbols apply to both automated and hand-drawn graphic displays and overlays. The symbols will apply to both the United Nations and belligerent forces and shall be distinguished by colour. Basic symbols are shown as follows in the table.

Size Indicator	Meaning
■	Installation
○	Team
●	Squad
●●	Section
●●●	Platoon/Troop
I	Company/Squadron
II	Battalion
III	Regiment/Group
X	Brigade
XX	Division
XXX	Corps
XXXX	Army
XXXXX	Army Group/Front

artillery   engineer   armor   infantry

2. Different names for same units of infantry, armour, artillery and engineer 相同单位在步兵、装甲兵、炮兵和工程兵中的不同名称

Different names are used in different branches of the Army.

What is called platoon in the infantry are named as "troops" in armour, artillery and engineer (including signal) units.

What is called companies in the infantry are named as "squadrons" in armour and engineer units. Moreover, in artillery companies are called "batteries", as shown in the following table.

Regiments are normally made up by battalion size units and company size units. Other than that, battlegroups and task forces are often formed by various units to complete specific missions.

platoon company   (1) troop squadron   troop (3) battery   (2) troop squadron

【词汇点拨】**Words and Expressions**

1. branch /brɑːntʃ/ *n.* an administrative division of some larger or more complex organization 部门；分支

He had a fascination for submarines and joined this branch of the service.

2. equivalent /ɪˈkwɪvələnt/ *adj.* having the same value, purpose, job etc as a person or thing of a different kind 等于

Many people believe that being wealthy is equivalent to gaining success.

3. regiment /ˈredʒəmənt/ *n.* a large group of soldiers, usually consisting of several battalions 军团

the 11th Armoured Cavalry Regiment 第十一装甲骑兵军团

4. troop /truːp/ *n.* a group of soldiers, especially on horses or in tanks（装甲兵、炮兵和工程兵的）排

We saw some troops coming and we ran away.

5. battery /ˈbætəri/ *n.* an artillery subunit of guns, men and vehicles 炮兵连

I was in the 10th battery before I came here.

6. squadron /ˈskwɒdrən/ *n.* a battalion-sized armoured cavalry grouping, consisting of three cavalry troops, one tank company and one battery（美）骑兵中队

The squadron flew on a reconnaissance mission.

7. battlegroup /ˈbætlgruːp/ *n.* a military force created to fight together, typically consisting of several different types of troops 战斗群

The death toll in the Battlegroup serving in Helmand has climbed to 27 since last October.
自去年10月以来，在赫尔曼德服役的战斗群的死亡人数已攀升至27人。

8. task force a military force sent to a place for a special purpose 特遣部队

The task force met for the first time in November.

【词汇点拨】Proper Names

1. 14th Signal Regiment 第十四信号团（美国陆军）
2. 3rd Artillery Regiment 第三炮兵团（美国陆军）

【长句解读】

1. Battalion size units and company size units are frequently grouped in regiments,...(2-6)
> 句子分析：该句中 sth. size units 指"与××体量相当的单位"，如 battalion size units 就是与营相当的单位；are grouped in 为固定搭配，表示"编（制）成为"。

> 翻译：营级和连级单位通常编制成为军团……

2. Units may be organised in a different way for combat. (2-6)
> 句子分析：该句中"be organised in a way for"表示"为……编成……"

> 翻译：单位可能会因为战斗的需要以不同方式进行编制。

【口语输出】

How are regiments and battalions organised in China's army?

【参考词汇】convert, be equipped with, company, armour, artillery, engineer, battlegroup, task force

Charlie Arms and services

【军事知识】

1. Army Air Corps 英国陆军航空团

The Army Air Corps (AAC) is the combat aviation arm of the British Army. Recognisable by their distinctive blue berets, AAC soldiers deliver firepower from battlefield helicopters and fixed wing aircraft to overwhelm and defeat enemy forces.

Army Air Corps soldiers have been wearing the blue beret with pride since 1957. They use Army aircraft, such as the Apache attack helicopter, to deliver hard-hitting and effective support to ground forces during the key stages of battle.

The AAC's role also includes reconnaissance. From high above the action, they observe enemy forces and pass information to troops on the ground.

This fearsome combination of manoeuvrability and firepower makes the AAC one of the most potent of the Army's combat arms.

2. Royal Army Medical Corps 英国皇家陆军医疗团

The Army Medical Service in the UK Army is a modern, inclusive, operationally proven organisation that is aligned totally with the National Health Service.

It is made up of four Corps (The Royal Army Medical Corps, The Royal Army Veterinary Corps, Royal Army Dental Corps and The Queen Alexandra's Royal Army Nursing Corps) of Regular and Reserve personnel delivering the very best patient care wherever the British Army can be found.

The RAMC was formed in 1898 and is the largest corps in the Army Medical Services (AMS). The Corps has a role which ranges from providing immediate first aid emergency care on front line and routine treatment or long term care at health centres and hospitals, as well as health promotion and disease prevention.

In the field RAMC surgeons, pharmacists, doctors and medics combine expert medical training with leadership skills to coordinate large-scale trauma situations and humanitarian operations.

3. Royal Artillery 英国皇家炮兵团

The Royal Artillery provides firepower to the British Army. It is responsible for finding the enemy using a variety of high-tech equipment. Under certain circumstances, it is responsible for striking the rivals using everything from explosive shells to advanced precision rockets.

4. The US Armed Forces 美国军队构成

The United States Armed Forces are the military forces of the United States of America. They consist of the Army, Navy, Marine Corps, Air Force, Space Force, and Coast Guard. All of the branches work together during operations and joint missions, under the Unified Combatant Commands and the authority of the Secretary of Defense with the exception of the Coast Guard. All five armed services are among the seven uniformed services of the United States; the others are the U.S. Public Health Service Commissioned Corps and the National Oceanic and Atmospheric Administration Commissioned Corps.

As of now, the US armed forces is divided into the United States Army, United States Marine, United States Marine Corps Forces Reserve, United States Army Reserve, United States Navy Reserve, United States Air National Guard, United States Air Force Reserve and United States Coast Guard Reserve.

【词汇点拨】Words and Expressions

1. aviation /ˌeɪviˈeɪʃən/ *n.*

① the design, development, production, operation, or use of aircraft, esp. heavier-than-air aircraft 航空技术（包括飞行器的设计、制造和维修）

② military aircraft（统称）军用飞机

the British Civil Aviation Authority 英国民航局

aviation fuel 航空燃料

2. combat /ˈkɒmˌbæt/ *n.* a fight, especially during a war（尤指战争中的）战斗，搏斗

There was a fierce combat between the two sides.

3. arm /ɑːm/ *n.* [C usually pl.] weapons and equipment used to kill and injure people 武器；军火；军备

The government's annual expenditure on arms has been reduced.

4. armour /ˈɑːmər/ *n.*

① strong covering that protects something, especially the body 盔甲；（尤指护身的）铠甲

Police put on body armour before confronting the rioters.

② military vehicles that are covered in strong metal to protect them from attack 装甲车辆；装甲部队

The troops were backed by tanks, artillery, and other heavy armour.

5. corps /kɔːr/ *n.* [pl. corps] a military unit trained to perform particular duties 特种部队；特殊兵种；军（团）

the Royal Army Medical Corps 皇家陆军医疗队

the intelligence corps 情报部队

6. quartermaster /ˈkwɔːtəˌmɑːstər/ *n.* an officer in the army who is in charge of providing food, uniforms and accommodation 军需官；军需主任

The rations were the responsibility of the quartermaster.

7. ordnance /ˈɔː(r)d(ə)nəns/ *n.* military supplies, especially weapons and bombs; large guns on wheels 军需供应（尤指武器和炸弹）；火炮，大炮

At least one person in Lebanon was killed by unexploded ordnance.

8. signal corps 通信队；【陆】（美）通信兵团

His father served in signal corps during the First World War.

9. intelligence /ɪnˈtelɪdʒ(ə)ns/ *n.*

① the ability to learn, understand and think in a logical way about things; the ability to do this well 智力；才智；智慧

② secret information that is collected, for example about a foreign country, especially one that is an enemy; the people that collect this information（尤指关于敌国的）情报；情报人

The intelligence gathered suggested the three men might be planning a vehicle-based attack.

10. civil affairs affairs and operations of the civil population of a territory that are supervised and directed by a friendly occupying power 民政事务

Their meetings with Afghan district officials gave the American civil affairs officers unique insights into local opinions.

11. logistic /ləˈdʒɪstɪk/ *n.* [C usually pl.] the process of planning and organizing to make sure that resources are in the places where they are needed, so that an activity or process happens effectively【军】后勤（学）

About 3% of his turnover went to paying for logistics and distribution.

【词汇点拨】Proper Names

1. Intelligence Corps 英国皇家情报团
2. Royal Engineers 英国皇家工兵团
3. Royal Signals 英国皇家信号部队
4. Royal Army Medical Corps 英国皇家军医部队
5. Royal Logistic Corps 英国皇家后勤部队
6. Royal Artillery 英国皇家炮兵团
7. Royal Electrical and Mechanical Engineers 英国皇家电气工程师部队
8. Royal Armoured Corps 英国皇家装甲兵团

【长句解读】

1. The Combat Arms are directly involved in fighting. (2-8)

   ➢ 句子分析：该句中 directly involved in 指"直接参与""直接参加"。固定搭配 be involved in fighting 为"参与战斗"。

   ➢ 翻译：战斗部队直接参与战斗。

2. Combat Support units provide operational assistance to the Combat Arms and also help with logistics and administration. (2-8)

   ➢ 句子分析：该简单句有两个谓语动词，表达战斗支援单位的两个作用。其中 operational assistance 指"作战帮助"；administration 指"行政（方面的帮助）"。

   ➢ 翻译：战斗支援单位为战斗部队提供作战、后勤与行政方面的帮助。

3. Welcome to this briefing about this British Army. (2-10)

   ➢ 句子分析：句中 briefing 指"简报；传达指示会；情况介绍会"。

   ➢ 翻译：欢迎参加此次关于英国军队的情况介绍会。

4. I'm going to begin with the arms and services of the army. (2-10)

   ➢ 句子分析：arms and services 在本单元内容中反复出现。其中 arms 指的是"军种"，services 指的是"兵种"。

   ➢ 翻译：首先介绍的是（英军）军兵种。

【口语输出】

1. Can you introduce the arms and services of the US Army to the class?

   【参考词汇】combat arms, infantry, armour, air defense artillery, field artillery, aviation, special forces, corps of engineers, combat support units, signal corps, military police corps, chemical corps military intelligence, combat service support branches, transportation, civil affairs, quartermaster, finance, army medical corps, ordnance

2. What does the intelligence corps do?

【参考词汇】collect and disseminate information about the enemy, present general assessments, engage in counter-intelligence

3. How to introduce the branches in the army of China?

【参考词汇】army, navy, air, artillery, strategic support force, logistic, medical, engineer, infantry, armour, special forces, combat support units, signal corps, military police, intelligence, quartermaster, finance, ordnance

Delta The Princess of Wales's Royal Regiment (PWRR)

【军事知识】

1. The Princess of Wales's Royal Regiment (PWRR) 英国威尔士王妃皇家军团

The Princess of Wales's Royal Regiment, also known as the Tigers, are a flexible, fighting Regiment. It is forward looking, yet fiercely proud of their forebear Regiments whose fighting spirit, values and traditions thrive in today's Regiment. The role of the infantry is to hold the ground and continue to take the fight forwards. Soldiers come from the South-East of England including London, Kent, Sussex, Surrey, Hampshire, Middlesex, the Isle of Wight and the Channel Islands.

As an infantry regiment, soldiers are trained in dismounted close combat or fighting on foot (1PWRR), specialised infantry trained to operate in small teams across the world (2PWRR) and it is at the forefront of the Army Reserves equipped with the very latest equipment (3 and 4PWRR).

2. Airborne infantry and airmobile infantry 空降步兵和空中机动步兵

Airborne forces are military units, usually light infantry, set up to be moved by aircraft and dropped into battle. The formations are limited only by the number and size of their aircraft, so given enough capacity a huge force can appear "out of nowhere" in minutes, an action referred to as vertical envelopment.

Airborne forces can be divided into three categories:

Paratrooper soldiers landed by parachute from aircraft and airlanding troops landed by aircraft (usually glider), which both belong to airborne infantry;

Air assault troops or airmobile infantry transported to the battle by helicopter or by aircraft.

3. Warrior Infantry Fighting Vehicle 步兵战车

The tracked Infantry Fighting Vehicle (IFV) known to the British Army as the Warrior was originally known as MCV-80. It was intended to replace the old FV432 armored personnel carrier. This vehicle was developed from the 1970s onwards. A series of pre-production prototypes built in the early 1980s. At the time of its introduction it was one of the best IFVs in the world. The Warrior has been in British Army service since 1988, seeing combat in the Gulf in 1991 and during recent military actions in Iraq and Afghanistan. Production of this armored vehicle ceased in 1995. A total of 384 Warriors were produced in the infantry section vehicle form. Another 105 Warriors were used as anti-tank guided missile team carriers. The British Army plans to operate these armored fighting vehicles until 2035. Version of the Warrior, the Desert Warrior, has been exported to Kuwait.

4. Saxon Armoured Personnel Carrier 撒克逊式装甲运兵车

The Saxon wheeled APC was developed by GNK Defence to provide a relatively low cost armored personnel carrier for the British Army. It was an economical supplement to the much more sophisticated and expensive Warrior tracked infantry fighting vehicle. The Saxon is a very basic infantry carrier. It is based on a revised Bedford M series 4×4 truck chassis and other commercially available components such as the engine and transmission. Developed from the earlier and less protected AT104, the base model was the AT105 which later became the Saxon before being ordered by the British Army in 1983 to provide United Kingdom-based infantry battalions to travel to North-West Europe in an emergency and still retain a measure of operational protection and mobility once arrived.

【词汇点拨】Words and Expressions

1. princess /ˈprɪnses/ *n.* an important female member of a royal family, especially a daughter or granddaughter of a king and queen, or the wife of a prince（除女王或王后外的）王室女成员；（尤指）公主

Lady Diana Spencer became Princess Diana when she married Prince Charles.

2. airborne /ˈeə(r)ˌbɔː(r)n/ *adj.* trained to jump out of aircraft onto enemy land in order to fight 空降的

airborne radar 机载雷达

airborne unit 空降部队

3. airmobile /eərˈməʊbaɪl/ *adj.* able to be transported into a combat zone by air, especially by helicopter 空中机动

an airmobile infantry regiment 空中机动步兵团

4. mechanised /ˈmekənaɪzd/ *adj. using a machine to do something that used to be done by hand* 机械化的

He reported that mechanised North Korean army units have been moving towards the border.

5. brigade /brɪˈɡeɪd/ *n. a large group of soldiers that forms a unit of an army* 旅（陆军编制单位）

There was a sigh of relief at the Pentagon Wednesday as the U. S. Army's final combat brigade crossed from Iraq into Kuwait.

6. assign /əˈsaɪn/ *v. to provide a person for a particular task or position* 指定；指派

Did you choose Russia or were you simply assigned there?

7. post /pəʊst/ *v. to send someone to a place to guard it or to watch who arrives and leaves; to send someone to another country or place to work for the government or a company, or for duty in the military* 使驻守

Guards have been posted along the border.

8. attached /əˈtætʃt/ *adj. working for or forming part of an organization* 隶属于；为……工作

This company is attached to a foreign enterprise.

9. station /ˈsteɪʃ(ə)n/ *v. to cause especially soldiers to be in a particular place to do a job* 驻扎

The regiment was stationed in Singapore for several years.

10. equip /ɪˈkwɪp/ *v. to provide yourself/sb./sth. with the things that are needed for a particular purpose or activity* 配备；装备

They try to equip their vehicles with gadgets to deal with every possible contingency.
他们尽量给他们的车辆配备各种小装置以应对任何可能的突发状况。

【词汇点拨】Proper Names

1. Tidworth /tidwɜ:(r)θ/（英国）蒂德沃思
2. 1PWRR 威尔士王妃皇家军团第 1 营（装甲步兵）
3. 2PWRR 威尔士王妃皇家军团第 2 营（轻步兵）
4. Warrior Infantry fighting vehicle "战士"步兵战车
5. Saxon Armoured personnel carrier 撒克逊式装甲运兵车

【长句解读】

1. The Regiment has two regular battalions. (2-11)

> 句子分析：该句中 the Regiment 指的就是 PWRR（威尔士王妃皇家军团）。这里 regular 指"常设部队"。

> 翻译：该军团拥有两个常设（步兵）营。

2. 1PWRR is assigned to the 1st Mechanised Brigade. (2-11)

> 句子分析：该句中 be assigned to 意为"隶属于"。

> 翻译：威尔士王妃皇家军团第 1 营隶属于第一机械化旅。

【口语输出】

1. Can you talk about the history of the Princess of Wales's Royal Regiment (PWRR)?

【参考词汇】the Queen, Princess of Wales, establishment, services, battles

2. Which unit do you come from? Can you introduce it to the class?

【参考词汇】assigned, posted, attached, based, stationed, equipped

3. What are the differences between the responsibilities of the 1PWRR, the 2PWRR, the 3PWRR and the 4PWRR?

【参考词汇】close combat or fighting, specialised infantry training, operation in small teams across the world, at the forefront of the Army Reserves, latest equipment

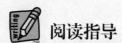

阅读指导

【文章导读】

This passage mainly focuses on British army organisation, namely the size and role of each organisation. The British army, in this sense, is divided into administrative organisation, operational organisation, and battlegroups. The groupings of units and terms also vary by the related branch, history and role.

【军事知识】

1. British Armed Forces 英国军队

Officially known as Her Majesty's Armed Forces, sometimes known as the King's

Armed Forces or Armed Forces of the Crown. The British monarch is the Commander-in-Chief of the British army, navy and air forces. It is administered by the Defence Council within the British Ministry of Defence.

2. Service branches of the U.K. armed forces 英军军种

1) Navy Service

Royal Navy（英国皇家海军）, also known as the British Royal Navy or the British Navy, is the UK's primary maritime combat force. Alfred the Great first used the sea army in the 9th century, and the English naval forces have been fought in it since the early 14th century. The modern Royal Navy dates from the early 16th century, which is the first service in the British armed forces. Because of this, the Royal Navy is also called the "Senior Service".

Corps of Royal Marines (shortened as RM, 英国皇家海军陆战队) light infantry in the British army, also serves as amphibious or sea, land and air combat forces, as well as snow operations, mountain operations special forces, forming Her Majesty's Navy together with the Royal Navy. As the main force of the British Rapid Response Force, the Royal Marines is an assault force capable of all-weather and independent operations.

2) Army Service

British Army（英国陆军）was the ground combat unit which had 80,040 active-duty troops, with 29,790 in the Homeland Army. Distinguished from the royal navy and the royal air force, the British Army has no "royal" title, since during the civil war it always stood with the British parliament. Besides, according to the British parliament, in the bill of rights passed in 1689, the country "in peace without the consent of the king shall not maintain a standing army." The British army does not belong to the jurisdiction of the royal family. However, it does not prevent the use of the title "royal" by several regiments within the army.

Territorial Army (shortened as TA, 英国地方自卫队) is a ground reserve unit under the jurisdiction of the British Army and the largest reserve force in the British military organization, now with 30,000-35,000 officers and men. The TA is not a full-time armed force. Its members are required to participate in a minimum of 19-27 days of military training outside the civilian occupation.

3) Royal Air Force (shortened as RAF, 英国皇家空军)

Founded on 1 April 1918, it was the world's first independent armed air service. The RAF currently has more than 800 aircraft, and 32,940 standing personnel.

【词汇点拨】Words and Expressions

1. administrative /ədˈmɪnɪstrətɪv/ *adj.* connected with organizing the work of a business or an institution 管理的；行政的

It is so easy to get bogged down in day-to-day administrative work that there is little time for strategic policymaking.

2. operational /ˌɒpəˈreɪʃ(ə)nəl/ *adj.* connected with a military operation 军事行动的

operational unit 作战单位

3. division /dɪˈvɪʒ(ə)n/ *n.* a unit of an army, consisting of several brigades or regiments 师

The average age of these officers was thirty in the regiment, thirty-five in a brigade, forty or over in a division.

4. militia /məˈlɪʃə/ *n.* a group of people who are not professional soldiers but who have had military training and can act as an army 民兵组织；国民卫队

The militia practise two or three times a week.

5. permanent /ˈpɜː(r)mənənt/ *adj.* lasting for a long time or for all time in the future; existing all the time 永久的；永恒的；长久的

permanent member of the United Nations security council 联合国安理会常任理事国

6. brigade /brɪˈɡeɪd/ *n.* a subdivision of an army, typically consisting of a small number of infantry battalions and/or other units and typically forming part of a division 旅

A third brigade is at sea, ready for an amphibious assault.

7. standing /ˈstændɪŋ/ *adj.* existing or arranged permanently, not formed or made for a particular situation 长期存在的；永久性的；常设的

Is there a standing review organization?

【参考译文】

英国陆军建制：单位规模和职能

英国陆军根据需要和局势被编成不同的单位。如果我们从行政部门、作战或者任务部门（需要执行特定任务）、战斗群（战时或解决冲突的单位）三者之间的区别来看，用于定义这些单位的术语更容易理解，然而，单位组织和相关一些术语也会随着军种、历史、职责而发生变化。

行政部门

师：由营或团组成的行政分组。师负责其单位的所有行政管理事务，从征兵、晋级、到制订长期计划。

团：由一个或多个正规营和相关的地方军队（包括预备役或民兵）营组成。

营：一般由五个连组成（总共约 700 人），由一名中校指挥。

作战部门

作战部门类似于行政部门，但是营和团的分组称为旅。一个旅通常由五个营或团组成。三到四个旅相应地组成一个作战师，执行特定的作战任务。这些作战单位的具体编制根据任务和参与军种的不同而有所不同。

战斗群

战斗群不是常设的或永久的单位。它们是根据执行的具体任务来编配的。战斗群是旅的下属机构，由中校指挥。战斗群由旅的不同军种组成，如步兵、装甲兵、炮兵、工兵，可能还有航空兵，其目的是完成特殊任务。

 词汇表

Alpha British Army organization and ranks

section 美军（分排）；（英军）班

platoon 排

company 连

battalion 营

officer commanding (OC) 指挥官

second in command (2IC) 副司令员，副指挥官，副中队长

commanding officer (CO) 指挥官，司令

Bravo Armour, artillery and engineer formations

branch 部门；分支

equivalent 等于

regiment 军团

troop （装甲兵、炮兵和工程兵的）排

battery 炮兵连

squadron （美）骑兵中队

battlegroup 战斗群

task force 特遣部队

Charlie Arms and services

aviation 航空技术（包括飞行器的设计、制造和维修）；（统称）军用飞机

combat （尤指战争中的）战斗，搏斗

arm [C usually pl.] 武器；军火；军备

armour 盔甲；（尤指护身的）铠甲；装甲车辆；装甲部队

corps [pl. corps] 特种部队；特殊兵种；军（团）

quartermaster 军需官；军需主任

ordnance 军需供应（尤指武器和炸弹）；火炮，大炮

signal corps 通信队；【陆】（美）通信兵团

intelligence 智力；才智；智慧；情报；情报人

civil affairs 民政事务

47

logistic【军】后勤（学）

Delta The Princess of Wales's Royal Regiment (PWRR)

princess（除女王或王后外的）王室女成员；（尤指）公主
airborne 空降的
airmobile 空中机动
mechanised 机械化的
brigade 旅（陆军编制单位）
assign 指定；指派
post 使驻守

attached 隶属于；为……工作
station 驻扎
equip 配备；装备

Echo Army organisation: unit size and role

administrative 管理的；行政的
operational 军事行动的
division 师
militia 民兵组织；国民卫队
permanent 永久的；永恒的；长久的
brigade 旅
standing 长期存在的；永久性的；常设的

 拓展学习

Military Services in China

The Chinese military services are composed of the Army, Navy, Air Force, Rocket Force, PLA Information Support Force, Armed Police Force and Reserve Force.

PLA Army

The PLAA plays an irreplaceable role in maintaining China's national sovereignty, security and development interests. It comprises maneuver operation, border and coastal defense, and garrison forces. Under the PLAA, there are 5 TC army commands, the Xinjiang military command, and the Xizang military command. The ETC Army has under it the 71st, 72nd, and 73rd group armies; the STC Army has the 74th and 75th group armies; the WTC Army has the 76th and 77th group armies; the NTC Army has the 78th, 79th and 80th group armies; the CTC Army has the 81st, 82nd and 83rd group armies.

PLA Navy

The PLAN has a very important standing in the overall configuration of China's national security and development. It comprises submarine, surface ship, aviation, marine, and coastal defense forces. Under the PLAN, there are the ETC Navy (Donghai Fleet), the STC Navy (Nanhai Fleet), the NTC Navy (Beihai Fleet), and the PLAN Marine Corps. Under the TC navies there are naval bases, submarine flotillas, surface ship flotillas and aviation brigades.

PLA Air Force

The PLAAF plays a crucial role in overall national security and military strategy. It comprises aviation, airborne, ground-to-air missile, radar, ECM, and communications forces. Under the PLAAF, there are 5 TC air force commands and one airborne corps. Under the TC air forces, there are air bases, aviation brigades (divisions), ground-to-air missile brigades (divisions) and radar brigades.

PLA Rocket Force

The PLARF plays a critical role in maintaining China's national sovereignty and security. It comprises nuclear missile, conventional missile and support forces, and subordinate missile bases. In line with the strategic requirements of having both nuclear and conventional capabilities and deterring wars in all battle spaces, the PLARF is enhancing its credible and reliable capabilities of nuclear deterrence and counterattack, strengthening intermediate and long-range precision strike forces, and enhancing strategic counter-balance capability, so as to build a strong and modernized rocket force.

PLA Information Support Force

The newly-established information support force of the Chinese People's Liberation Army (PLA) is a brand-new strategic arm of the PLA. With this round of reform, the PLA now features a new system of services including the army, the navy, the air force and the rocket force, and arms including the aerospace force, the cyberspace force, the information support force and the joint logistic support force, under the leadership and command of the Central Military Commission. The establishment of the information support force is a major decision that the CPC Central Committee and the CMC has made in light of the overall need of building a strong military.

The information support force is a new, strategic branch of the military and a key pillar in coordinating the construction and application of the network information system. It will play a crucial role in advancing the Chinese military's high-quality development and competitiveness in modern warfare.

The force is urged to integrate deeply into the Chinese military's joint operation system, carry out information support operations in a precise and effective manner, and facilitate military operations in various directions and fields. It is also urged to build a network information system that fulfills the requirements of modern warfare and features the Chinese military's own characteristics, as well as efforts to accelerate the development of integrated combat capabilities more effectively.

PLA Armed Police Force

The PAP shoulders important responsibilities in safeguarding national security, social stability and public well being. China has adopted a CMC-PAP-Troops leadership and command system with the basic duties and nature of the PAP unchanged. The PAP is not in the force structure of the PLA. The PAP border defense, firefighting and security guard forces have been decommissioned. The coast guard under the leadership of State Oceanic Administration has been transferred to the PAP. PAP goldmine, forest and hydroelectricity forces have been reorganised into specialized forces of non-active service under corresponding state authorities. Meanwhile, the PAP customs guard forces have been withdrawn. In this way, the leadership, management, command and employment of the PAP has become more coherent. Following adjustment and reorganization, the PAP is mainly composed of the internal security corps, the mobile corps, and the coast guard.

PLA Reserve Force

The Reserve Force is an armed organization composed of the people not released from their regular work. As an assistant and backup force of the PLA, the Reserve Force is tasked with participating in the socialist modernization drive, performing combat readiness support and defensive operations, helping maintain social order and participating in emergency rescue and disaster relief operations.

(*Selected from http://eng.mod.gov.cn/services/index.htm*)

【Notes】

1. coastal 海岸的，沿海的

2. garrison 要塞，驻防

3. TC（Theatre Command）战区

4. flotilla 小型舰队

5. long-range（飞机、火箭等）远程的

6. strategic counter-balance capability 战略平衡能力

7. endeavor 努力，尽力

8. CMC-PAP-Troops leadership（CMC: Central Military Commission, PAP: People's Armed Police）"中央军委—武警部队—部队"领导指挥体制

9. decommission 使退役

10. emergency rescue 紧急救援

11. disaster relief operation 救灾行动

Unit 2　Military Organisation

单元练习

I. Label the picture

platoon company | (1) _____ squadron | (2) _____ squadron | troop (3) _____

(4) | (5) | (6) | (7)

1. _____ 2. _____ 3. _____ 4. _____ 5. _____
6. _____ 7. _____

II. Multiple choice

1. The _____ is the smallest element in the army.

   A. section　　　B. company　　　C. battery　　　D. battalion

2. The section commander is a _____ and the second in command (2IC) is a _____.

   A. lieutenant, second lieutenant　　B. private, corporal

   C. major, captain　　D. corporal, lance corporal

3. An infantry _____ has five companies. The commander is called the CO (commanding officer).

   A. division　　B. battalion　　C. squadron　　D. corps

4. The Intelligence Corps _____ .

   A. collect information about the enemy　　B. build roads and bridges

   C. provide medical support　　D. provide fire support

5. The Royal Engineers _____ .

   A. collect information about the enemy　　B. build roads and bridges

   C. provide medical support　　D. provide fire support

6. The Royal Army Medical Corps _____ .

   A. collect information about the enemy　　B. build roads and bridges

   C. provide medical support　　D. provide fire support

7. The Royal Artillery _____.

A. collect information about the enemy B. build roads and bridges

C. provide medical support D. provide fire support

8. Armour, artillery and engineer units equivalent to infantry platoons are called _____.

A. batteries B. squadrons C. troops D. regiments

9. Armour units and engineer units equivalent to infantry companies are called _____.

A. batteries B. squadrons C. troops D. regiments

10. Battalion size units and company size units are frequently grouped in_____.

A. batteries B. squadrons C. troops D. regiments

III. Translation: English to Chinese

1. A platoon has three sections. The platoon commander is a second lieutenant or lieutenant. A sergeant is second in command. An infantry platoon has between 24 and 30 officers and men.

2. Regiments are commanded by a colonel. Units may be organised in a different way for combat. These groups are called battlegroups or task forces.

3. The US Army classifies the different branches of the Army as Combat Arms, Combat Support units and Combat Service Support branches.

4. The Combat Service Support branches include Transportation, Civil Affairs, Quartermaster, Finance, Army Medical Corps and Ordnance.

5. The mission of the Royal Electrical and Mechanical Engineers is to maintain the Army's vehicles and equipment.

IV. Translation: Chinese to English

英国陆军将陆军的不同军团分为战斗团、战斗武器支援团和战斗服务支援团。战斗团直接参与战斗。战斗团包括皇家装甲兵团、步兵团（INF）和陆军航空团（AAC）。战斗武器支援团的任务是为战斗团提供近距离的支援。战斗武器支援团包括皇家炮兵团（RA）、皇家工兵团（RE）、皇家信号部队（R SIGNALS）和皇家情报团（INT CORPS）。战斗服务支援团包括皇家后勤部队（RLC）、皇家军医部队（RAMC）和皇家电气工程师部队（REME）。

V. Oral practice

What do you know about the branches in the U.K. armed forces? Can you introduce them?

附录：Unit 2 练习答案

I. Label the picture

1. troop   2. battery   3. troop   4. artillery

5. engineer   6. armour   7. infantry

II. Multiple choice

1-5 ADBAB   6-10 CDCBD

III. Translation: English to Chinese

1. 一个排分为三个部分。排长是少尉或中尉。副手是中士。一个步兵排有24~30名官兵。

2. 兵团由一名上校指挥。（其中的）单位可以以不同的方式组织起来进行战斗。这些小组被称为战斗小组或特遣部队。

3. 美国陆军分为战斗部队、战斗支援单位和战斗服务支援部门三个不同的分支。

4. 战斗服务支援部门包括运输、民政、军需、财政、医疗和军械部门。

5. 皇家电气和机械工程师部队的任务是维护陆军的车辆和设备。

IV. Translation: Chinese to English

The British Army classifies the different corps and regiments of the Army as Combat Arms, Combat Arms Support and Combat Service Support. The Combat Arms are directly involved in fighting. The Combat Arms include the Royal Armoured Corps, the Infantry (INF) and the Army Air Corps (AAC). The mission of Combat Arms Support corps is to provide close support to the Combat Arms. The Combat Arms Support corps include the Royal Artillery (RA), the Royal Engineers (RE), the Royal Signals (R SIGNALS) and the Intelligence Corps (INT CORPS). Combat Service Support corps include the Royal Logistic Corps (RLC), the Royal Army Medical Corps (RAMC) and the Royal Electrical and Mechanical Engineers (REME).

V. Oral practice

There are altogether 3 service branches of the U.K. armed forces, namely Navy, Army and Royal Air Force. The Navy Service includes Royal Navy, also known as the British Royal Navy or the British Navy, is the UK's primary maritime combat force. It is also called the "Senior Service".

It also has Corps of Royal Marines, light infantry in the British army, also serves as amphibious or sea, land and air combat forces, as well as snow operations, mountain operations special forces, forming Her Majesty's Navy together with the Royal Navy.

The Army Service includes British Army and Territorial Army. The former was the ground combat unit of the British Army. The British Army has no "royal" title. However, it does not prevent the use of the title "royal" by several regiments within the army. Territorial Army is a ground reserve unit under the jurisdiction of the British Army and the largest reserve force in the British military organization. It is not a full-time armed force. The last is the Royal Air Force, the world's first independent armed air service.

Unit 3 Military Technology

单元导学清单

模块	主题	学习目标	任务	军事知识	核心词汇
Alpha	军事发明	1）掌握军事发明相关词汇与表达；2）识别并匹配图片中的军事发明；3）了解、记忆并谈论某些军事发明（机关枪、雷达、坦克、潜水艇、直升机、原子弹）的背景知识（时间、地点、人物等）。	Task 1 Task 2 Task 3	■ Matabele War ■ Battle of Cambrai ■ Battle of Flers-Courcelette ■ Turtle	aircraft carrier, atomic, jet fighter, machine gun, radar, submarine, portable
Bravo	飞机、车辆和军舰	1）了解固定翼飞机、旋翼飞机、装甲车辆和军舰的分类、定义和功能；2）掌握与飞机、车辆和军舰相关的词汇与表达；3）介绍我军轰-6K型轰炸机。	Task 1 Task 2 Task 3 Task 4 Task 5	■ Fixed-wing aircraft ■ Rotary-wing aircraft ■ H-6K bomber	bomber, destroyer, frigate, transport aircraft, fixed-wing, rotary-wing, armoured, naval, underwater, vertical, column, stealth
Charlie	过去与现在的军事科技	1）了解英尺、英寸、英里、海里、节等距离和速度单位；2）掌握距离、速度、温度的单位及其换算；3）介绍并对比过去和现在潜水艇、战斗机的技术参数。	Task 1 Task 2 Task 3 Task 4	■ Holland 1 ■ Seawolf ■ ME 262 ■ Eurofighter	knot, nautical mile, convert, multiply, divide, dive, armament, crew, torpedo, tube, cannon, Mach
Delta	装甲车辆	1）了解不同装甲车辆及其特点；2）熟练掌握与车辆性能相关的词汇和表达；3）运用所学知识介绍不同的装甲车辆。	Task 1 Task 2 Task 3 Task 4 Task 5	■ Leopard 2 Main Battle Tank(MBT) ■ Warrior Infantry Fighting Vehicle(IFV) ■ FV432 Armoured Personnel Carrier(APC)	hull, main gun, tracks, turret, specification

续表

模块	主题	学习目标	任务	军事知识	核心词汇
Echo	综合技能：描述与定义	1）了解描述与定义物品的方法；2）熟练掌握相关的词汇和表达；3）运用所学知识描述常见的军事装备，如"黑鹰"直升机；4）介绍我军的新型直升机。	Task 1 Task 2 Task 3 Task 4 Task 5 Task 6	■ The Black Hawk helicopter ■ Hellfire anti-armour missile ■ Identification Friend or Foe (IFF)	fabric, helmet, mortar, tracksuit, cruise, equip, rotor blade, logistical, assault, evacuation, calibre, mount
Foxtrot (Review)	飞机、车辆和海军舰艇	1）了解飞机、舰船和地面车辆的类型；2）熟练掌握与各类军事装备相关的词汇和表达；3）运用所学知识介绍国内外军事装备。	Task 1 Task 2	■ The US B-2 Spirit ■ The Eurofighter 2000/Typhoon ■ The US C-130 Hercules ■ The US Spruance-class destroyer	fixed-wing aircraft, rotary-wing aircraft, surface vessels, submarines, manoeuvrable, reconnaissance, humanitarian, relief, aerial, versatile, escort, engage, ballistic

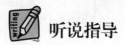 听说指导

Alpha Military inventions

【军事知识】

1. Matabele War 马塔贝莱战争（1893 年马塔贝莱人抗击英国殖民者入侵的战争）

The first Matabele War was fought between 1893 and 1894 in the country today called Zimbabwe. It pitted the British South Africa Company against the Ndebele (Matabele) Kingdom. Lobengula, king of the Ndebele, had tried to avoid outright war with the company's pioneers because he and his advisors were mindful of the destructive power of European-produced weapons on traditional Matabele impis (units of Zulu warriors) attacking in massed ranks.

2. Battle of Cambrai 康布雷战役

Battle of Cambrai, military engagement in northern France, took place during World War I from September 27 to October 11, 1918. It was part of a series of connected battles at the start of the "hundred days" campaign, which began with the Battle of Amiens in August and led to the defeat of Germany and the end of the war. The battle was among the Canadian

Corps' most impressive tactical victories of the war, particularly because of the Canadians' skillful use of military engineers.

3. Battle of Flers-Courcelette 弗莱尔—库尔瑟莱特之战

The Battle of Flers-Courcelette, was a battle within the Franco-British Somme Offensive which took place in the summer and autumn of 1916. Launched on 15 September 1916 the battle went on for one week. Flers-Courcelette began with the objective of cutting a hole in the German line by using massed artillery and infantry attacks. This hole would then be exploited with the use of cavalry. It was the third and final general offensive mounted by the British Army during the Battle of the Somme. By its conclusion on 22 September, the strategic objective of a breakthrough had not been achieved; however tactical gains were made in the capture of the villages of Courcelette, Martinpuich and Flers. The battle is significant for the first use of the tank in warfare. It also marked the debut of the Canadian and New Zealand Divisions on the Somme battlefield.

4. Turtle "海龟号"潜艇

Turtle, one-man submarine, the first to be put to military use, was built and designed by the American inventor David Bushnell in 1775 for use against British warships. The pear-shaped vessel, made of oak reinforced with iron bands, measured about 2.3 m (7.5 feet) long by 1.8 m (6 feet) wide. It was equipped with a mine that was to be attached to the hull of an enemy ship. In 1776, in New York harbour, the Turtle tried to sink the British warship HMS Eagle but failed; none of its succeeding missions was successful.

【词汇点拨】Words and Expressions

1. aircraft carrier a large ship that carries aircraft which use it as a base to land on and take off from 航空母舰

An aircraft carrier is like an airport on the sea.

2. atomic /əˈtɒmɪk/

① *adj.* connected with atoms or an atom 原子的；与原子有关的

② *adj.* related to the energy that is produced when atoms are split; related to weapons that use this energy 原子能的；原子武器的

atomic structure 原子结构

atomic energy/power/bomb 原子能 / 原子动力 / 原子弹

3. jet /dʒet/ *n.* a plane driven by jet engines 喷气式飞机

a jet aircraft/fighter/airliner 喷气式飞机 / 战斗机 / 客机

The accident happened as the jet was about to take off.

4. fighter /ˈfaɪtə(r)/ *n.* a fast military plane designed to attack other aircraft 战斗机；歼击机

fighter pilot 战斗机飞行员

Fighter pilots searched out and attacked enemy aircraft.

5. machine gun a gun that automatically fires many bullets one after the other very quickly 机关枪；机枪

a burst/hail of machine-gun fire 一阵猛烈的机关枪扫射

6. radar /ˈreɪdɑː(r)/ *n.* a system that uses radio waves to find the position and movement of objects, for example planes and ships, when they cannot be seen 雷达

They located the ship by radar.

7. submarine /ˌsʌbməˈriːn/ *n.* a ship that can travel underwater 潜艇

a nuclear submarine 核潜艇

a submarine base 潜艇基地

8. portable /ˈpɔːtəbl/ *adj.* that is easy to carry or to move 便携式的；手提的；轻便的

Portable laser pointing devices are less expensive, more powerful, and more readily available than ever.

【词汇点拨】Proper Names

1. Hiram Maxim 海勒姆·马克沁（美国发明家，第一支"马克沁"机枪的发明者）

2. Matabele War 马塔贝莱战争

3. Battle of Cambrai 康布雷战役

4. Battle of Flers-Courcelette 弗莱尔－库尔瑟莱特之战

5. Leonardo da Vinci 列奥纳多·达·芬奇

6. Paul Cornu 保罗·科尔尼

【长难句解读】

1. The machine gun wasn't used in combat until 1893. (3-1)

   ➢ 句子分析：该句中 not until 结构表示"直到……才"。combat 意为"战斗"。

   ➢ 翻译：直到 1893 年，机关枪才用于战斗。

2. Leonardo da Vinci made drawings of helicopters in about 1485 but the French pilot Paul Cornu flew the first helicopter on 13 November, 1907. (3-1)

   ➢ 句子分析：该句中"drawing"意为"素描"，此处的 drawings of helicopter 是指达·芬奇在 1485 年设计的飞行器手稿。

   ➢ 翻译：
   莱昂纳多·达·芬奇在 1485 年左右绘制了直升机的图纸，但法国飞行员保罗·科尔尼在 1907 年 11 月 13 日实现了直升机的首次飞行。

【口语输出】

1. Can you introduce one of the military inventions that you are familiar with? Please talk about when and by whom it was invented.

   【参考词汇】aircraft carrier, submarine, machine gun, helicopter, radar

2. What military inventions are introduced in the recording?

   【参考词汇】portable machine gun, radar, tank, submarine, helicopter, and atomic bomb

3. What have you learned from the recording about the first portable machine gun?

   【参考词汇】Hiram Maxim, 1883, combat, 1893, British troops, Matabele War

4. What have you learned from the recording about the first helicopter?

   【参考词汇】Leonardo da Vinci, drawings of helicopters, French pilot, Paul Cornu, 13 November, 1907, twenty seconds

Bravo Aircraft, vehicles and naval Ships

【军事知识】

1. Fixed-wing aircraft 固定翼飞机

A fixed-wing aircraft is a heavier-than-air flying machine, which is capable of flight using wings that generate lift caused by the aircraft's forward airspeed and the shape of the wings. The wings of a fixed-wing aircraft are not necessarily rigid. Gliding fixed-wing aircraft use moving air to gain altitude while powered fixed-wing aircraft gain forward thrust from an engine.

2. Rotary-wing aircraft 旋转翼飞机

A rotary-wing aircraft, also known as rotorcraft, is a heavier-than-air aircraft with rotary wings or rotor blades, which generate lift by rotating around a vertical mast. They generally include aircraft where one or more rotors provide lift throughout the entire flight, such as helicopters, autogyros, and gyrodynes.

3. H-6K bomber 轰-6K 型轰炸机

The H-6K praised as the "Almighty of War" is a new generation of medium and long-range bomber independently developed by China. As an important combat force of the Air Force, the H-6K is able to carry out stand off attack and conduct wide-range patrol and long-distance strike. Its main mission objective is to strike key strategic points deep within enemy territories. It is equipped with very advanced weapons including air-launched cruise missiles, enabling it to attack targets far away. Leading experts predicted that the warplane could eventually be armed with hypersonic weapons that can destroy hostile military hubs 3,000 kilometers away within minutes. Together with J-20 stealth fighters, H-6Ks could devastate the enemy's fighting capability even before a war gets fully underway.

【词汇点拨】Words and Expressions

1. bomber /ˈbɒmə(r)/ *n.* a plane that carries and drops bombs 轰炸机

The bomber scored a direct hit on the bridge.

2. destroyer /dɪˈstrɔɪə(r)/ *n.* a small fast ship used in war, for example to protect larger ships 驱逐舰

The latest Royal Navy Type 45 destroyer, HMS Defender, has been formally commissioned into the fleet.

3. frigate /ˈfrɪɡət/ *n.* a small fast ship in the navy that travels with other ships in order to protect them（小型）护卫舰

The crew from the Type 23 frigate was undergoing combat training at the time.

4. transport aircraft 【航】运输机

It is a version of the C-130 Hercules transport aircraft and flies at less than 300 mph (482 kph).

5. fixed-wing /ˌfɪkstˈwɪŋ/ *adj.* used to describe aircraft with wings that remain in the same position, rather than helicopters, etc. 固定机翼的

Fixed-wing aircraft may be built with many wing configurations.

6. rotary-wing /ˈrəʊtəriˈwɪŋ/ *adj.* used to describe aircraft with an airfoil that rotates about an approximately vertical axis, as that supporting a helicopter or autogiro in flight 旋翼的

Pakistan Army Aviation Corps flies a mixture of fixed and rotary-wing aircraft in close support of ground operations.

7. armoured /ˈɑːməd/ *adj.* (especially of a military vehicle 尤指军用交通工具) protected by metal covers 有装甲的

an armoured car 装甲车

The cruiser was heavily armoured.

8. naval /ˈneɪvl/ *adj.* connected with the navy of a country 海军的

a naval base/battle 海军基地/海战

He accepted a commission as a naval officer.

9. underwater /ˌʌndəˈwɔːtə(r)/ *adj.&adv.* situated, occurring, or done beneath the surface of the water 水下的；供水下用的；水下生长的；水下操作的；在水下，在水中

There are underwater volcanoes in the region.

The seal spent a lot of time underwater.

10. vertical /ˈvɜːtɪkl/ *adj.* going straight up or down from a level surface or from top to bottom in a picture, etc. 竖的；垂直的；直立的

a plane with vertical take-off and landing capabilities

11. column /ˈkɒləm/ *n.* a long, moving line of people or vehicles（人或车辆排成行移动的）长列，纵队

a long column of troops and tanks

12. stealth /stɛlθ/ *n.* the act or characteristic of moving with extreme care and quietness, esp so as to avoid detection 秘密行动；暗中活动

stealth bomber 隐形轰炸机

【词汇点拨】Proper Names

1. **IFV (Infantry Fighting Vehicle)** 步兵战斗车辆；步战车
2. **APC (Armoured Personnel Carrier)** 装甲运兵车
3. **H-6K bomber** 轰-6K 型轰炸机

【长难句解读】

1. A submarine is a type of ship. It can travel underwater. (3-2)

> 句子分析：该句子中 underwater 为副词，意为"在水下"，travel 在此处指

"（以某速度、朝某方向或在某距离内）行进，转送，传播"。

> **翻译**：潜艇是船舶的一种，可以在水下行进。

2. An IFV is a type of armoured vehicle. It can carry a section of soldiers. (3-2)

> **句子分析**：该句中 IFV 为 Infantry Fighting Vehicle 的缩写，翻译为"步兵战斗车辆"或"步战车"，a section of soldiers 在此处意为一个班的士兵，armoured vehicle 中 armoured 为形容词，修饰 vehicle，意为"装甲车辆"。

> **翻译**：步战车是一种装甲车辆，它能够运载一个班的士兵。

【口语输出】

1. Can you list some examples of fixed-wing aircraft and naval ships?

【参考词汇】bomber, fighter, transport aircraft, aircraft carrier, destroyer, frigate, submarine

2. How would you define aircraft carriers, submarines and helicopters?

【参考词汇】naval ship, take off, land, travel, underwater, rotary-wing aircraft, take off, vertically, fly high

3. What are the functions of the following aircraft and vehicles: destroyer, fighter-bomber, fighter, stealth bomber, transport aircraft?

【参考词汇】naval vessel, support, protect, destroy enemy aircraft, detect, transport

Charlie　Military technology now and then

【军事知识】

1. Holland 1　英制"荷兰一号"潜艇

Holland 1 (or HM submarine Torpedo Boat No.1) was the first submarine commissioned by the Royal Navy, the first in a six-boat batch of the Holland-class submarine. In 1901 she was ordered from John Philip Holland and built at Barrow-in-Furness. On 2 October 1901, she was launched and on 20 March 1902 she dived for the first time.

In 1913 the submarine was decommissioned and sold to Thos W Ward for £410. While being towed to the scrapyard she encountered very severe weather and sank. She was located in 1981 at a depth of 63 meters and salvaged in 1982. Between 1995 and 2000 extensive conservation measures were applied. In 2001 she opened to the public in a climate-controlled gallery.

2. Seawolf 美国"海狼"级潜艇

The Seawolf-class is a class of nuclear-powered fast attack submarines (SSN) in service with the United States Navy. The class was the intended successor to the Los Angeles class. Design work began in 1983. At one time, an intended fleet of 29 submarines was to be built over a ten-year period, later reduced to twelve submarines. Finally, only 3 were finished, the Seawolf (SSN21, commissioned in July 1997), Connecticut (SSN22, commissioned in December 1998) and Jimmy Carter (SSN23, launched in June 2004 and commissioned in February 2005).

The Seawolf-class is about 107 meters long, and 12 meters wide with a displacement of 8,600 tonnes (surfaced) and a displacement of 9,138 tonnes (submerged).

Compared to previous Los Angeles class submarines, the Seawolf-class submarines are larger, faster, and significantly quieter; they also carry more weapons and have twice as many torpedo tubes, for a total of 8. The boats are able to carry up to 50 UGM-109 Tomahawk cruise missiles for attacking land and sea surface targets.

3. ME 262 德制 ME 262 型喷气式战斗机

The Messerschmitt ME 262 was the first operational fighter aircraft powered by jet in the world. It started to develop before the Second World War and first flew in April 1941. It was introduced in 1944, and was majorly used by the Luftwaffe and Czechoslovak Air Force and was retired in 1945 and 1951, respectively. A total of 1,430 aircraft were built.

The ME 262 was designed with higher speed and equipped with better armaments compared with other fighter aircraft of the Allied forces. It was one of the most innovative aviation concepts that were used in the Second World War and served as a light bomber, reconnaissance aircraft, and an experimental night fighter. Several versions of the ME 262 were produced including a reconnaissance version designated as A-1a/U3, a bomber-destroyer version designated as A-1a/U4, a heavy jet fighter designated as A-1a/U5, and a few more.

4. Eurofighter 欧制"台风"战斗机

The Eurofighter Typhoon is the world's most modern swing-role fighter. Once conceived as a multi-national programme to modernise the Air Forces of the European industrial partner nations, the Eurofighter Typhoon has in the meantime been sold to five additional customers (Austria, the Kingdom of Saudi Arabia, Sultanate of Oman, Kuwait and Qatar) and thus doubled the number of its original user nations.

It is powered by two EJ200 engines that give the Eurofighter Typhoon its impressive thrust-to-weight ratio and maneuverability. The core of this state-of-the-art weapon system is its identification capability and sensor fusion, based on the CAPTOR-E AESA radar and the PIRATE FLIR sensor while being protected by the PRAETORIAN Electronic Defensive Aid Sub System (DASS).

【词汇点拨】Words and Expressions

1. knot /nɒt/ *n.* a unit for measuring the speed of boats and aircraft; one nautical mile per hour 节（速度单位，等于 1 海里 / 小时）

Some days the vessel logged 12 knots.

2. nautical mile a unit for measuring distance at sea; 1,852 metres 海里（合 1852 米）

The record-breaking 9,000 nautical mile (16,668 km) trip took the PacX Wave Glider just over a year to achieve.

3. convert /kənˈvɜːt/ *v.* ~ (sth.) (from sth.) (into/to sth.) to change or make sth. change from one form, purpose, system, etc. to another （使）转变，转换，转化

What rate will I get if I convert my dollars into euros?

4. multiply /ˈmʌltɪplaɪ/ *v.* to add a number to itself a particular number of times 乘；乘以

2 multiplied by 4 is/equals/makes 8.

Multiply 2 and 6 together and you get 12.

5. divide /dɪˈvaɪd/ *v.* to find out how many times one number is contained in another 除；除以

Measure the floor area of the greenhouse and divide it by six.

6. dive /daɪv/ *v.* to go to a deeper level underwater 下潜；潜到更深的水下

The fish dive down to about 1400 feet and then swim south-west.

7. armament /ˈɑːməmənt/ *n.* [C, usually pl.]

① a general term for a weapon 武器；装备

② the process of equipping with weapons 武装；战备

armaments factory 军工厂

The main innovation in the new vehicle was the turret and its armament.

8. crew /kruː/ *n.* a team of people who man a ship, aircraft or vehicle; a team of people who operate a weapon or equipment 机组人员；炮班（自行火炮等）；车组（人员）

None of the passengers and crew were injured.

9. torpedo /tɔːˈpiːdəʊ/ *n.* a long narrow bomb that is fired under the water from a ship or submarine and that explodes when it hits a ship, etc. 鱼雷

But most observers believe that the most likely explanation was that a torpedo misfired, stuck in the tube and exploded.

10. tube /tjuːb/ *n.* a long hollow pipe made of metal, plastic, rubber, etc., through which liquids or gases move from one place to another（金属、塑料、橡胶等制成的）管，管子

torpedo tube 鱼雷发射管

11. cannon /ˈkænən/ *n.*

① an old type of large heavy gun, usually on wheels, that fires solid metal or stone balls（通常装有轮子并发射铁弹或石弹的旧式）大炮

② an automatic gun that is fired from an aircraft（飞机上的）自动机关炮

Police turned water cannon on the rioters.

Twenty people were killed when German planes attacked Newport with bombs and cannon fire in April 1943.

12. Mach /mɑːk/ *n.* a measurement of speed, used especially for aircraft. Mach 1 is the speed of sound. 马赫，马赫数（常用于表示飞行速度，马赫数 1 即 1 倍声速）

a fighter plane with a top speed of Mach 3 (= 3 times the speed of sound) 最高速度为马赫数 3 的歼击机

【词汇点拨】Proper Names

1. Royal Navy（英）皇家海军

2. German /ˈdʒɜːmən/ 德国的；德国人；德语

【长难句解读】

1. To convert Centigrade into Fahrenheit, multiply by 9, divide by 5 and add 32. (3-4)

➢ **句子分析**：该句中 Centigrade 和 Fahrenheit 为两个温度单位，分别表示"摄氏度"和"华氏温度"，multiply、divide 和 add 意为"乘""除"和"加"。

➢ **翻译**：将摄氏度转换为华氏度，乘以 9，除以 5，再加 32。

2. It had seven crew and carried one 18-inch torpedo tube. (3-5)

➢ **句子分析**：该句中 It 指上文的 Holland 1，英国皇家海军的"荷兰一号"潜艇。crew 意为潜艇上的机组人员，torpedo tube 则指鱼雷发射管。

➢ **翻译**："荷兰一号"潜艇上有七名机组人员，携带一根 18 英寸鱼雷发射管。

3. Today's jet, like the Eurofighter can fly at Mach 2–that's twice the speed of sound. (3-5)

➢ **句子分析**：句中以 Eurofighter 为例说明当今喷气式战斗机的飞行速度。破折号后的句子用于解释 Mach 2，帮助读者更好地理解该速度单位。

➢ **翻译**：今天的喷气式战斗机，比如欧制"台风"战斗机，可以以马赫数 2 的速度飞行——这是声速的两倍。

【口语输出】

1. How to convert inches into centimetres, feet into meters, miles into kilometres, nautical miles into kilometres and Centigrade into Fahrenheit?

【参考词汇】convert into, multiply by, divide by, add

2. What are the differences between 1901 Holland 1 and Seawolf in terms of speed, operating depth, armament and crew?

【参考词汇】knots, feet, 18-inch torpedo tubes, 30-inch torpedo tubes

3. What are the differences between 1942 ME 262 and Eurofighter in terms of maximum speed, crew and armament?

【参考词汇】Mach, speed of sound, 30/27 mm cannon, missiles

Delta Armoured vehicles

【军事知识】

1. Leopard 2 Main Battle Tank (MBT) 德制"豹"2型主战坦克

Leopard 2 main battle tank was developed by Krauss-Maffei in the early 1970s for the West German Army. The tank first entered service in 1979 and succeeded the earlier Leopard 1 as the main battle tank of the German Army. Various versions have served in the armed forces of Germany and twelve other European countries,

as well as several non-European nations, including Canada, Chile, Indonesia and Singapore. More than 3,480 Leopard 2s have been manufactured. It is armed with a 120 mm smoothbore cannon, and is powered by a V-12 twin-turbo diesel engine. The tank has the ability to engage moving targets while moving over rough terrain. It has a crew of four, including commander, gunner, loader and driver.

2. Warrior Infantry Fighting Vehicle (IFV) 英制"勇士"步兵战斗车辆

Warrior is one of the most widely deployed vehicles in the British Army. The Warrior family includes several variants such as the Infantry Fighting Vehicle, Infantry Command Vehicle, Repair & Recovery Vehicle and Observation Post Vehicle. Warrior is currently in service with the British Army and the Desert Warrior with the Kuwait Land Forces. It has the speed and performance to keep up with Challenger 2 main battle tanks over the most difficult terrain, and the firepower and armour to support infantry in the assault. They provide excellent mobility, lethality and survivability for the infantry and have enabled key elements from the Royal Artillery and Royal Electrical and Mechanical Engineers to operate effectively within the battlegroup.

3. FV432 Armoured Personnel Carrier (APC) 英制 FV432 型（履带式）装甲人员输送车辆

The FV432 Armoured Personnel Carrier is a variant of the British Army's FV430 series of armoured fighting vehicles. Since its introduction in the 1960s, it has been the most common variant, being used for transporting infantry on the battlefield. In the 1980s, almost 2,500 vehicles were in use, with around 1,500 now remaining in operation, mostly in supporting arms rather than front-line infantry service.

The FV432 is an all-steel construction. The FV432 chassis is a conventional tracked design with the engine at the front and the driving position to the right. Directly behind the driver position is the vehicle commander's hatch. There is a large round opening in the passenger compartment roof, which has a split hatch, and a side-hinged door in the rear for loading and unloading.

【词汇点拨】Words and Expressions

1. hull /hʌl/ *n.* the outer covering of a ship or boat; the lower part of an armoured fighting vehicle（坦克、舰艇等）壳体；车体

The explosion penetrated the hull.

2. main gun 主炮

This photo is of a parade with the main gun reversed to face the rear of the vehicle.

3. track /træk/ *n.* a moving band of metal links fitted around the wheels of a tank or other armoured vehicle, enabling it to move over soft or uneven ground 履带

The tank came off the road when it lost tracks.

4. turret /ˈtʌrɪt/ *n.* a small metal tower on a ship, plane or tank that can usually turn around and from which guns are fired（战舰、飞机或坦克的）回转炮塔；旋转枪架

She began to talk to the man in the turret of the car.

5. specification /ˌspesɪfɪˈkeɪʃn/ *n.* a detailed description of how sth. is, or should be, designed or made 规格；技术参数；技术指标

The barracks have been built exactly to our specifications.

【词汇点拨】Proper Names

1. **Leopoard 2** /ˈlepəd/ 德制"豹"2 型主战坦克
2. **Austria** /ˈɒstriə/ 奥地利
3. **Germany** /ˈdʒɜːməni/ 德国
4. **Greece** /griːs/ 希腊
5. **Holland** /ˈhɒlənd/ 荷兰
6. **Poland** /ˈpəʊlənd/ 波兰
7. **Spain** /speɪn/ 西班牙
8. **Sweden** /ˈswiːd(ə)n/ 瑞典
9. **Switzerland** /ˈswɪtsələnd/ 瑞士

【长难句解读】

1. Leopard 2 is currently in service with Austria, Germany, Greece, Holland, Poland, Spain, Sweden and Switzerland. (3-6)

➢ 句子分析：该句中的 service 指的是"兵役"的意思。常见搭配 be in service with sth. 意思为"在……服役"。

➢ 翻译：德制"豹"2 型主战坦克目前装备国家包括奥地利、德国、希腊、荷兰、波兰、西班牙、瑞典和瑞士。

2. It is 2.48 m high and 3.7 m wide. (3-6)

➢ 句子分析：英语中常用"be+ 基数词 + 单位词 + 形容词（high/ wide/ long/ deep）"或者"be+ 基数词 + 单位词 +in+ 长度或重量等的名词（height/width/length/depth/weight）"描述物体的高度、宽度、长度、深度、重量等。表示重量时，只能用 in weight 结构。

➢ 翻译：它高 2.48 米，宽 3.7 米。

【口语输出】

1. What do you know about Leopard 2?

【参考词汇】main battle tank, one of the best tanks in the world, be in service with, main armament, machine gun, a crew of four, weigh 55,150 kg

2. How many crew do Leopard 2, Warrior IFV and FV 432 have?

【参考词汇】a crew of four men, a crew of three men and seven men, a crew of two men and eight men

3. Which armoured vehicle is the fastest, Leopard 2, Warrior IFV or FV 432?

【参考词汇】a maximum road speed, 72 km/h, 75 km/h, 64 km/h

Echo Integrated skills: describing and defining

【军事知识】

1. The Black Hawk helicopter 美制"黑鹰"直升机

The UH-60 Black Hawk helicopter manufactured by Sikorsky Aircraft is the U.S. Army's primary medium-lift utility transport and air assault aircraft. It is a twin-engined medium-lift utility helicopter. It is equipped with a single 4-bladed rotor and a single 4-bladed tail rotor.

The UH-60A entered service with the U.S. Army in 1979, to replace the Bell UH-1 Iroquois as the Army's tactical transport helicopter. This was followed by the fielding of electronic warfare and special operations variants of the Black Hawk. Improved UH-60L and UH-60M utility variants have also been developed. Modified versions have also been developed for the U.S. Navy, Air Force, and Coast Guard. In addition to U.S. Army use, the UH-60 family has been exported to several nations.

The basic crew complement for the UH-60A is three: pilot, co-pilot and crew chief. Its rotor blades are resistant to AAA (anti-aircraft artillery) fire up to 23mm and are equipped with pressurized sensors capable of detecting loss of rotor pressurization (damage).

2. Hellfire anti-armour missile 美制"地狱火"反坦克导弹

Hellfire is an air-to-ground, laser-guided, subsonic tactical missile with significant anti-tank capacity. It can also be used as an air-to-air weapon. The missile is used to target armored vehicles, including tanks, bunkers, radar systems and antennas, communications equipment, soft targets, or hovering helicopters.

The Hellfire was first developed to target tanks for the AH-64 Apache attack helicopter, but it is now used by fixed-wing aircraft, helicopters, UAVs, ground and sea vessels, and land-based sites.

The Hellfire family of missiles includes Hellfire II (AGM-114) and Hellfire Longbow (AGM-114L), as well as different variants of these missiles.

3. Identification Friend or Foe (IFF) 敌友识别

Identification, Friend or Foe (IFF) was first developed during World War II, with the arrival of radar, and several friendly fire incidents. It is a radar-based identification system designed for command and control. It uses a transponder that listens for an interrogation signal and then sends a response that identifies the broadcaster. It enables military and civilian air traffic control interrogation systems to identify aircraft, vehicles or forces as friendly and to determine their bearing and range from the interrogator.

【词汇点拨】Words and Expressions

1. fabric /ˈfæbrɪk/ *n.* material made by weaving wool, cotton, silk, etc., used for making clothes, curtains, etc. and for covering furniture 织物；布料

The military uniform is made of fabric.

2. helmet /ˈhelmɪt/ *n.* a protective head covering 头盔

He locked his helmet into position with a click.

3. mortar /ˈmɔːtə(r)/ *n.* a simple indirect-fire weapon, which is designed to fire projectiles at very high trajectories 迫击炮

The 120 mm mortar has a range of 18,000 yards.

4. tracksuit /ˈtræksuːt/ *n.* a warm loose pair of trousers/pants and matching jacket worn for sports practice or as informal clothes 运动服

The soldier reached into the wardrobe and extracted another tracksuit.

5. cruise /kruːz/

① *n.* a journey by sea, visiting different places, especially as a holiday/vacation 乘船游览；航行

The aim of the cruise was to awaken an interest in and an understanding of military cultures.

② *v.* [usually + adv./prep.] (of a car, plane, etc.) to travel at a steady speed 以平稳的速度行驶；巡航

This light aircraft is cruising at 4,000 feet.

6. equip /ɪˈkwɪp/ *v.* to provide yourself/sb./sth. with the things that are needed for a particular purpose or activity 配备；装备

to be fully/poorly equipped 装备齐全／简陋

The centre is well equipped for canoeing and mountaineering.

7. rotor blade the long airfoil that rotates to provide the lift that supports a helicopter in the air 旋翼叶片；旋翼桨叶

The Black Hawk helicopter has four rotor blades.

8. logistical /ləˈdʒɪstɪk(ə)l/ *adj.(or logistic)* of or relating to logistics 后勤的

WHO provided technical and logistic support.

9. assault /əˈsɔːlt/

① *n.* the act of attacking an enemy position in order to take control of it 攻击；袭击；突击

A third brigade is at sea, ready for an amphibious assault.

② *v.* to use force in order to occupy an enemy position 攻击；袭击；突击

The enemy assaulted the small town.

10. evacuation /ɪˌvækjuˈeɪʃn/ *n.* the act of removing people from their homes because of danger and making them stay elsewhere until that danger is over 撤离；疏散

The evacuation is being organized at the request of the United Nations Secretary General.

11. calibre /ˈkælɪbə(r)/ *n.* the internal diameter of a gun barrel; the external diameter of a projectile 口径

The soldier was hit in the head by a 22-calibre bullet.

12. mount /maʊnt/ *v.* to fix sth. into position on sth., so that you can use it, look at it or study it 安置

The machine gun was mounted to the top of the truck.

【词汇点拨】Proper Names

1. **GPS (Global Positioning System)** 全球定位系统
2. **Sikorsky Aircraft Corporation** 西科斯基飞行器公司
3. **Connecticut** /kəˈnetɪkət/ 康涅狄格州（美国）
4. **UHF (Ultra High Frequency)** 超高频
5. **VHF (Very High Frequency)** 甚高频

【长难句解读】

1. It is designed to carry troops into battle and to serve as a logistical support aircraft, but all missions are possible—troop assault, combat support, combat service support. (3-9)

> 句子分析：该句中有两个并列分句，由表示转折的并列连词 but 连接，介绍了"黑鹰"直升机的设计用途。其中，第一个分句中使用了被动语态，有两个并列的不定式结构。

> 翻译：它的设计目的是运送部队参与战斗，并作为后勤支援飞机，但是也能够执行各类任务——突击作战、战斗支援、战斗勤务支援。

2. The helicopter can carry a weight of 9,000 pounds outside and can be armed with a variety of missiles, including the Hellfire anti-armour missile, and also rockets, machine guns and 20 mm cannons. (3-9)

> 句子分析：该句中的 be armed with sth. 意思为"用……武装""装备了……"。

> 翻译：该型直升机可在机舱外携带9000磅的物资装备，并装备多种型号导弹，包括"地狱火"反坦克导弹、火箭弹、机枪和20毫米口径自动机关炮。

3. That is why it is used to transport troops into the combat zone. It can also fly in almost any weather condition. (3-9)

> 句子分析：该句是主系表结构，表语为 why 引导的从句。

> 翻译：这就是它被用于向战区运送部队的原因。它几乎能在任何天气条件下飞行。

4. Finally, it is equipped with voice, satellite, UHF and VHF communication systems, and also IFF (which means identification friend or foe). (3-9)

> 句子分析：该句中的 be equipped with sth. 意思为"装备了……"。

> 翻译：最后，它还配备了语音、卫星、超高频和甚高频通信系统，以及敌我识别系统。

【口语输出】

1. How do you describe and define a beret?

【参考词汇】an item of uniform, be made of fabric, to cover one's head

2. What do you know about the Black Hawk helicopter?

【参考词汇】a crew of three, carry up to eleven fully equipped troops, weigh 11,780 pounds, a height of 16 feet and 9 inches, 162 knots, a range of 304 nautical miles, close to ground, a height of 19,000 feet

3. What are the main functions of the Black Hawk helicopter?

【参考词汇】carry troops into battle, serve as a logistical support aircraft, troop assault, combat support, combat service support, medical evacuation, search and rescue, command and control

4. What is the Black Hawk helicopter armed and equipped with?

【参考词汇】a variety of missiles, the Hellfire anti-armour missile, rockets, machine guns, 20 mm cannons, 7.62 mm or 50 calibre machine guns, voice, satellite, UHF and VHF communication systems, IFF

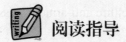

阅读指导

【文章导读】

This passage introduces the common armaments in the air force, the navy and the army, such as aircraft, naval vessels and ground vehicles. Aircraft can primarily be divided into two categories, that is, fixed-wing aircraft and rotary-wing aircraft. Naval vessels mainly include surface vessels and submarines. Finally, there are various ground vehicles for the army, including jeep, truck, APC, IFV and MBT.

【军事知识】

1. The US B-2 Spirit 美制"幽灵"B-2 型轰炸机

The Northrop B-2 Spirit, also known as the Stealth Bomber, is an American heavy strategic bomber, capable of delivering both conventional and nuclear munitions. Designed during the Cold War, it is a flying wing design with a crew of two, a pilot in the left seat and mission commander in the right. The first B-2 was publicly displayed on Nov. 22, 1988.

The B-2 Spirit is capable of all-altitude attack missions up to 50,000 feet, with a range of more than 6,000 nautical miles unrefuelled and over 10,000 nautical miles with one refuelling, giving it the ability to fly to any point in the world within hours.

2. The Eurofighter 2000/Typhoon 欧制"台风"2000型战斗机

The Eurofighter 2000/Typhoon (EF-2000) is a single-seat, twin-engined, canard delta wing, multirole fighter which was designed for the air-to-air, air-to-ground and tactical reconnaissance roles.

The Eurofighter 2000/Typhoon is a result of joint efforts on the part of UK, Germany, Italy and Spain to develop a 5th Generation fighter capable of engaging all contemporary threats. The aircraft was designed with Air Dominance as its primary mission, and ground attack being the secondary. To achieve better combat performance, the Eurofighter 2000/Typhoon had to be endowed with features like unstable design, carefree handling, advanced sensor suite and defensive counter-measures, ultra long-range missiles and a top-end cockpit.

3. The US C-130 Hercules 美制"大力神"C-130型运输机

The Lockheed C-130 Hercules is an American four-engine turboprop military transport aircraft designed and built by Lockheed. Capable of using unprepared runways for takeoffs and landings, the C-130 was originally designed as a troop, medevac, and cargo transport aircraft. The versatile airframe

has found uses in other roles, including as a gunship, for airborne assault, search and rescue, scientific research support, weather reconnaissance, aerial refueling, maritime patrol, and aerial firefighting. It is now the main tactical airlifter for many military forces worldwide. More than 40 variants of the Hercules, including civilian versions marketed as the Lockheed L-100, operate in more than 60 nations.

The C-130 Hercules has been in service with the US Air Force for more than half a century, fulfilling a wide range of operational missions in both peace and war situations. The C-130 Hercules is appropriately the most reliable military air transport vehicle in US military history.

4. The US Spruance-class destroyer 美军"斯普鲁恩斯"级驱逐舰

The Spruance-class destroyer was developed by the United States to replace a large number of World War II-built destroyers, and was the primary destroyer built for the U.S. Navy during the 1970s and 1980s. It was named in honor of United States Navy Admiral

Raymond A. Spruance, who successfully led major naval battles in the Asiatic-Pacific Theater during World War II.

First commissioned in 1975, the class was designed with gas-turbine propulsion, all-digital weapons systems, automated 5-inch guns. All 31 vessels of the Spruance-class entered service between 1975 and 1983.

The Spruance-class destroyer is 171 meters long, and 16.8 meters wide with a full load displacement of 8,040 tonnes.

Serving for three decades, the Spruance class was designed to escort a carrier group with a primary anti-submarine warfare mission, though in the 1990s, 24 members of the class were upgraded with the Mark 41 Vertical Launching System (VLS) for the Tomahawk surface-to-surface missile. And the last ship of the class was decommissioned in 2005. The Spruance-class destroyer was succeeded by the Arleigh Burke-class destroyer.

【词汇点拨】Words and Expressions

1. rotary /ˈrəʊtəri/ *adj.* moving in a circle around a central fixed point 旋转的；绕轴转动的

Invented in 1951, the rotary engine is a revolutionary concept in internal combustion.

2. blade /bleɪd/ *n.* one of the flat parts that turn around in an engine or on a helicopter（机器上旋转的）叶片；桨叶

rotor blades on a helicopter 直升机的旋翼

3. rotate /rəʊˈteɪt/ *v.* to move or turn around a central fixed point; to make sth. do this（使）旋转；转动

Stay well away from the helicopter when its blades start to rotate.

4. manoeuvrable /məˈnuːvərəbl/ *adj.* capable of moving easily over all types of terrains 可调动的；机动的

These jet fighters were extremely manoeuvrable.

5. reconnaissance /rɪˈkɒnɪsns/ *n.* an examination or survey of ground or a specific location, in order to plan an operation or task 侦查

The helicopter was returning from a reconnaissance mission.

6. paratrooper /ˈpærətruːpə(r)/ *n.* soldiers who are trained to be dropped by parachute into battle or into enemy territory 伞兵

The paratroopers descended slowly through the air.

7. humanitarian /hjuːˌmænɪˈteəriən/ *adj.* concerned with reducing suffering and improving the conditions that people live in 人道主义的

They are calling for the release of the hostages on humanitarian grounds.

8. relief /rɪˈliːf/ *n.* an act of providing assistance or support 救济；救助

A UN spokesman said that the mission will carry 20 tons of relief supplies.

9. aerial /ˈeəriəl/

① *n.* a metal wire, rod, mast or structure used in the receiving and transmission of radio signals 天线

② *adj.* in the air; existing above the ground 空中的

The aerial doesn't look very secure to me.

Weeks of aerial bombardment had destroyed factories and motorways.

10. refuel /ˌriːˈfjuːəl/ *v.* to fill sth., especially a plane, with fuel in order to continue a journey; to be filled with fuel（尤指给飞机）补充燃料，加燃料；加油

His plane stopped in Hawaii to refuel.

11. patrol /pəˈtrəʊl/

① *n.* an act of walking or driving around an area on a regular basis in order to deter or prevent illegal or hostile activity 巡逻；巡查

② *v.* to carry out a patrol 巡逻；巡查

Highway patrol officers closed the road.

Guards patrol the perimeter of the estate.

12. versatile /ˈvɜːsətaɪl/ *adj.* having many different uses 多用途的；多功能的

He's a versatile actor who has played a wide variety of parts.

13. escort

① /ˈeskɔːt/ *n.* a person, vehicle or aircraft of ship which accompanies an individual or group in order to protect them 护送者；护卫队；护卫舰（或车队、飞机）

Armed escorts are provided for visiting heads of state.

② /ɪˈskɔːt/ *v.* to act as an escort 护卫；护送

Policemen had to escort the referee from the football field.

14. engage /ɪnˈɡeɪdʒ/ *v.* to begin fighting with sb. 与（某人）交战；与（某人）开战

It could engage the enemy beyond the range of hostile torpedoes.

15. ballistic /bəˈlɪstɪk/ *adj.* connected with ballistics 弹道（学）的

Ballistic tests have matched the weapons with bullets taken from the bodies of victims.

16. warhead /ˈwɔːhed/ *n.* the explosive part of a missile（导弹的）弹头

The missile warhead hit the target, effecting a nuclear explosion.

17. configuration /kənˌfɪɡəˈreɪʃn/ *n.* an arrangement of the parts of sth. or a group of things; the form or shape that this arrangement produces 结构；构造

Configuration and functions of the digital TV receiver are introduced.

18. bulldozer /ˈbʊldəʊzə(r)/ *n.* a powerful vehicle with a broad steel blade in front, used for moving earth or knocking down buildings 推土机

The bulldozer leveled the mountain.

【词汇点拨】Proper Names

1. **The US B-2 Spirit** 美制"幽灵"B-2型轰炸机
2. **The Eurofighter 2000/Typhoon** 欧制"台风"2000型战斗机
3. **The US C-130 Hercules** 美制"大力神"C-130型运输机
4. **The US Spruance-class destroyer** 美军"斯普鲁恩斯"级驱逐舰

【参考译文】

飞机、车辆和海军舰船

空军、陆军和海军在执行任务时使用的飞机、地面车辆和军舰多达几十种。下面是一些主要的分类。

飞机

飞机主要分为固定翼飞机和旋转翼飞机两大类。固定翼飞机的机翼不能移动，比如客机和喷气式飞机；旋转翼飞机通过飞机的桨叶或机翼的旋转升空，比如直升机。飞机也可以根据执行任务的类型进行分类。

固定翼飞机

轰炸机：通常能够长距离飞行，并携带重型武器装备。最著名的现代轰炸机之一是美制"幽灵"B-2型轰炸机。因为其较难被雷达发现，也被称为"隐形轰炸机"。

战斗机：具有快速及高机动性的特点，通常用于执行作战和侦查任务。比如由英国、德国、意大利和西班牙联合制造的欧制"台风"2000型战斗机。

运输机：用于执行多样运输任务，从敌后空投伞兵到提供人道主义救援。比如美制"大力神"C-130型运输机。

空中加油机：用于飞机空中加油。

旋转翼飞机

攻击直升机：具有快速及高机动性的特点，通常装备机枪和导弹。

运输直升机：能够运送部队，有时运送大型车辆。

海军舰船

海军舰船可以分为水面舰船和水下潜艇。

水面舰船

水面舰船的形状和大小各不相同，大到航母，小到河流巡逻艇。

航空母舰：设计目的是用于运输喷气式战斗机和直升机，并为舰载机提供起降平台。

护卫舰：多用途、多功能舰船，可执行反潜作战、护卫和巡逻任务。

驱逐舰：主要用于发现和打击高速潜艇，但是也可以与其他舰船和岸基目标交战，比如美军"斯普鲁恩斯"级驱逐舰。

潜艇

弹道导弹潜艇：可携带带有核弹头的远程导弹。

攻击型潜艇：可携带常规导弹，用于发现、击毁敌方潜艇。比如美国"海狼"级潜艇。

地面车辆

地面车辆在形状、大小和构造上千差万别，比如装甲或非装甲、攻击型或运输型、轮式或履带式（比如推土机和拖拉机上的传送带）。下面是一些常见的类型：

吉普：轻型、非装甲、四人运输车辆，通常配有车载式机枪。

卡车（truck 为美式英语，lorry 为英式英语）：用于运送较重型的货物或部队。

装甲运兵车：战斗中用于运送部队（通常被叫作"战场出租车"），比如英制 FV432 型（履带式）装甲人员输送车辆。装甲运兵车有装甲车身和车载式机枪，通常可以运送八到十人。

步兵战车：装甲步兵营使用，被认为是轻型坦克，也可以运送少量士兵，比如英制"勇士"步兵战斗车辆。

主战坦克：装备较大型火炮的重型坦克，火力更为强悍且可以承受较严重的损伤，比如德制"豹"2 型主战坦克。

 词汇表

Alpha Military inventions

aircraft carrier 航空母舰

atomic 原子的；与原子有关的；原子能的；原子武器的

jet 喷气式飞机

fighter 战斗机；歼击机

jet fighter 喷气式战斗机

machine gun 机关枪；机枪

radar 雷达
submarine 潜艇
portable 便携式的；手提的；轻便的
explode 爆炸；爆破；爆裂

Bravo Aircraft, vehicles and naval ships
bomber 轰炸机
destroyer 驱逐舰
frigate （小型）护卫舰
transport aircraft 【航】运输机
fixed-wing aircraft 固定翼飞机
rotary-wing aircraft 旋转翼飞机
armoured 有装甲的
armoured vehicle 装甲车辆
naval 海军的
naval ship 军舰
underwater 水下的；在水下、在水中
vertical 竖的；垂直的；直立的
vertically 垂直地
column （人或车辆排成行移动的）长列，纵队
stealth 秘密行动；暗中活动
stealth bomber 隐形轰炸机
Infantry Fighting Vehicle (IFV) 步兵战斗车辆；步战车
Armoured Personnel Carrier (APC) 装甲运兵车

Charlie Military technology then and now
foot 英尺
inch 英寸
knot 节（速度单位，等于 1 海里 / 小时）
nautical mile 海里（合 1852 米）

convert （使）转变，转换，转化
multiply 乘；乘以
divide 除；除以
dive 下潜；潜到更深的水下
armament 武器；装备
torpedo 鱼雷
tube （金属、塑料、橡胶等制成的）管，管子
cannon 大炮；（飞机上的）自动机关炮
Mach 马赫，马赫数（常用于表示飞行速度，马赫数 1 即 1 倍声速）

Delta Armoured vehicles
hull （坦克、舰艇等）壳体；车体
main gun 主炮
tracks 履带
turret （坦克的）炮塔
specification 规格；技术参数；技术指标

Echo Integrated skills
fabric 织物；布料
helmet 头盔
mortar 迫击炮
tracksuit 运动服
cruise 航行；巡航
equip 配备；装备
rotor blade 旋翼叶片；旋翼桨叶
logistical 后勤的
assault 攻击；袭击；突击
evacuation 撤离；疏散
calibre 口径
mount 安置

Foxtrot Aircraft, vehicles and naval ships

rotary 旋转的；绕轴转动的
blade （机器上旋转的）叶片；桨叶
rotate （使）旋转；转动
manoeuvrable 可调动的；机动的
reconnaissance 侦查
paratrooper 伞兵
humanitarian 人道主义的
relief 救济；救助
aerial 空中的；天线

refuel （尤指给飞机）补充燃料；加燃料；加油
patrol 巡逻；巡查
versatile 多用途的；多功能的
escort 护卫；护送；护送者；护卫队
engage 与（某人）交战；与（某人）开战
ballistic 弹道的
warhead （导弹的）弹头
configuration 结构；构造
bulldozer 推土机

 拓展学习

Artificial Intelligence, Machine Learning and Big Data

AI, ML and Big Data have numerous applications in security and defence, including providing decision support in contexts such as nuclear security. AI, ML and Big Data may provide several potential general opportunities for defence:

- The control of information flows and access to information may provide a significant advantage to any actor in the context of the future battlefield. As such, AI, ML and Big Data systems present significant opportunities for EU MS to establish a strategic advantage in data access and management over a potential adversary as well as breaking through an adversary's Observe, Orient, Decide and Act (OODA) loop.

- At the strategic level, the capacity of AI, ML and Big Data for rapid data processing and inference may produce a qualitative edge for decision-makers. For example, strategic-level AI may support decision-making through challenging accepted wisdom concerning relationships between various factors relevant to strategy-making or identifying key vulnerabilities in an adversary.

- Through enabling autonomy, AI and ML bring opportunities for defence through increased stealth and rapid analysis for defensive action on the battlefield.

- AI can provide opportunities for improving organisational innovativeness and the adaptability of public sector organisations, including defence, to fast-paced changes in the socio-economic, geo-political and security environment.

In addition to these general opportunities, specific examples of AI, ML and Big Data military applications can be identified, particularly in the areas of ISR, decision-support, command-and-control and logistics.

The ability, however, of AI, ML and Big Data systems to rapidly process and make inferences from large amounts of data also indicate various challenges for future defence. The speed of decision-making may substantially increase in the event of wide-spread uses of AI and ML-enabled systems on the future battlefield. In particular, the employment of networked agents with autonomous decision-making abilities may enable extremely rapid sequential action, even in uncertain operating environments.

AI and ML are also being integrated across many offensive and defensive weapons and cyber-physical systems. This includes the integration of AI into offensive cyber capabilities, as well as advanced remote sensing, precision-guided munitions and hypersonic weapons. The integration of AI with advanced weapons systems, including systems based on biotechnologies and systems for the delivery of non-kinetic effects, could significantly advance the potency of weapons systems as well as their lethality and speed of decision-making concerning their deployment.

While AI may improve the ability of EU MS forces to establish an understanding of adversary actions and capabilities, AI may also challenge the attribution of attacks or hostile action. AI applications to programming automation may, for example, lead to uses of AI-generated malicious code, which would be significantly more difficult to trace and attribute. When being used for covert purposes or influence operations, AI and ML may be used to create "deep fake" images and videos or operation of sophisticated social media "bots", challenging the ability of armed forces to navigate an increasingly complex information environment.

Future trends with regards to uses of AI on the future battlefield may be underpinned by the potential of AI to spark unmitigated arms race-like escalations in military AI, including AI-enabled weapons systems. Existing literature points to the inherent nature of military technology investments as creating situations of a security dilemma, fostering uncertainty and incentivizing increased investment in increasingly autonomous systems resulting in an armament spiral.

While its development and adoption remain highly unlikely in the 2040 timeframe, if actualised, AGI would present additional and far-reaching challenges in the battlefield context as well as for societal resilience more broadly. This is due particularly to the uncertainty as to the alignment of an AGI agent with established human values and intentions, including ethical standards and commonly accepted norms of conduct. The ability of AGI to replicate human-like learning abilities may also drastically magnify the challenges associated with inexplicability of AI models, significantly reducing predictability and increasing the risk of accidental or maliciously-induced conflicts or collisions involving artificial agents. (651 words)

(*Selected from Innovative technologies shaping the 2040 battlefield - Publications Office of the EU*)

【Notes】

1. **strategic** 战略性的；战略上的
2. **adversary**（辩论、战斗中的）敌手；对手
3. **stealth** 秘密行动
4. **logistics** 后勤；物流；组织工作
5. **substantially** 非常；大大地
6. **sequential** 按次序的；顺序的；序列的
7. **munition** 军火（包括弹药、装备及库藏）
8. **hypersonic** 高超声速的
9. **non-kinetic** 非动力学
10. **potency** 威力；影响
11. **lethality** 致命性
12. **malicious** 怀有恶意的；恶毒的
13. **covert** 秘密的；隐蔽的；暗中的
14. **unmitigated** 完全的；十足的，彻底的（通常指坏事）
15. **escalation**（战争等事态）升级
16. **incentivize** 刺激；鼓励；使人有积极性
17. **spiral** 逐渐加速上升（或下降）
18. **drastically** 猛烈地；力度大地
19. **magnify** 扩大；增强
20. **inexplicability** 无法说明；费解
21. **EU (European Union)** 欧盟
22. **MS (Member State)** 成员国
23. **ISR (Intelligence, Surveillance, Reconnaissance)** 情报、监视和侦查

24. Observe, Orient, Decide and Act (OODA): The OODA loop (an acronym that stands for Observe, Orient, Decide, Act) is a four-step approach developed by military strategist John Boyd. This looping concept referred to the ability possessed by fighter pilots that allowed them to succeed in combat. It is now used by the U.S. Marines and other organizations. Nation-states around the world use the OODA Loop as part of their military strategy. 观察、调整、决策以及行动循环（也称包以德循环）

单元练习

I. Label the picture

1._____ 2._____ 3._____

4._____ 5._____ 6._____

II. Multiple choice

1. _____ is a type of ship. It can travel underwater.

A. Fighter   B. Submarine   C. Tank   D. Frigate

2. A(n) _____ is a type of armoured vehicle. It can carry a section of soldiers.

A. IFV   B. MBT   C. jeep   D. truck

3. A _____ is a type of aircraft. It can take off vertically, but it can't fly very high.

A. bomber   B. fighter   C. helicopter   D. radar

4. To convert nautical miles into kilometres, multiply by _____.

A. 9   B. 2.54   C. 1.609   D. 1.852

5. Which of the following is the unit of speed?

A. Nautical mile　　B. Fahrenheit　　C. Knot　　D. Inch

6. Which part is the main armament of a tank?

A. Hull　　B. Tracks　　C. Turret　　D. Main gun

7. A _____ is a type of military aircraft. It is designed to carry out a variety of transport missions.

A. fighter　　B. bomber　　C. tanker　　D. cargo

8. Which of the following aircraft is used for aerial refuelling?

A. Fighter　　B. Tanker　　C. Cargo　　D. Bomber

9. Which of the following isn't the function of the frigates?

A. To transport aircraft.　　B. To carry out anti-submarine combat.

C. To escort other ships.　　D. To patrol.

10. Which is called "a battle taxi"?

A. Jeep　　B. MBT　　C. APC　　D. IFV

III. Translation: English to Chinese

1. Some people think that the Battle of Cambrai in 1917 was the first battle with tanks, but the first tanks were used in the Battle of Flers-Courcelette in September 1916.

2. An aircraft carrier is a type of naval ship. Aircraft can take off and land on the ship.

3. The German ME 262 was the world's first operational jet fighter. It could fly at 870 km/h, but it couldn't fly at the speed of sound.

4. The Leopard 2's main armament is a 120 mm gun. It also has a 7.62 mm machine gun. It has a crew of four. Leopard has a maximum road speed of 72 km/h.

5. The Black Hawk can fly very close to the ground and can tolerate small arms fire. And it can't be detected very easily.

IV. Translation: Chinese to English

高超声速武器主要是指以高超声速飞行技术为基础、飞行速度超过5倍声速的武器，包括导弹、轰炸机、侦查机等。这种以快见长的新型武器，被诸多军事专家认为将从一定程度上改变未来战争模式。近年来，美俄等世界军事大国在高超声速武器领域展开激烈角逐，启动了多个高超声速武器研发项目，一些项目取得重大突破。

V. Oral practice

1. What are the functions of carriers, frigates and destroyers?

2. What are the main categories of aircraft, naval vessels and ground vehicles?

附录：Unit 3 练习答案

I. Label the picture

1. bomber　　　　2. tank　　　　　3. radio

4. aircraft carrier　5. radar　　　　6. submarine

II. Multiple choice

1-5 BACDC　　　　　6-10 DDBAC

III. Translation: English to Chinese

1. 有人认为1917年的康布雷战役是首次使用坦克参战的战役，但实际上坦克首次投入战斗是1916年9月的弗莱尔－库尔瑟莱特战役。

2. 航空母舰是一种海军舰艇。飞机可以在舰上起降。

3. 德制ME 262型喷气式战斗机是世界上第一架实战喷气式战斗机。它能以每小时870千米的速度飞行，但它不能以声速飞行。

4. 德制"豹"2型主战坦克的主要武器是一门120毫米大炮。它还配有一挺7.62毫米机关枪，可携带四名乘员。其最大行驶速度为每小时72千米。

5. "黑鹰"直升机可以近地飞行，能承受轻武器攻击，且不易被发现。

IV. Translation: Chinese to English

Hypersonic weapons mainly refer to weapons based on hypersonic flight technology and flying at a speed of more than 5 times the speed of sound, including missiles, bombers, reconnaissance planes, etc. This new type of weapon, which is known for its speed, is believed by many military experts to change the future war mode to a certain extent. In recent years, the world's military powers such as the United States and Russia have launched a fierce competition in the field of hypersonic weapons, and launched a number of research and development projects of hypersonic weapons, some of which have made major breakthroughs.

V. Oral practice

1. Carriers, frigates and destroyers are the main warship types for modern navies.

Carriers are designed to transport fighter jets and helicopters and provide a platform for their operations. Frigates are versatile multi-purpose ships that can be used for anti-submarine combat, escort and patrol operations. Destroyers are designed primarily to find and destroy high-speed submarines but can also engage (enter into combat with) other ships and shore targets, e.g. the US Spruance-class destroyer.

2. The air force, navy and army are equipped with various aircraft, naval vessels and ground vehicles. Aircraft can be divided into two main categories: the fixed-wing aircraft and the rotary-wing aircraft. Considering their function, the fixed-wing aircraft include bombers, fighters, cargo and tankers, while the other category includes attack helicopters and cargo helicopters.

Naval vessels can be divided between surface vessels and submarines. Carriers, frigates, destroyers cruise on the sea and belong to the surface vessels. Submarines include ballistic submarines and fleet or attack submarines because of their different armament.

Ground vehicles vary greatly in shape, size and configuration, such as armoured or unarmoured, attack or transport, wheels or tracks. The common types are jeep, truck, Armoured Personnel Carrier (APC), Infantry Fighting Vehicle (IFV) and Main Battle Tank (MBT).

Unit 4　War Games

单元导学清单

模块	主题	学习目标	任务	军事知识	核心词汇
Alpha	军演的基本概念	1）了解军事演习的基本定义和类型；2）掌握描述军演的相关词汇。	Task 1	■ Military Exercise	joint, multinational, parachutist, airborne, amphibious
	"闪亮之星"联合军演	1）了解联合军演"闪亮之星"的基本情况；2）掌握描述军演的相关句型和基本结构。	Task 2 Task 3 Task 4	■ Military Exercise	staff, terrain, interior
Bravo	地形地貌	1）了解常见的地形地貌；2）掌握描述地形地貌的相关词汇；3）能够通过图例准确识别地形地貌。	Task 1 Task 2 Task 3 Task 4	■ Map symbols	ford, marsh, saddle, spur, jungle, ravine, scrub, oasis, draw, pass, ridge, dune, vegetation, mosque
	地图定位	1）了解军事地图的基本概念和使用方法；2）掌握地图上定位的相关词汇和表达；3）能够准确描述地图上的某一定位。	Task 5 Task 6 Task 7	■ Military maps	grid, vicinity, minefield, objective, deploy, flank, rear
Charlie	连队指挥官下达任务1	1）了解下达任务的基本流程；2）熟练掌握相关的词汇与术语；3）能够运用所学知识描述敌我双方的位置部署。	Task 1 Task 2 Task 3 Task 4 Task 5	■ Military briefings	recce, protective, aerial, brief, briefing
Delta	连队指挥官下达任务2	1）了解下达任务的主要内容和结构；2）熟练掌握相关的词汇和表达；3）能够运用所学知识进行简单的口头任务下达。	Task 1 Task 2 Task 3 Task 4 Task 5	■ Mission briefings	assault, reserve, execute, assembly, secure, column, standby

续表

模块	主题	学习目标	任务	军事知识	核心词汇
Echo	坠落事故简报	1）了解事故简报的基本结构；2）熟练掌握相关的词汇和表达。	Task 1 Task 2 Task 3 Task 4 Task 5	■ Landing zone	travel, crew, co-pilot, shaken, bleed, unconscious, senior, compass
Foxtrot (Review)	识图	1）了解识图的重要性、军事地图的基本结构和使用方法；2）熟练掌握相关的词汇和表达；3）能够根据军事地图熟练描述位置信息。	Task 1 Task 2	■ The grid system ■ Terrain features	rank, impassable, landscape, co-ordinate, reference, specify, precision, subsection, terminology, crest, horizontal

 听说指导

Alpha　Exercise Bright Star

【军事知识】

Military Exercise 军事演习

Exercises are important tools through which the alliance tests and validates its concepts, procedures, systems and tactics, either exploring the effects of warfare or testing strategies without actual combat. More broadly, they enable militaries and civilian organisations deployed in theatres of operation to test capabilities and practise working together efficiently in a demanding crisis situation. They serve the purpose of ensuring the combat readiness of garrisoned or deployable forces prior to deployment from a home base.

军事演习（military exercise）简称演习。部队在导演部组织和想定情况诱导下进行的作战指挥和行动的演练，是军事训练的高级形式。按层次，分为战略演习、战役演习、战术演习；按对象，分为首长机关演习、实兵演习；按场所，分为室内演习、现地演习；按形式，分为单方演习、对抗演习；按目的，分为检验性演习、研究性演习和示范性演习；还可以按演习内容划分。（中国人民解放军军语（全本）2011.12）

【词汇点拨】Words and Expressions

1. joint /dʒɔɪnt/ *adj.* [only before noun] involving two or more people together　联合的；共同的

The joint exercise follows a series of smaller steps to break the ice, including a joint mountaineering expedition and joint naval exercises.

2. multinational /ˌmʌltiˈnæʃnəl/ *adj.* including or involving several countries or individuals of several nationalities 多国的

1,500 troops were sent to join the multinational force.

3. parachutist /ˈpærəʃuːtɪst/ *n.* a person who uses a parachute 跳伞者；跳伞运动员；伞兵

Parachutist Training Camp 伞兵训练营

4. airborne /ˈeəbɔːn/ *adj.* [only before noun] (of soldiers) trained to jump out of aircraft onto enemy land in order to fight 空降的

an airborne division 空降师

5. amphibious /æmˈfɪbiəs/ *adj.* (of military operations) involving soldiers landing at a place from the sea 两栖作战的；登陆的

an amphibious assault 两栖突袭

6. staff /stɑːf/ *n.* [treated as sing. or pl.] a group of officers assisting an officer in command of an army formation or administration headquarters 工作班子；（军队的）全体参谋人员

a staff officer 参谋

7. terrain /təˈreɪn/ *n.* [mass noun] a stretch of land, especially with regard to its physical features 地面；地形、地势；地带

digital terrain model 数字地形模型

They were delayed by rough terrain.

8. interior /ɪnˈtɪəriə(r)/ *n.* the inland part of a country or region（国家或地区的）内地，内陆

the plains of the interior 内陆平原

【词汇点拨】**Proper Names**

1. Alexandria /ˌælɪɡˈzɑːndriə/ 亚历山大（埃及海滨城市）

2. Mediterranean /ˌmedɪtəˈreɪniən/ 地中海

【长句解读】

1. Bright Star is a joint exercise with the participation of ground forces, air forces and naval forces. (4-1)

> 句子分析：该句中 with the participation of 表示"有……，由……参与"。ground

原意为"地、地面"，ground forces 指地面部队或陆军，air forces 指空军，naval forces 指海军。

➤ **翻译**："明亮之星"是陆军、空军、海军参加的联合演习。

2. In Command Post Exercises (CPXs) there are no troops - commanders and their staff practice command and communications using computers. (4-1)

➤ **句子分析**：该句中 practice 为及物动词，表示"练习、演练"。

➤ **翻译**：在指挥所演习 (CPXs) 中，没有部队参演——指挥官和参谋使用计算机进行指挥、通信演练。

【口语输出】

1. Could you explain what the multinational joint exercise is? Please give an example.

【参考词汇】with the participation of, ground forces, air forces, naval forces, terrains

2. What are the differences between CPXs and FTXs?

【参考词汇】real operation, troops, staff, military skills, command, communication

Bravo Location and terrain

【军事知识】

1. Map symbols【测】地图符号、图例

A map symbol or cartographic symbol is a graphical device used to visually represent a real-world feature on a map, working in the same fashion as other forms of symbols. Map symbols represent the physical features of land and help the map's reader gain an acute awareness of his surroundings. The most common map symbols include contour lines, buildings, water features, forests, etc. Contour lines are indicators of elevation and each contour line has a number beside it that indicates "feet above sea level".

2. Military maps 军事地图

Military maps have grids, which are essentially two sets of parallel lines. One set is called Eastings and runs from the South to the North with numbers increasing Eastwards. The other set is called Northings and goes from the West to the East with numbers rising Northwards. The sets of lines form small squares called grid squares. The squares are numbered, which starts from the bottom left-hand corner. The numbers make up the digits in coordinates, and this is what specifies a point location.

【词汇点拨】Words and Expressions

1. ford /fɔ:d/ *n.* a shallow place in a river where it is possible to drive or walk across 浅滩，可涉水而过的地方

The hikers carefully crossed the ford.

2. marsh /mɑ:ʃ/ *v.* [C,U] an area of low land that is always soft and wet because there is nowhere for the water to flow away to 湿地；沼泽；草本沼泽

Cows were grazing on the marshes.

3. saddle /ˈsædl/ *n.* something resembling a saddle in appearance, function, or position, in particular a low part of a ridge between two higher points or peaks 鞍状物，（山的）鞍部

We crossed the saddle of the mountain to continue our hike.

4. spur /spɜ:(r)/ *n.* an area of high ground that sticks out from a mountain or hill 山嘴；尖坡；支脉

The trail led us to the spur of the mountain.

5. jungle /ˈdʒʌŋgl/ *n.* [C,U] an area of land overgrown with dense forest and tangled vegetation, typically in the tropics （尤指热带的）丛林地带

The area is covered in dense jungle.

6. ravine /rəˈvi:n/ *n.* a deep and narrow valley with steep sides 沟壑；溪谷

The bus overturned and fell into a ravine.

7. scrub /skrʌb/ *n.* (also scrubland) [U] an area of dry land covered with small bushes and trees 硬叶灌丛带；低矮灌木丛林地

There is an area of scrub beside the railway line.

8. oasis /əʊˈeɪsɪs/ *n.* an area in the desert where there is water and where plants grow （沙漠中的）绿洲

Suddenly, an oasis started up before us.

9. draw /drɔ:/ *n.* a gully shallower than a ravine 凹地，山沟，冲沟

The hikers decided to set up camp in the draw.

10. pass /pɑ:s/ *n.* a road or way over or through mountains 关口；关隘；山路

The pass over the mountain was open again after the snows.

11. ridge /rɪdʒ/ *n.* a narrow area of high land along the top of a line of hills; a high pointed area near the top of a mountain 山脊；山脉

We were walking along the ridge for a whole day.

12. dune /djuːn/ *n.* (**sand dune**) a small hill of sand formed by the wind, near the sea or in a desert（风吹积成的）沙丘

A dune field covers some of the ground.

13. vegetation /ˌvedʒəˈteɪʃn/ *n.* [U] plants in general, especially the plants that are found in a particular area or environment（统称）植物；（尤指某地或环境的）植被

The chalk cliffs are mainly sheer with little vegetation.

14. mosque /mɒsk/ *n.* a building in which Muslims worship 清真寺

Hundreds of people packed into the mosque.

15. grid /ɡrɪd/ *n.* a pattern of squares on a map that are marked with letters or numbers to help you find the exact position of a place（地图上的）坐标方格

The grid reference is C8.

16. vicinity /vəˈsɪnəti/ *n.* [sing.] the area near or surrounding a particular place 周围地区；邻近地区；附近

Crowds gathered in the vicinity of the square.

17. minefield /ˈmaɪnfiːld/ *n.* an area planted with explosive mines 雷区

They were skirmishing close to the minefield now.

18. objective /əbˈdʒektɪv/

① *n.* a thing aimed at or sought; a goal 目的，目标

Our objective must be to secure a peace settlement.

② *adj.* not influenced by personal feelings or opinions; considering only facts 客观的；就事论事的；不带个人感情的

I find it difficult to be objective where he's concerned.

19. deploy /dɪˈplɔɪ/ *v.* move (troops) into position for military action 部署（部队）

Forces were deployed at strategic locations.

20. flank /flæŋk/ *n.* the right or left side of a body of people such as an army, a naval force, or a soccer team（陆军、海军或足球队的）侧翼，翼侧

The assault element, led by Captain Ramirez, opened up from their right flank.

21. rear /rɪə(r)/

① *adj.* [only before noun] at or near the back of sth. 后面的；后部的

She was shepherded by her guards up the rear ramp of the aircraft.

② *n.* the space or position at the back of something or someone 后方

They set off, two men out in front as scouts, two behind in case of any attack from the rear.

【长句解读】

1. There are some friendly forces about 10 km north of here. (4-3)
   > 句子分析：该句中 friendly forces 意为"友军"，north of 意为"在……以北"。
   > 翻译：友军位于此处以北约 10 千米处。

2. Our current location is about 10 km south of the objective. (4-3)
   > 句子分析：该句中 current 作形容词，此处意为"目前的，现在的"，south of 意为"在……以南"。
   > 翻译：我们现在的位置在目标以南约 10 千米处。

3. 1st platoon will deploy in firing position at grid 675798. (4-4)
   > 句子分析：该句中的 firing position 意为"开火、射击位置"，at grid... 意为"在某个坐标位置"。
   > 翻译：一排部署到坐标 675798 的射击位置。

4. 2nd platoon will assault the position from the right flank and 3rd platoon will remain in reserve 100 meters to the rear of the LZ. (4-4)
   > 句子分析：该句中的 from the right flank 表示"从右翼（进攻）"，reserve 此处意为"后备部队"，缩写 LZ 代表 landing zone 意为"（飞机）着陆地区"。
   > 翻译：2 排从右翼攻击该阵地，3 排在离登陆区后方 100 米做后备力量。

【口语输出】

Please describe the features on the map on the page 301.

【参考词汇】wood, hill, bridge, river, road, at grid, south of, north of

Charlie　Company commander's briefing 1

【军事知识】

Military briefings　军事会议

Military briefings are designed to present selected information to commanders, staffs and other audiences in a clear, concise and expedient manner. The types of military briefings are dictated by purpose. There are four basic types: the information briefings, the decision briefings, the staff briefings, and the mission briefings.

【词汇点拨】Words and Expressions

1. recce /ˈreki/ *n.* (BrE) (informal) = reconnaissance 侦查

a quick recce of an area 对一个区域的快速侦查

2. protective /prəˈtektɪv/ *adj.* [only before noun] providing or intended to provide protection 保护的；防护的

Workers should wear full protective clothing.

3. aerial /ˈeəriəl/

① *n.* a rod, wire, or other structure by which signals are transmitted or received as part of a radio or television transmission or receiving system 天线

② *adj.* [attrib.] existing, happening, or operating in the air; coming or carried out from the air, especially using aircraft 在空中（存在、发生或进行）的；（尤指用飞机）来自空中的，在空中进行的

an aerial battle 空战

aerial bombardment of civilian targets 对民用目标的空中轰炸

4. brief /bri:f/ *v.* ~ sb. (on/about sth.) to give sb. information about sth. so that they are prepared to deal with it 给（某人）指示；向（某人）介绍情况

The officer briefed her on what to expect.

5. briefing /ˈbri:fɪŋ/ *n.* a meeting for giving information or instructions; the information or instructions given 情况介绍会，指令发布会；简报，情况介绍

Captain Trent gave his men a full briefing.

【词汇点拨】Proper Names

Communications Center 通信中心

【长句解读】

1. I'm here to brief you on the operation order for our task. (4-6)

   ➤ 句子分析：该句中 operation 意为"行动"。

   ➤ 翻译：我来讲一下这次任务的行动指令。

2. Take out pen and paper and hold all questions until the end. (4-6)

   ➤ 句子分析：该句中 hold all questions until the end 意为"有问题留到讲完再问"。

   ➤ 翻译：拿出纸和笔，有任何问题等讲完再问。

3. A platoon-size unit is located at hill 120, in the vicinity of grid 1637. (4-6)

➤ **句子分析**：句中 a platoon-size unit 表示"一支排规模的部队"，in the vicinity of 表示"在……附近"。

➤ **翻译**：一支排规模的部队位于 120 山地，在坐标网格 1637 附近区域。

【口语输出】

Work in pairs to create dialogue with the location information. To get familiar with the description of specific location on the military map.

【参考词汇】is located at..., at grid..., in the vicinity of..., south of...

Delta　Company commander's briefing 2

【军事知识】

Mission briefing　任务部署会

The mission briefings are used under operational conditions to provide information, to give specific instructions, or to instill an appreciation of a mission. It is usually presented by a single briefing officer, who may be the commander, an assistant, a staff officer, or a special representative. Mission briefings serve to convey critical mission information not provided in the plan or order to individuals or small units. The mission briefing reinforces orders, provides more detailed requirements and instructions for each individual, and ensure participants know the mission objective, their contribution to the operation, problems they may confront, and ways to overcome them.

【词汇点拨】Words and Expressions

1. assault /əˈsɔːlt/

① *v.* [with obj.] make a physical attack on 武力攻击，袭击

He has been charged with assaulting a police officer.

② *n.* a military attack or raid on an enemy position（对敌方的）军事攻击

A group of 80 attackers launched an all-out assault just before dawn.

2. reserve /rɪˈzɜːv/ *n.* an extra military force, etc. that is not part of a country's regular forces, but is available to be used when needed 预备役部队；后备部队

in reserve means available to be used in the future or when needed 储备；备用

200 police officers were held in reserve.

3. execute /ˈeksɪkjuːt/ *v.* [with obj.] carry out or put (a plan, order, or course of action) into effect 执行；实行，实施（计划、命令、行动方针）

Check that the computer has executed your commands.

4. assembly /əˈsembli/ *n.* [mass noun] the action of gathering together as a group for a common purpose 集合，集会

an assembly point 集合地点

5. secure /sɪˈkjʊə(r)/

① *v.* protect against threats; make safe 保护；使安全

The government is concerned to secure the economy against too much foreign ownership.

② *adj.* fixed or fastened so as not to give way, become loose, or be lost; not subject to threat, certain to remain or continue safe and unharmed 牢固的，结实的；无威胁的，安全的

a secure job/income 稳定的工作/收入

6. column /ˈkɑ:ləm/ *n.* one or more lines of people or vehicles moving in the same directio（人或交通工具的）队；行

A column of tanks moved north-west.

7. standby /ˈstændbaɪ/ *n.* [mass noun] readiness for duty or immediate deployment 随时待命

We downgraded the military alert from emergency to standby.

【长句解读】

1. Our forces will be organized into three elements. (4-8)

➢ 句子分析：该句中 three element 表示"三个部分"。

➢ 翻译：我们这支部队将分为三部分。

2. We will move to the attack position and firing position on foot and in a column formation. (4-9)

➢ 句子分析：该句中 on foot 意为"步行"，in a column formation 表示"成纵队（行进）"。

➢ 翻译：我们成纵队步行到达进攻地点和射击地点。

3. A recce party with the company commander will check the area south of the target. (4-9)

➢ 句子分析：句中 a recce party 表示"侦查队"。

➢ 翻译：由连长带领的侦查队检查目标以南区域。

4. 3rd platoon will remain in a standby position 100 m to the rear of the attack position. (4-9)

➢ 句子分析：句中 in a standby position 表示"待命（位置）"。

➢ 翻译：三排在攻击阵地后方 100 米处待命。

5. We will then establish a defensive position on the objective. (4-9)

➢ 句子分析：句中 defensive position 表示"防御阵地"。

➢ 翻译：然后我们在目标位置建立防御阵地。

【口语输出】

What are the basic elements of a mission briefing?

【参考词汇】mission, organization, execution, brief you on..., be located at..., in the rear of..., position, move

Echo Crash landing

【军事知识】

Landing zone（飞机）着陆区

In military terminology a landing zone (LZ) is an area where aircraft can land. A helicopter requires a relatively level, cleared, circular area from 25 to 100 meters in diameter for landing. This depends on the type of helicopter. The area around the landing point must be cleared of all trees, brush, stumps, or other obstacles that could damage the helicopter. Generally, a helicopter requires more landing area during darkness than during daylight. Considerations such as helicopter type, nature of load, climate, and visibility affect what size landing point is used for a particular landing site.

【词汇点拨】**Words and Expressions**

1. travel /ˈtrævl / v. to go or move at a particular speed, in a particular direction, or a particular distance（以某速度、朝某方向或在某距离内）行进，转送，传播

The troops travel at 50 miles an hour.

2. crew /kruː/ n. [sing. or pl.] a group of people who work on and operate a ship, boat, aircraft, or train（轮船、飞机等上面的）全体工作人员

None of the passengers and crew were injured.

3. co-pilot n. a second pilot in an aircraft（飞机）副驾驶员

The co-pilot was at the controls when the plane landed.

4. shaken /ˈʃeɪkən / adj. [not usually before noun] shocked, upset or frightened by sth. 震惊；烦恼；恐惧

The survivors were ashen-faced and visibly shaken.

5. bleed /bliːd / v. [no obj.] lose blood from the body as a result of injury or illness 出血，流血

He was bleeding from a gash on his head.

6. unconscious /ʌnˈkɒnʃəs/ *adj.* in a state like sleep because of an injury or illness, and not able to use your senses 无知觉的；昏迷的；不省人事的

They found him lying unconscious on the floor.

7. senior /ˈsiːniə(r)/ *n.* a person who is higher in rank or status 级别（或地位）较高者；上级；上司

She felt unappreciated both by her colleagues and her seniors.

8. compass /ˈkʌmpəs/ *n.* [C] an instrument for finding direction, with a needle that always points to the north 罗盘；罗经；指南针；罗盘仪

a map and compass 地图和指南针

【词汇点拨】**Proper Names**

1. Captain Garcia 加西亚上尉
2. SKALE 虚构地名
3. Camp WHISKY（代号）W 营地
4. Camp YANKEE（代号）Y 营地

【长句解读】

1. From here, you will fly over the village of SKALE first and then about 10 km further north you will see the landing zone on the hill, right in front of you. (4-10)

➤ 句子分析：该句中 fly over 意为"飞过……"。

➤ 翻译：从这起飞，首先飞越 SKALE 村，然后再向北 10 千米就会看到降落区，位于你面前的山上。

2. You are travelling in a military helicopter from Camp WHISKY to Camp YANKEE, located 150 km to the south. (4-10)

➤ 句子分析：该句中过去分词短语 located 150 km to the south 做定语，修饰 Camp YANKEE。

➤ 翻译：你乘坐军用直升机从 W 营地飞往 Y 营地，Y 营地位于向南 150 千米处。

3. You are the most senior and you take command. (10-10)

➤ 句子分析：句中 take command 意思为"（接过）指挥权，指挥；挂帅"。

➤ 翻译：你的级别最高，你来指挥。

【口语输出】

1. What happened to the speaker's aircraft? And where did it crash according to the

description?

【参考词汇】smoke, crash, at grid of..., in the vicinity of...

2. If you were the speaker, what should you do?

【参考词汇】equipment, compass, radio, recue, on foot, injured

 阅读指导

【文章导读】

This passage mainly focuses on the important military skill of map reading, including the grid system and the vocabulary of terrain features. The definition and function of the grid system are elaborated, followed by the detailed explanation of how to use the four-figure references and six-figure references to locate a specific position on the military map.

【军事知识】

1. The grid system 坐标系网格

A map grid enables you to find or give any position on a map by the use of a grid reference. The grid is the name given to the two sets of parallel lines on a map which together form grid squares. One set of parallel lines runs south to north with numbers increasing eastwards (EASTINGS) and the other, at right angles, runs west to east with numbers increasing northwards (NORTHINGS). As in the diagram below, grid lines and grid squares are numbered from the bottom left-hand corner of the map.

To read a grid reference, first read the figures left to right along the EASTINGS lines, then the figures bottom to top up the NORTHINGS lines. Remember: In reading a grid reference, it is always EASTINGS figures before NORTHINGS figures-EASTINGS before NORTHINGS.

2. Terrain features 【测】地形 / 地貌特征

Terrain features refer to the characteristics of the land, such as hills, ridges, valleys, saddles, depressions, and so forth. Maps represent these features in specific ways. U.S. Army divides terrain features into three groups: major, minor, and supplementary terrain features. Major terrain features include hills, saddles, valleys, ridges, and depressions. Minor terrain features include draws, spurs, and cliffs. Supplementary terrain features include cuts and fills.

【词汇点拨】Words and Expressions

1. rank /ræŋk/ *v.* ~ **(sb.) (as sth.)** to give sb./sth. a particular position on a scale according to quality, importance, success, etc.; to have a position of this kind 把……分等级；属于某等级

Last year, he was ranked second in his age group.

2. impassable /ɪmˈpɑːsəbl/ *adj.* impossible to travel along or over 不能通行的，不可逾越的

Snow and ice made the road impassable.

3. landscape /ˈlændskeɪp/ *n.* everything you can see when you look across a large area of land, especially in the country（陆上，尤指乡村的）风景，景色

The woods and fields that are typical features of the English landscape.

4. co-ordinate /kəʊˈɔːdɪneɪt/ *n.* the two sets of numbers or letters on the map or graph that you need in order to find that point 坐标

5. reference /ˈrefrəns/ *n.*

① use of a source of information in order to ascertain something 印证；资料引用

popular works of reference 常用参考书

② [C] a number, word or symbol that shows where sth. is on a map, or where you can find a piece of information（为方便查询所用的）标记，标识，编号

The map reference is Y4.

6. specify /ˈspesɪfaɪ/ *v.* [with obj.] identify clearly and definitely 具体说明，具体指定

Remember to specify your size when ordering clothes.

7. precision /prɪˈsɪʒn/ *n.* [U] the quality of being exact, accurate and careful 精确；准确；细致

precision instruments/tools 精密仪器/工具

8. subsection /ˈsʌbsekʃn/ *n.* a division of a section 分区，分部，分段，分支

a subsection of a text 文本后的一个字部分

9. terminology /ˌtɜːmɪˈnɒlədʒi/ *n.* [mass noun] the body of terms used with a particular technical application in a subject of study, theory, profession, etc. 术语

medical terminology 医学术语

10. crest /krest/ *n.* [usually sing.] the top part of a hill or wave 山顶；顶峰；波峰；浪尖

She reached the crest of the hill.

11. horizontal /ˌhɒrɪˈzɒntəl/ *adj.* parallel to the plane of the horizon; at right angles to the vertical 水平的，横向的

horizontal lines 横线

【参考译文】

识图

识图是一项重要的军事技能，通常被军队指挥官视为与武器使用技能同等重要。识图错误不仅会浪费宝贵的时间，还可能导致部队进入无法通过的地形或是不必要地暴露在敌人面前。熟练的识图者看到任何地形都能在脑海里呈现相应地图，同样，看到详细地图就知道实际的地形地貌。然而，这一技能只有通过实践、训练、并充分理解地图的工作原理才能获得。

坐标系网格

军事地图将地形划分为方格，用网格线表示。这样，就可以使用网格坐标数字轻松准确定位。

四位数坐标：坐标数字类似于大多数街道地图使用的字母加数字组合，这样就可以找到特定的街道（例如，贝克街位于G4方格）。军用地图使用四位数字：前两位数字表示东西向（或左右）坐标；后两位数字表示南北向（或上下）坐标。因此，四位数坐标 8040 指的是东西向坐标 80，南北向坐标 40 的方格。

六位数坐标：为了更加精确，可以将单个网格进一步划分为 10×10 的细分网格。此时使用六位数坐标数字。数字中的第三位表示东西向细分坐标，第六位表示南北向细分坐标。因此，数字 809403 指的是与上面（8040）相同位置的网格，但网格中的定位更加精确：在第九个东西向细分坐标和第三个南北向细分坐标的交叉点。

重点词汇：地形特征

以下是一些常见地图特征（地貌特征）术语，你可能不太熟悉：

- 冲沟：水流冲出的沟壑或深沟；
- 隘口：狭窄的开口或缺口，尤指两座山之间；
- 沟壑：在陆地上长而深的沟，通常由水流冲刷形成；
- 山脊：狭长的高地，通常指山顶；
- 鞍部：两个山峰或峰顶之间山脊较低的部分；
- 山坡：从山的主峰（水平）延伸出去的一部分。

 词汇表

Alpha Exercise Bright Star
joint 联合的；共同的

multinational 多国的
parachutist 跳伞者；跳伞运动员；伞兵

airborne 空降的
amphibious 两栖作战的；登陆的
staff（军队的）全体参谋人员
terrain 地面；地形、地势；地带
interior（国家或地区的）内地，内陆

Bravo Location and terrain
ford 浅滩，可涉水而过的地方
marsh 湿地；沼泽；草本沼泽
saddle（山的）鞍部
spur 山嘴；尖坡；支脉
jungle（尤指热带的）丛林地带
ravine 沟壑；溪谷
scrub 硬叶灌丛带；低矮灌木丛林地
oasis（沙漠中的）绿洲
draw 凹地，山沟，冲沟
pass 隘口；关隘
ridge 山脊；山脉
dune（风吹积成的）沙丘
vegetation 植物；植被
mosque 清真寺
grid（地图上的）坐标方格
vicinity 周围地区；邻近地区；附近
minefield 雷区
objective 目的，目标；客观的，就事论事的，不带个人感情的
deploy 部署（部队）
flank（陆军、海军或足球队的）侧翼，翼侧
rear 后面的，后部的；后方

Charlie Military commander's briefing 1
recce 侦查
protective 保护的；防护的

aerial 天线；（尤指用飞机）来自空中的，在空中进行的
brief 给（某人）指示；向（某人）介绍情况
briefing 情况介绍会，指令发布会；简报，情况介绍

Delta Military commander's briefing 2
assault 武力攻击，袭击；（对敌方的）军事攻击
reserve 预备役部队，后备部队；储备，备用
execute 执行；实行，实施（计划、命令、行动方针）
assembly 集合，集会
secure 保护；使安全
column（人或交通工具的）队；行
standby 随时待命

Echo Crash landing
travel 行进，转送，传播
crew（轮船、飞机等上面的）全体工作人员
co-pilot（飞机）副驾驶员
shaken 震惊；烦恼；恐惧
bleed 出血，流血
unconscious 无知觉的；昏迷的；不省人事的
senior 级别（或地位）较高者；上级；上司
compass 罗盘；罗经；指南针；罗盘仪

Foxtrot Map reading
rank 把……分等级；属于某等级

impassable 不能通行的，不可逾越的
landscape 风景，景色
co-ordinate 坐标
reference 标记，标识，编号
specify 具体说明，具体指定

precision 精确；准确；细致
subsection 分区，分部，分段，分支
terminology 术语
crest 山顶；顶峰；波峰；浪尖
horizontal 水平的，横向的

 拓展学习

Carrying out Military Training in Real Combat Conditions

Military training is the basic practice of the armed forces in peacetime. China's armed forces put military training in an important position and take combat effectiveness as the sole and fundamental criterion. In order to enhance realistic training, they optimize the policy framework and criteria in this respect, establish and improve the relevant supervision system, conduct supervision on military training for emergencies and combat across the services, implement the responsibility system for training and readiness, and organize extensive contests and competitions to encourage officers and soldiers to step up military training.

Military training in real combat conditions across the armed forces is in full swing. Since 2012, China's armed forces have carried out extensive mission-oriented training tailored to the specific needs of different strategic directions and exercises of all services and arms, including 80 joint exercises at and above brigade/division level.

The TCs have strengthened their leading role in joint training and organized serial joint exercises codenamed the Bast, the South, the West, the North and the Central, to improve joint combat capabilities.

The PLA Army (PLAA) has organized training competitions and conducted live exercises codenamed Stride and Fire-power. The PLA Navy (PLAN) has extended training to the far seas and deployed the aircraft carrier task group for its first far seas combat exercise in the West Pacific. It has organized naval parades in the South China Sea and the waters and air space near Qingdao, and conducted a series of live force-on-force exercises codenamed Mobility and systematic all-elements exercises. The PLA Air Force (PLAAF) has strengthened systematic and all-airspace training based on operational plans. It has conducted combat patrols in the South China Sea and security patrols in the East China Sea, and operated in the

West Pacific. It has completed a series of regular system-vs.-system exercises such as Red Sword. The PLA Rocket Force (PLARF) has organized force-on-force evaluation-oriented training and training based on operational plans at brigade and regiment levels, strengthened training for joint strikes, and completed regular exercises such as Heavenly Sword. The PLA Strategic Support Force (PLASSF) has made active efforts to integrate into the joint operations systems. It has carried out confrontational training in new domains and trained for emergencies and combats. The PLA Joint Logistic Support Force (PLAJLSF) has striven to align itself with the joint operations systems, and conducted exercises such as Joint Logistics Mission 2018. The PAP has developed to meet the requirements of nationwide coverage, effective connectivity, all-area response and integrated functions, and conducted a series of exercises including Guard.

(Selected from China's National Defense in the New Era, July 2019)

【Notes】

1. **real combat** 实战
2. **criteria** 标准，准则
3. **optimize** 优化，充分利用
4. **framework** 构架，结构
5. **supervision** 监督，管理
6. **implement** 执行，贯彻
7. **readiness** 准备就绪状态
8. **in full swing** 在热烈进行中；处于兴盛阶段
9. **mission-oriented** 以任务为导向的
10. **TC (Theatre Command)** 战区
11. **live exercises** 实弹演习
12. **far seas combat exercise** 远海作战演练
13. **live force-on-force exercises** 实兵对抗演习
14. **systematic** 系统的
15. **strive** 努力，力争
16. **align** 参加，加入
17. **PAP (People's Armed Police)** 武警部队
18. **coverage** 覆盖范围
19. **integrated** 综合的

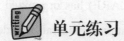 单元练习

I. Label the pictures

1. _____   2. _____   3. _____   4. _____

5. _____   6. _____   7. _____   8. _____

II. Multiple choice

1. There are always more than one country and different services participated in the _____.

A. multinational exercise

B. joint exercise

C. multinational joint exercise

D. PLA's training exercise

2. There are no troops in the _____.

A. multinational joint exercise

B. joint combat training exercise

C. Field Training Exercise

D. Command Post Exercise

3. On the map, the bridge is at grid _____.

A. 182372 B. 172368

C. 180377 D. 170368

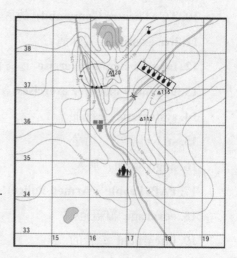

4. There is a _____ at grid 155335, and a _____ at grid 165385 on the map.

A. lake; forest B. river; jungle

C. desert; hill D. mosque; vegetation

5. The protective minefield is at grid _____.

A. 182378 B. 180380 C. 178375 D. 175369

6. According to the map, the south-north road is built along a _____.

A. draw B. valley C. ridge D. spur

7. The four basic types of military briefings refer to information briefings, decision briefings, _____ briefings and staff briefings.

A. mission B. deployment

C. convoy D. patrol

8. _____ refers to an exploration of an area in order to gather military information.

A. Check B. Move C. Security D. Recce

9. The lower part of a ridge between two peaks or summits is usually called _____.

A. spur B. saddle C. depression D. pass

10. _____ usually refers to an attack launched by land and sea forces.

A. Airborne assault B. Ambush

C. Amphibious assault D. Air to ground attack

III. Translation: English to Chinese

1. In military terminology a landing zone (LZ) is an area where aircraft can land.

2. Military training is the basic practice of the armed forces in peacetime.

3. Military training in real combat conditions across the armed forces is in full swing.

4. The PLA Strategic Support Force (PLASSF) has made active efforts to integrate into the joint operations systems and it has carried out confrontational training in new domains and trained for emergencies and combats.

5. Since 2012, China's armed forces have carried out extensive mission-oriented training tailored to the specific needs of different strategic directions and exercises of all services and arms, including 80 joint exercises at and above brigade/division level.

IV. Translation: Chinese to English

地形特征是指陆地地表特征，如丘陵、山脊、山谷、鞍部、洼地等。地图以特定的方式表示这些特征。美国陆军将地形特征分为三组：主要地形特征、次要地形特征和辅助地形特征。主要的地形特征包括丘陵、鞍部、山谷、山脊和洼地。次要地形特征包括凹陷、山坡和悬崖。

V. Oral practice

1. What is the definition and purpose of military exercises?

2. What is the purpose and structure of mission briefing?

附录：Unit 4 练习答案

I. Label the picture

1. hill   2. draw   3. saddle   4. cliff

5. depression   6. ridge   7. valley   8. spur

II. Multiple choice

1-5   CDBAC   6-10   BADBC

III. Translation: English to Chinese

1. 在军事术语中，着陆区（LZ）是指飞机可以降落的区域。

2. 战略支援部队积极融入联合作战体系，扎实开展新兴领域对抗演练和应急应战训练。

3. 全军兴起大抓实战化军事训练的热潮。

4. 军事训练是和平时期军队的基本实践活动。

5. 2012年以来，全军部队广泛开展各战略方向使命课题针对性训练和各军兵种演训，师旅规模以上联合实兵演习80余场。

IV. Translation: Chinese to English

Terrain features refer to the characteristics of the land, such as hills, ridges, valleys, saddles, depressions, and so forth. Maps represent these features in specific ways. U.S. Army divides terrain features into three groups: major, minor, and supplementary terrain features. Major terrain features include hills, saddles, valleys, ridges, and depressions. Minor terrain features include draws, spurs, and cliffs.

V. Oral practice

1. The rationale for planning and executing military exercises is to prepare commands and forces for operations in peace, crisis and conflict. Therefore, the aims and objectives of military exercises must mirror current operational requirements and priorities. It is carried out in order to practice and evaluate collective training of staffs, units and forces to enable them to operate effectively together, to demonstrate Military Capability, or to provide improvements to the capability. In addition, in today's international affairs, military diplomacy has emerged

a major tool in the interest of Nations. It is an indication of the highest level of trust and confidence between the member nations. Now looking at the operational side. Military exercises certainly enable militaries to understand each other's drills and procedures, overcome language barriers, and facilitate familiarization with equipment capabilities. And this is very necessary whether in war or in operations other than war (OOTW)—humanitarian aid, disaster relief, anti-piracy, etc. Last but not least, military exercises can show military strength and military quality. Conducting military exercises during special periods can achieve the purpose of deterring unfriendly countries. For instance, PLA continues military exercises in the air and waters off the Taiwan island. The drills in the area of the Taiwan Strait are a necessary measure to safeguard national sovereignty. It is prompted by the "seriously incorrect words and actions of relevant countries over the issue of Taiwan" and the actions of those advocating the self-governing island's independence.

2. The mission briefings are used under operational conditions to provide information, to give specific instructions, or to instill an appreciation of a mission. Mission briefings serves to convey critical mission information not provided in the plan or order to individuals or small units. The mission briefing reinforces orders, provides more detailed requirements and instructions for each individual, and ensure participants know the mission objective, their contribution to the operation, problems they may confront, and ways to overcome them.

Unit 5 Peacekeeping

单元导学清单

模块	主题	学习目标	任务	军事知识	核心词汇
Alpha	联合国维和任务	1）了解联合国维和任务基本知识；2）掌握联合国关于维和任务机构的常用词汇和表达；3）能厘清并描述以上机构之间的关系及其在维和任务中的作用。	Task 1 Task 2 Task 3	■ United Nations Security Council ■ UN Secretary General	Peacekeeping, headquarter, head of mission, comprise, sector, contingent
Bravo	维和任务的发展历史	1）掌握与维和任务相关的词汇和表达；2）辨析传统的维和任务与当代维和任务的区别；3）讨论中国军队在联合国维和行动中发挥的重要作用。	Task 1 Task 2 Task 3 Task 4 Task 5	■ Blue Berets ■ Buffer Zone ■ United Nations Military Observer	truce, demobilize, disarm, refugee, monitor, patrol, ceasefire, supervise
Charlie	UNTAC维和任务	1）了解UNTAC维和任务的背景知识；2）掌握UNTAC维和任务中主要职责。	Task 1 Task 2 Task 3 Task 4	■ UNTAC	assign, repatriation, landmine, duration, for good
Charlie	维和人员注意事项	1）掌握维和任务中使用的词汇；2）了解执行维和任务过程中的文化禁忌及交战规则；3）讨论维和人员在执行任务前所需要掌握知识。	Task 5 Task 6 Task 7 Task 8 Task 9	■ Rules of Engagement	typical, punctual, prohibited, authorisation, hostile, detain
Delta	地雷与排雷	1）了解地雷的基本知识与使用历史；2）熟练掌握与排雷任务相关的词汇和表达；3）掌握排雷、扫雷任务流程。	Task 1 Task 2 Task 3 Task 4	■ AT mine ■ AP mine ■ American Civil War	body armour, deminer, visor, tripwires, booby traps, probe

续表

模块	主题	学习目标	任务	军事知识	核心词汇
Echo	战区总部	1）了解战区总部相关的词汇和表达；2）熟练掌握传递信息的书面表达方式。	Task 1 Task 2 Task 3	■ Kalashnikov AK-47 ■ RPG ■ Chain of command	compound, subordinate, superior
	安全检查与营区秩序	1）了解通过检查站的基本流程；2）熟练掌握与营区秩序相关的词汇和表达。	Task 4 Task 5 Task 6 Task 7 Task 8	■ Checkpoint ■ GPS	smuggle, manifest, custody, issue, stand by
	安防布置	1）了解安全防护相关的词汇和表达；2）掌握一个既定区域的安防布置方式和注意事项。	Task 9 Task 10 Task 11		blast wall, barbed wire, sangar, searchlights, sentry
Foxtrot (Review)	联合国维和任务	1）了解联合国维和任务的相关知识（促进和平和维持和平的区别，维持和平任务的组成部分）；2）熟练掌握相关词汇和表达；3）运用所学知识介绍一次联合国维和任务（任务目的及其组成部分等）。	Task 1 Task 2	■ UN Charter ■ NGOs ■ Humanitarian aid	come into effect, resolve, dispute, negotiation, sanction, diplomatic, electoral

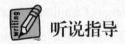

听说指导

Alpha　United Nations peacekeeping missions

【军事知识】

1. United Nations Security Council 联合国安理会

The United Nations Security Council (UNSC) is one of the six principal organs of the United Nations (UN) and is charged with ensuring international peace and security, recommending the admission of new UN members to the General Assembly, and approving any changes to the UN Charter. Its powers include establishing peacekeeping operations, enacting international sanctions, and authorizing military action. The UNSC is the only UN body with the authority to issue binding resolutions on member states.

2. UN Secretary General 联合国秘书长

The secretary-general of the United Nations (UNSG or SG) is the chief administrative officer of the United Nations and head of the United Nations Secretariat, one of the six principal organs of the United Nations. The role of the secretary-general is described as combining the functions and responsibilities of an advocate, diplomat, civil servant, and chief executive officer. The UN Charter designates the secretary-general as the "chief administrative officer" of the UN and allows them to perform "such other functions as are entrusted" by other United Nations organs. The Charter also empowers the secretary-general to inform the Security Council of "any matter which in his opinion may threaten the maintenance of international peace and security". These provisions have been interpreted as providing broad leeway for officeholders to serve a variety of roles as suited to their preferences, skill set, or circumstances.

【词汇点拨】Words and Expressions

1. peacekeeping /ˈpiːskiːpɪŋ/ *n.* a deployment, usually by the United Nations, of a neutral military force into an area where two sides are, or have recently been, engaged in armed conflict, in order to prevent or deter further military action by either side 维护和平

The UN is deploying a peacekeeping force in the region.

2. headquarters /ˈhedˈkwɔrtərz/ *n.* the place from which military action is controlled; used for referring to the people in charge of a military operation 总部；司令部

A tactical and administrative military unit consisting of a headquarters, two or more corps, and auxiliary forces.

3. Head of Mission the person in charge of a national or international mission to a particular country 特派团团长

The Head of Mission reports to the Secretary-General, who in turn reports to the Security Council.

4. comprise /kəmˈpraɪz/ *v.* to consist of two or more things; to form something 组成；构成

The armies were expanded and directed from Yenan, and there Mao wrote most of the articles that comprise the Selected Works of Mao Tse-tung.

5. sector /ˈsektər/ *n.* a part of a particular area, especially an area under military control（尤指军事管制的）区域，地带

Officers were going to retake sectors of the city.

Sector Commander 战区司令

6. contingent /kənˈtɪndʒənt/ *n.* a small military force which forms part of a larger grouping 代表团；（军队）分遣队

The British contingent is made up of marines and reconnaissance units.

【词汇点拨】Proper Names

1. United Nations Security Council 联合国安理会
2. United Nations Secretary General 联合国秘书长
3. Head of Mission 特派团团长
4. Force Commander 部队指挥官
5. Sector Commander 战区司令
6. Contingent Commander 代表团团长

【长难句解读】

1. The FC is responsible to the Head of Mission... The Secretary General is responsible for all UN peacekeeping operations and he reports to the UN Security Council. (5-1)

➢ 句子分析：该句中两个相近词组 responsible to 和 responsible for 的含义有明显区别：be responsible to 指"对某人负责"，后面一般接人物，或者接动词，表示"对做某事负有责任"；be responsible for 指"为某事承担责任"，后接名词。

➢ 翻译：军事指挥官向特派团团长负责。秘书长负责联合国所有维和行动，并向联合国安理会报告。

2. Each AO is under the command of a Sector Commander. (5-1)

➢ 句子分析：该句中 AO 指代的是 areas of operation，意思是"作战区域、行动区"。词组 under the command of 意为"由……指挥"，和前文中的"be commanded by"意思一致。

➢ 翻译：每个作战区都是由一个战区司令所指挥。

【口语输出】

1. How many UN departments do you know? Which departments are in charge of peacekeeping mission?

【参考词汇】UN General Assembly, UN Security Council, The Economic and Social Council, The Trusteeship Council, The International Court of Justice, The Secretariat

2. There are different positions in the United Nations Security Council. Please describe the chart on page 51 using the phrases in the box.

【参考词汇】under the command of, commanded by, responsible to, in charge of, report to

Bravo　Peacekeeping then and now

【军事知识】

1. Blue Berets　蓝盔部队

A blue beret is a blue-colored beret used by various (usually special) military and other organizations, notably the United Nations peacekeepers who are sometimes referred to as the Blue Berets. UN peacekeepers (often referred to as Blue Berets or Blue Helmets because of their light blue berets or helmets) can include soldiers, police officers, and civilian personnel. Peacekeepers monitor and observe peace processes in post-conflict areas and assist ex-combatants in implementing the peace agreements they may have signed. Such assistance comes in many forms, including confidence-building measures, power-sharing arrangements, electoral support, strengthening the rule of law, and economic and social development.

2. Buffer Zone　缓冲区

A buffer zone is a neutral zonal area that lies between two or more bodies of land, usually pertaining to countries. Depending on the type of buffer zone, it may serve to separate regions or conjoin them. Common types of buffer zones are demilitarized zones, border zones and certain restrictive easement zones and green belts. Such zones may be comprised by a sovereign state, forming a buffer state.

Buffer zones have various purposes, politically or otherwise. They can be set up for a multitude of reasons, such as to prevent violence, protect the environment, shield residential and commercial zones from industrial accidents or natural disasters, or even isolate prisons. Buffer zones often result in large uninhabited regions that are themselves noteworthy in many increasingly developed or crowded parts of the world.

3. United Nations Military Observer 联合国军事观察员

United Nations Military Observer (UNMO) is a military official deployed by the United Nations to provide support to a UN mission or peace operation. Described as the "eyes and ears" of the UN Security Council, observers fulfill a variety of roles depending on scope, purpose, and status of the UN mission to which they are attached. An UNMO is generally tasked with monitoring and assessing post-conflict agreements, such as a ceasefire or armistice; the withdrawal of military forces; or the maintenance of a neutral buffer zone. Observers usually undergo special training to ensure neutrality, diplomacy, and de-escalation techniques.

4. United Nations Truce Supervision Organization 联合国停战监督组织

United Nations Truce Supervision Organization (UNTSO) was set up in May 1948, which was the first ever peacekeeping operation established by the United Nations. Since then, UNTSO military observers have remained in the Middle East to monitor ceasefires, supervise armistice agreements, prevent isolated incidents from escalating and assist other UN peacekeeping operations in the region to fulfil their respective mandates. UNTSO personnel have also been available at short notice to form the nucleus of some other peacekeeping operations worldwide. The ability of UNTSO's military observers to deploy almost immediately after the Security Council has authorized a new mission, has been a significant factor in the success of those operations.

5. United Nations Disengagement Observer Force 联合国脱离接触观察员部队

The United Nations Disengagement Observer Force (UNDOF) was established on 31 May 1974 by Security Council resolution 350 (1974), following the agreed disengagement of the Israeli and Syrian forces in the Golan. Since then, UNDOF has remained in the area to maintain the ceasefire between the Israeli and Syrian forces and to supervise the implementation of the disengagement agreement. During the Syrian conflict, however, there were violations of the ceasefire with the escalation of military activity in the area of separation patrolled by UNDOF peacekeepers.

【词汇点拨】Words and Expressions

1. truce /truːs/ *n.* an agreement between enemies or opponents to stop fighting for an agreed period of time; the period of time that this lasts 休战；停战协定；停战期

An Israeli mayor on the front lines of the conflict with the Palestinians is calling for a truce with Islamic militants.

2. demobilize /diːˈmoʊbəlaɪz/ *v.* to release sb. from military service, especially at the end of a war 使退伍，使复员

All peace agreements should include specific measures to demobilize and reintegrate child soldiers into society.

3. disarm /dɪsˈɑːrm/ *v.* ① to take a weapon or weapons away 缴械；解除（某人）的武装； ② to reduce the size of an army or to give up some or all weapons, especially nuclear weapons 裁军；裁减军备（尤指核武器）

The troops will not attempt to disarm the warring militias.

The North has since all but disowned the promise at the six-party talks aimed at getting it to disarm.

4. refugee /ˌrefjuˈdʒiː/ *n.* a person who has been forced to leave their country or home, because there is a war or for political, religious or social reasons 避难者；难民

Somali government spokesperson Omar Osman says the point of the operation was to protect displaced people living in refugee camps.

5. monitor /ˈmɑːnɪtər/

v. ① ~ **sth.** ~ **what, how, etc.** to watch and check sth. over a period of time in order to see how it develops, so that you can make any necessary changes 监视；检查；跟踪调查

② ~ **sth.** to listen to telephone calls, foreign radio broadcasts, etc. in order to find out information that might be useful 监听（电话、外国无线电广播等）

n. ① a television screen used to show particular kinds of information; a screen that shows information from a computer 显示屏；监视器；（计算机）显示器

② a person whose job is to check that sth. is done fairly and honestly, especially in a foreign country（尤指派往国外的）监督员，核查员

Officials had not been allowed to monitor the voting.

UN observers are monitoring the ceasefire.

This monitor is not working.

6. patrol /pəˈtroʊl/

v. ~ **sth.** to go around an area or a building at regular times to check that it is safe and that there is no trouble 巡逻；巡查

n. ① [C,U] the act of going to different parts of a building, an area, etc. to make sure that there is no trouble or crime 巡逻；巡查

② [C] a group of soldiers, vehicles, etc. that patrol an area 巡逻队；巡逻车队

a garrison of 5000 troops 有 5000 士兵驻守的防地

Patrol the areas under charge regularly, keep high alertness and report abnormal cases timely.

He said that military action against Libya would be "several days" to combat the transition to the no-fly zone by the military on patrol.

7. ceasefire /ˈsiːsfaɪər/ *n.* a time when enemies agree to stop fighting, usually while a way is found to end the fighting permanently（通常指永久性的）停火，停战

He said the situation was the most serious it had been since the ceasefire in 2002.

8. supervise /ˈsuːpərvaɪz/ *v.* to be in charge of sb./sth. and make sure that everything is done correctly, safely, etc. 监督；管理；指导；主管

The UN pledged to help supervise the clearance of mines.

【词汇点拨】Proper Names

1. **Iran** /ɪˈrɑːn/ 伊朗（亚洲国家）

2. **Iraq** /ɪˈræk/ 伊拉克（西南亚国家）

3. **Israeli forces** 以色列军队

4. **Syrian forces** 叙利亚军队

5. **Cambodia** /kæmˈboʊdiə/ 柬埔寨（位于亚洲）

6. **Lebanon** /ˈlebənɑːn/ 黎巴嫩（西南亚国家，位于地中海东岸）

7. **UNIFIL** (UN Interim Force in Lebanon) 联合国驻黎巴嫩临时部队

8. **Pakistan** /ˈpækɪstæn/ 巴基斯坦（南亚国家名）

9. **UNMOGIP** (UN Military Obsewer Group in India and Pakistan) 联合国印度和巴基斯坦观察组

10. **Cyprus** /ˈsaɪprəs/ 塞浦路斯（地中海东部一岛）

11. **UNFICYP** (UN Peacekeeping Force in Gyprus) 联合国驻塞浦路斯维持和平部队

12. **UNMIK** (United Nations Interim Administration Mission in Kosovo) 联合国科索沃临时行政当局特派团

13. **MINURSO** (UN Mission for the Referendum in Western Sahara) 西撒哈拉公民投票特派团

14. **Democratic Republic of the Congo** 刚果民主共和国

15. **MONUSCO** (UN Organisation Stabilisation Mission in the Democratic Republic of the Congo) 联合国刚果民主共和国稳定特派团

16. **Abyei** 阿卜耶伊地区（连接苏丹南北部的地区）

17. **UNISFA** (UN Interim Security Force for Abyei) 联合国阿卜耶伊临时安全部队

18. **Republic of South Sudan** 南苏丹共和国

19. **UNMISS** (UN Mission in the Republic of South Sudan) 联合国南苏丹共和国特派团

20. **MINUSMA** (The United Nations Multidimensional Integrated Stabilization Mission in Mali) 联合国马里多层面综合稳定特派团

21. **Central African Republic** 中非共和国（非洲中部的一个内陆国家）

22. **MINUSCA** (Multidimensional Integrated Stabilisation Mission in the Central African Republic) 联合国中非共和国多层面综合稳定特派团

【长难句解读】

1. For example, UN troops disarm and demobilize soldiers, help refugees return to their homes and clear mines. (5-2)

➢ 句子分析：该句子中几个动词"disarm、demobilize、help、clear"并列使用，主语同为 UN troops，表示联合国部队同时执行该任务。其中 disarm 和 demobilize 的宾语都是 soldiers; help sb. do sth. 结构中，接不定式作 return 的宾语; clear 在这个句子中表示"清除，排除"的意思。

➢ 翻译：例如，联合国部队（的任务是）解除士兵的武装、复员士兵、帮助难民返回家园以及清除地雷。

2. In today's peacekeeping missions there are many more civilian peacekeepers, including police officers, engineer, medical personnel and drivers. (5-2)

➢ 句子分析：该句中"many more"表示相当多的，该词组后面一般接复数可数名词，不可修饰形容词和副词，若不接名词，则 more 本身就被看作是一个复数名词，表示复数概念。相近词组 much more 后可跟形容词和副词，表示"更加"，如果接名词，则是不可数名词。句子中 including 引导独立主格结构，对前句进行补充说明。

➢ 翻译：在今天的维持和平特派团中，有更多的民事维和人员，包括警察、工程师、医务人员和司机。

【口语输出】

1. What are the differences between the missions in traditional peacekeeping and

peacekeeping today?

【参考词汇】time, personnel, situation, buffer zone, refugees, clear mines, supervise, monitor ceasefire

2. How much do you know about China's troop contribution to UN peacekeeping operations?

【参考词汇】ceasefire supervision, stabilizing the situation, protecting civilians, providing force protection, deploying, sowing the seeds of hope

Charlie　We are United Nations peacekeepers

【军事知识】

1. UNTAC 联合国柬埔寨过渡政权

The United Nations Transitional Authority in Cambodia (UNTAC) was a United Nations peacekeeping operation in Cambodia in 1992–1993 formed following the 1991 Paris Peace Accords. This was the first occasion in which the UN directly assumed responsibility for the administration of an outright independent state (the UN did the administration of the former Dutch territory of Netherlands New Guinea between 1962–1963 prior), rather than simply monitoring or supervising the area. The UN transitional authority organised and ran elections, had its own radio station and jail, and was responsible for promoting and safeguarding human rights at the national level.

2. Rules of Engagement 交战规则

Rules of engagement (ROE) are the internal rules or directives afforded military forces (including individuals) that define the circumstances, conditions, degree, and manner in which the use of force, or actions which might be construed as provocative, may be applied. They provide authorization for and/or limits on, among other things, the use of force and the employment of certain specific capabilities. In some nations, articulated ROE have the status of guidance to military forces, while in other nations, ROE constitute lawful command. Rules of engagement do not normally dictate how a result is to be achieved, but will indicate what measures may be unacceptable.

【词汇点拨】Words and Expressions

1. assign /əˈsaɪn/ *v.*

① to give sb. sth. that they can use, or some work or responsibility　分配（某物）；分

派，布置（工作、任务等）

② to provide a person for a particular task or position 指定；指派

~ sb. to sb./sth. 委派；派遣

You can decide who to assign tasks to, who is best suited to take on certain tasks, and what tasks get priority.

2. repatriation /ˌriːpeɪtriˈeɪʃn/ *n.* the act of returning to the country of origin 遣送回国

It warns it will be forced to shut down the repatriation operation this year if it does not get the money.

3. landmine /ˈlændmaɪn/ *n.* the bomb placed on or under the ground, which explodes when vehicles or people move over it 地雷

Yet the number of reported casualties from landmine explosions was eleven percent higher than in two thousand four.

4. duration /duˈreɪʃn/ *n.* the length of time that sth. lasts or continues 持续时间；期间

for the duration (informal) 直到……结束；在整个……期间

It could also increase the potential range and duration of infantry operations as it would not need to be recharged at base.

5. for good 永久地；一劳永逸地

back for good 永久停留；重归于好

here for good 一心做好；始终如一

for good and all 永远；永久地

The war is over, then, for good maybe, and I've missed it all.

6. typical /ˈtɪpɪkl/ *adj.*

① having the usual qualities or features of a particular type of person, thing or group 典型的；有代表性的 **~ (of sb./sth.)** 不出所料；特有的

② happening in the usual way; showing what sth. is usually like 一贯的；平常的

Since reserve, a show of modesty and a sense of humor are part of his own nature, the typical Englishman tends to expect them in others.

I am a really typical teenager, the kind that people often see on televisions.

7. punctual /ˈpʌŋktʃuəl/ *adj.* happening or doing sth. at the arranged or correct time; not late 按时的；准时的；守时的

He has always been punctual in attendance, diligent in the discharge of his duties, and careful and practical in dealing with everything.

8. prohibit /prəˈhɪbɪt/ *v.* to stop sth. from being done or used especially by law（尤指以

法令）禁止

If lawmakers do not do something soon to prohibit spam, the problem will certainly get much worse.

9. authorisation /ˌɔːθərəˈzeɪʃn/ *n*. [C,U] official permission or power to do sth.; the act of giving permission 批准；授权

Any reproduction or use of copyrighted material for commercial purposes without authorisation is strictly prohibited.

10. hostile /ˈhɑːstaɪl/ *adj.* very unfriendly or aggressive and ready to argue or fight; belonging to a military enemy 敌意的；敌对的；敌军的；敌人的

This may be enough to deter hostile governments, but insurgents and militias might be less worried.

11. detain /dɪˈteɪn/ *v.* (~ **sb.**) to keep sb. in an official place, such as a police station, a prison or a hospital, and prevent them from leaving 拘留；扣押

Soldiers should not unlawfully arrest and detain civilians in military facilities.

【词汇点拨】**Proper Names**

1. Mohammed /moʊˈhɑːmɪd/ 默罕默德（人名）

2. Captain Sanders 上尉桑德斯

3. Major Thompson 少校汤普森

【长难句解读】

1. So UNTAC troops cleared mines and we set up training programmes in mine clearance and mine awareness. (5-4)

➢ 句子分析：该句中 UNTAC troops 指的是执行联合国柬埔寨过渡政权任务的部队，set up 指"发起，开办，创立"；词组 clear mines 和 mine clearance 都是指清扫地雷；mine awareness 指的是"防雷宣传"。

➢ 翻译：因此，联合国柬埔寨过渡政权机构的部队负责扫雷，我们也设立了关于扫雷和防雷培训的计划。

2. This was important because there were villages between the two sides—on both sides of the border we were patrolling—and these people were often attacked. (5-5)

➢ 句子分析：句中主句是 this was important, 后由 because 引导一个较长的原因状语从句，中间 on both sides of the border we were patrolling 作为插入语，补充说明前面的 villages，插入语中又包含一个由 border 作为先行词的定语从句，其中引导词 that 被

省略掉。

> **翻译**：这很重要，因为（冲突的）双方之间都有村庄——我们在边境的两边巡逻——这些村民经常受到袭击。

3. This was very tiring—it was very hot out in the midday sun—and also very stressful—several of my men died as a result of this daily routine looking for mines everywhere and others were severely injured. (5-5)

> **句子分析**：该句中 it was very hot out in the midday sun, and also very stressful 作为并列插入语，补充说明前面 tiring，其中 midday sun 指的是"正午烈日"。后面 as a result of 引导原因状语，现在分词 looking for mines 作状语进一步表达 daily routine 的任务内容。

> **翻译**：这是一项很累人的工作——正午的阳光下非常热——人们也很有压力——我手下的几个战士因为这种每天到处寻找地雷的日常工作而牺牲，其他人则严重受伤。

4. My name is Captain Sanders and I'm going to brief you on some of the ground rules for this mission. (5-6)

> **句子分析**：句中 Captain 指的是军衔中的上尉；词组 brief sb. on sth. 指是"就某事向某人简要汇报"；ground rules 指的是"基本原则、基本规则"。

> **翻译**：我是桑德斯上尉，接下来我将向你们简要介绍这次任务的一些基本规则。

5. As you all know, the first rule to remember is that all military operations should be conducted according to the laws of war. (5-7)

> **句子分析**：该句中 as you all know 是一个口语表达方式，意为"正如你们大家所知"，不定式 to remember 作为后置定语，修饰 the first rule；词组 according to 意思是"按照、根据"。

> **翻译**：大家都知道，要记住的第一条规则是，所有军事行动都应按照战争法进行。

【口语输出】

1. Can you retell the exact tasks in UNTAC mission?

【参考词汇】clear, demobilise, deploy, provide, support, monitor, organize, patrol

2. What are the dos and don'ts when carrying out a peacekeeping mission?

【参考词汇】respect, religion, culture, laws, environment, polite, punctual, alcohol, prohibited, press

3. What are the rules of engagement for the peacekeeping mission?

【参考词汇】laws of war, CO, self-defence, open fire, minimun force, treatment of the local people, respect, civilians, detain

Delta   Mines and demining

【军事知识】

1. AT mine 反坦克地雷

An anti-tank mine (abbreviated to "AT mine") is a type of land mine designed to damage or destroy vehicles including tanks and armored fighting vehicles. Compared to anti-personnel mines, anti-tank mines typically have a much larger explosive charge, and a fuze designed to be triggered by vehicles or, in some cases, remotely or by tampering with the mine.

2. AP mine 防步兵地雷

Anti-personnel mines are a form of mine designed for use against humans, as opposed to anti-tank mines, which are designed for use against vehicles. Anti-personnel mines may be classified into blast mines or fragmentation mines; the latter may or may not be a bounding mine. The mines are often designed to injure, not kill, their victims to increase the logistical (mostly medical) support required by enemy forces that encounter them. Some types of anti-personnel mines can also damage the tracks on armoured vehicles or the tires of wheeled vehicles.

3. American Civil War 美国内战

The American Civil War (April 12, 1861–May 26, 1865) was a civil war in the United States between the Union (states that remained loyal to the federal union, or "the North") and the Confederacy (states that voted to secede, or "the South"). The central cause of the war was the status of slavery, especially the expansion of slavery into territories acquired as a result of the Louisiana Purchase and the Mexican–American War. On the eve of the Civil War in 1860, four million of the 32 million Americans were enslaved black people, almost all in the South.

【词汇点拨】Words and Expressions

1. body armour clothing worn by the police, etc. to protect themselves 防弹服，胸甲，（警察等穿的）防弹背心

After all, body armour may be an expected part of an army, police or security services uniform, but civilians need more discretion.

2. deminer /diːˈminər/ *n.* specialized persons whose job is to remove, deactivate, or safely detonate land mines in an area 排雷人员

The deminer carefully removed the landmine from the ground.

3. visor /ˈvaɪzər/ *n.* a part of a helmet that can be pulled down to protect the eyes and face（头盔上的）面甲，面罩，护面

Oxygen is circulated around the helmet to stop the visor misting.

4. tripwires /ˈtrɪpwaɪər/ *n.* a wire that is stretched close to the ground as part of a device for catching sb./sth. if they touch it 绊索；绊网

A border guard grumbles that the only tourist in weeks has set off her tripwire.

5. booby traps a hidden bomb that explodes when the object that it is connected to is touched 饵雷；诡雷

A spate of attacks more frequent and reckless than before has seen the deployment of booby-trap and car bombs, as well as other devices.

6. probe /proʊb/

v. ① to ask questions in order to find out secret or hidden information about sb./sth. 盘问；追问；探究 ② [T](~ sth.) to touch, examine or look for sth., especially with a long thin instrument（用细长工具）探查，查看

n. ① ~ **(into sth.)** a thorough and careful investigation of sth. 探究；详尽调查 ② a small device put inside sth. and used by scientists to test sth. or record information 探测仪；传感器；取样器

He has not been charged or found guilty of wrongdoing and the probe is continuing, according to people familiar with the matter.

7. UXO unexploded ordnance 未爆弹药；未爆炸物

【词汇点拨】Proper Names

1. American Civil War 美国内战（南北战争 1861 年 4 月 12 日至 1865 年 4 月 9 日）
2. World War II 第二次世界大战

【长难句解读】

1. After he checks for tripwires, he clears the vegetation. When the lane is clear of vegetation, he checks the area again for tripwires. (5-9)

> 句子分析：该句中 check sth. 表示一般的检查的对象，check for sth. 指的是检查的目标、目的；vegetation 泛指地面上的植被，is clear of vegetation 指地面干净，没有植被。

> 翻译：在检查（地表上）绊网之后，他会清除地面植被。当车道上没有植被时，他再次检查该区域是否有绊网。

2. Then, he probes the ground where the metal detector found metal. He keeps the probe at a 30° angle. He continues until he finds a mine or UXO. (5-9)

> 句子分析：该句中第一个 probe 是动词，意为"探查"，第二个 probe 是名词，指代"探针、探测器"。第一句中是由 where 引导的定语从句，补充前面中心词 ground 的位置状语；第二句中 UXO 指代没有爆炸的弹药。

> 翻译：然后，他探测金属探测器发现金属的地面。他将探针保持在 30°的角度，他继续前进，直到发现地雷或未引爆武器。

【口语输出】

1. Can you introduce the history of landmines?

【参考词汇】first use, American Civil War, World War II, eighty countries, casualties, victims, clear, cost, produce

2. What are the necessary equipment when demining?

【参考词汇】body armour, metal detector, prodder, red triangle, tripwires, visor

3. What is the procedure of demining?

【参考词汇】before, put on, first of all, study, looks for, clear, next, check, then, probe, locate, mark, finally

Echo   Sector headquarters

【军事知识】

1. Kalashnikov AK-47 卡拉什尼科夫 AK-47 突击步枪

The AK-47, officially known as the Avtomat Kalashnikova (Russian: Автомат Калашникова, also known as the Kalashnikov or just AK), is a gas-operated assault rifle that is chambered for the 7.62mm×39mm cartridge. Developed in the Soviet Union by Russian small-arms designer Mikhail Kalashnikov, it is the originating firearm of the Kalashnikov (or "AK") family of rifles. After more than seven decades since its creation, the AK-47 model and its variants remain the most popular and widely used rifles in the world. The number "47" refers to the year the rifle was finished. Design work on the AK-47 began in 1945. It was presented for official military trials in 1947, and, in 1948, the fixed-stock version was introduced into active service for selected units of the Soviet Army. In early 1949, the AK was officially accepted by the Soviet Armed Forces and used by the majority of the member states of the Warsaw Pact. The model and its variants owe their global popularity to their reliability under harsh conditions, low production cost (compared to contemporary weapons), availability in virtually every geographic region, and ease of use. The AK has been manufactured in many countries, and has seen service with armed forces as well as irregular forces and insurgencies throughout the world.

2. RPG 火箭榴弹

A rocket-propelled grenade (RPG) is a shoulder-fired missile weapon that launches rockets equipped with an explosive warhead. Most RPGs can be carried by an individual soldier, and are frequently used as anti-tank weapons. These warheads are affixed to a rocket motor which propels the RPG towards the target and they are stabilized in flight with fins. Some types of RPG are reloadable with new rocket-propelled grenades, while others are single-use. RPGs are generally loaded from the front. RPGs with high-explosive anti-tank (HEAT) warheads are very effective against lightly armored vehicles such as armored personnel carriers (APCs) and armored cars. However, modern, heavily-armored vehicles, such as upgraded APCs and main battle tanks, are generally too well-protected (with thick composite or reactive armor) to be penetrated by an RPG, unless less armored sections of the vehicle are exploited. Various warheads are also capable of causing secondary damage to vulnerable systems (especially sights, tracks, rear and roof of turrets) and other unarmored targets.

3. Chain of Command 指挥系统，指挥链

In a military context, the chain of command is the line of authority and responsibility along which orders are passed within a military unit and between different units. In simpler terms, the chain of command is the succession of leaders through which command is exercised and executed. Orders are transmitted down the chain of command, from a responsible superior, such as a commissioned officer, to lower-ranked subordinate(s) who either execute the order personally or transmit it down the chain as appropriate, until it is received by those expected to execute it. "Command is exercised by virtue of office and the special assignment of members of the Armed Forces holding military rank who are eligible to exercise command."

4. GPS 全球卫星定位系统

The Global Positioning System (GPS), originally Navstar GPS, is a satellite-based radionavigation system owned by the United States government and operated by the United States Space Force. It is one of the global navigation satellite systems (GNSS) that provides geolocation and time information to a GPS receiver anywhere on or near the Earth where there is an unobstructed line of sight to four or more GPS satellites. It does not require the user to transmit any data, and operates independently of any telephonic or Internet reception, though these technologies can enhance the usefulness of the GPS positioning information. It provides critical positioning capabilities to military, civil, and commercial users around the world. Although the United States government created, controls and maintains the GPS system, it is freely accessible to anyone with a GPS receiver.

【词汇点拨】Words and Expressions

1. compound /ˈkɑːmpaʊnd/ *n.* an area surrounded by a fence or wall in which a factory or other group of buildings stands 有围栏（或围墙）的场地（内有工厂或其他建筑群）

He said the Afghan government can not enter, because this compound, "a NATO coalition forces," the police can not enter.

2. subordinate /səˈbɔːrdɪnət/

① *n.* a person who has a position with less authority and power than sb. else in an organization 下级；部属

② *v.* ~ **sb./sth. (to sb./sth.)** to treat sb./sth. as less important than sb./sth. else 使从属于

③ *adj.* ~ **(to sb.)** having less power or authority than sb. else in a group or an organization 隶属的；从属的；下级的

Subordinate commanders who might have once used the compound also must have long since established alternate command sites.

3. superior /suːˈpɪriər/

① *n.* a person of higher rank, status or position 级别（或地位、职位）更高的人；上级；上司

② *adj.* of very good quality; better than other similar things 质量卓越的；出类拔萃的；超群的

The superior was suspicious of his survival,— he thought the soldier might be a traitor.

4. smuggle /ˈsmʌɡl/ *v.* ~ **sth./sb. (+ adv./prep.)** to take, send or bring goods or people secretly and illegally into or out of a country, etc. 走私；私运；偷运

Traditional human spies risk arrest or execution by trying to smuggle out copies of documents.

5. manifest /ˈmænɪfest/

① *n.* a list of goods or passengers on a ship or an aircraft（船或飞机的）货单，旅客名单

② *v.* ~ **sth. (in sth.)** to show sth. clearly, especially a feeling, an attitude or a quality 表明，清楚显示（尤指情感、态度或品质）

The enemies have manifested their intention to attack the city.

6. custody /ˈkʌstədi/ *n.* ① the state of being in prison, especially while waiting for trial（尤指在候审时的）拘留，拘押，羁押 ② the legal right or duty to take care of or keep sb./sth.; the act of taking care of sth./sb. 监护权；保管权；监护；保管

The magistrate decided to keep him in custody to avoid a risk of relapse or tampering with evidence.

7. issue /ˈɪʃuː/

① *n.* [C] a problem or worry that sb. has with sth.（有关某事的）问题，担忧

② *v.* to give sth. to sb., especially officially（正式）发给，供给

The police were issued with firearms.

8. stand by be available or ready for a certain function or service 待命

The US 48th fighter wing based in England keeps peregrines and other falcons on stand by.

9. blast wall a barrier designed to protect vulnerable buildings or other structures and the people inside them from the effects of a nearby explosion, whether caused by industrial accident, military action or terrorism 防爆墙

Today he is a contractor, building the Aynak mine's electric fence, blast wall, workers' dormitories and a road to Kabul.

10. sangar /ˈsʌŋgər/ *n.* a small low temporary defensive work, usually built of stone around an existing hollow in the ground 散兵壕；（常用于围挡洼地的）石砌矮墙

It is much safer to place more sangars made with sandbags on the roof of the building.

11. searchlight /ˈsɜːrtʃlaɪt/ *n.* a powerful lamp that can be turned in any direction, used, for example, for finding people or vehicles at night 探照灯

Smoke rose into the night sky from one of the attacked compounds, and a helicopter circled overhead, scanning the ground with a searchlight.

12. sentry /ˈsentri/ *n.* a soldier whose job is to guard sth. 哨兵

During the Cuban missile crisis, an air force sentry in Minnesota shot at a fence-climbing intruder and sounded the sabotage alarm.

【词汇点拨】**Proper Names**

1. **Lieutenant Jarvis** 中尉贾维斯
2. **Major Stanton** 少校斯坦顿
3. **Sergeant Denton** 中士登顿
4. **Sergeant Clay** 中士克莱
5. **Captain Robertson** 上尉罗伯逊
6. **Lieutenant Turner** 中尉特纳
7. **Sergeant Crossley** 中士克罗斯利
8. **Sergeant Thomas** 中士托马斯
9. **Sergeant Harris** 中士哈里斯
10. **Briggs** /brɪgz/ 布里格斯（人名）
11. **Corporal Hayne** 下士海恩
12. **Sergeant Brooks** 中士布鲁克斯

13. **Corporal Burns** 下士彭斯
14. **Corporal Taylor** 下士泰勒
15. **Geoff** /dʒɪɔf/ 杰夫（人名）
16. **Claude** /klɔːd/ 克劳德（人名）

【长难句解读】

1. I'm calling to inform you that the Alert State will change from red to yellow at 2359 hours tonight. (5-10)

> 句子分析：该句中 to inform 作目的状语，inform 后接直接宾语 you，和 that 所引导的宾语从句；Alert State 指军事警戒状态，一般由五种颜色代指不同的警戒状态："严重（Severe，红色）：严重的恐怖袭击风险；高（High，橙色）：高恐怖袭击风险；较高（Elevated，黄色）：显著的恐怖袭击风险；警戒（Guarded，蓝色）：一般的恐怖袭击风险；低（Low，绿色）：低恐怖袭击风险"。

> 翻译：我打电话是想通知你，警戒状态将在今晚23点59分由红色变为黄色。

2. We have a truck here at the VCP loaded with weapons hidden under a cargo of wood. (5-11)

> 句子分析：上句中 VCP 全称是 vehicle check point，指汽车检查站；loaded with weapons 作后置定语，修饰前句中的 truck；hidden under a cargo of wood 也作后置定语，修饰前句面的 weapons。

> 翻译：我们在汽车检查站发现有一辆满载武器的卡车，这些武器都藏在一堆木头下面。

3. When we searched the back of the truck, hidden under the wood we found twelve boxes of Kalashnikov AK-47 rifles and magazines, about fifty rocket-propelled grenades and another twenty boxes containing probably as many as 500 hand grenades. (5-11)

> 句子分析：该句中 hidden under the wood 作为地点状语，放到了主句之前，强调了主句中一系列武器的发现地；containing... 补充说明前面的 another twenty boxes；as many as 后跟可数名词，用于数字之前表示数目极大。

> 翻译：当我们搜查卡车后部时，发现了藏在木头下面的12箱卡拉什尼科夫AK-47步枪和弹匣，大约50枚火箭榴弹，还有20箱可能多达500枚手榴弹的箱子。

4. The Ops Room is manned twenty-four hours a day and that means we work in shifts. (5-13)

> 句子分析：句中 Ops Room 指 operations room，意为作战指挥室；is manned 意思是"有人值守"；that 指代前句"指挥室需要24小时有人值守"，work in shifts 意为

"轮班作业、分班轮换"。

> **翻译**：作战指挥室一天 24 小时都有人值守，这意味着我们要轮班工作。

5. We placed four sangars made with sandbags on the roof.... There's a wire fence with an anti-sniper screen that goes around the perimeter of the building. (5-15)

> **句子分析**：前句中 place 为动词，表示"放置"，made with sandbags 作后置定语，修饰前面的 four sangars；后句中 with an anti-sniper screen 补充说明前面 wire fence，而 that 引导的定语从句用来修饰 anti-sniper screen，引导词在句中作主语。

> **翻译**：我们在屋顶设置了四个用沙袋做成的散兵壕……这栋建筑物周边有一个带防狙击屏的铁丝栅栏。

【口语输出】

1. What is Geoff's duty according to task 6?

【参考词汇】Ops Room, military operations, 24 hours, shifts, weapons, compound, load, magazine, clean, Rules of Engagement, training, Commander's Brief, work parties, maintenance, uniform

2. Can you retell the Camp Orders?

【参考词汇】weapons, unload, Commander's Brief, dress, combat uniform, alcohol, work parties

3. Can you give a quick briefing on what to do with the campus security?

【参考词汇】main gate, sangars, sand bags, corner, buildings, guards, searchlights, wire fence, anti-sniper screen, perimeter of the building, compound, blast wall, checkpoint, tank stops, sentries, patrol, defence positions

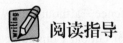 阅读指导

【文章导读】

This passage mainly focuses on United Nations peacekeeping operations (PKOs). In the beginning, the article elaborates the duties of UN when it comes to both international and regional conflicts, it also explains the differences between peacemaking and peacekeeping. In the latter part, followed by the components of a peacekeeping operation, the article further introduced the personnel involved, and their respective roles in the operation.

【军事知识】

1. UN Charter 联合国宪章

The Charter of the United Nations (UN) is the foundational treaty of the UN, an intergovernmental organization. It establishes the purposes, governing structure, and overall framework of the UN system, including its six principal organs: the Secretariat, the General Assembly, the Security Council, the Economic and Social Council, the International Court of Justice, and the Trusteeship Council.

The UN Charter mandates the UN and its member states to maintain international peace and security, uphold international law, achieve "higher standards of living" for their citizens, address "economic, social, health, and related problems", and promote "universal respect for, and observance of, human rights and fundamental freedoms for all without distinction as to race, sex, language, or religion". As a charter and constituent treaty, its rules and obligations are binding on all members and supersede those of other treaties.

2. NGOs 非政府组织

A non-governmental organization (NGO) is an organization that generally is formed independent from government. They are typically nonprofit entities, and many of them are active in humanitarianism or the social sciences; they can also include clubs and associations that provide services to their members and others. Surveys indicate that NGOs have a high degree of public trust, which can make them a useful proxy for the concerns of society and stakeholders. However, NGOs can also be lobby groups for corporations, such as the World Economic Forum. The term as it is used today was first introduced in Article 71 of the newly-formed United Nation's Charter in 1945. While there is no fixed or formal definition for what NGOs are, they are generally defined as nonprofit entities that are independent of governmental influence—although they may receive government funding. According to the UN Department of Global Communications, an NGO is "a not-for profit, voluntary citizen's group that is organized on a local, national or international level to address issues in support of the public good". The term NGO is used inconsistently, and is sometimes used synonymously with civil society organization (CSO), which is any association founded by citizens. In some countries, NGOs are known as nonprofit organizations, and political parties and trade unions are sometimes considered NGOs as well.

3. Humanitarian aid 人道主义救助

Humanitarian aid is material and logistic assistance to people who need help. It is usually short-term help until the long-term help by the government and other institutions replaces it. Among the people in need are the homeless, refugees, and victims of natural disasters, wars, and famines. Humanitarian relief efforts are provided for humanitarian purposes and include natural disasters and man-made disasters. The primary objective of humanitarian aid is to save lives, alleviate suffering, and maintain human dignity. It may, therefore, be distinguished from development aid, which seeks to address the underlying socioeconomic factors which may have led to a crisis or emergency. There is a debate on linking humanitarian aid and development efforts, which was reinforced by the World Humanitarian Summit in 2016. However, the conflation is viewed critically by practitioners.

【词汇点拨】Words and Expressions

1. come into effect 开始生效；开始实施

A new law came into effect in Alabama that cracks down on illegal immigration, after a judge ruled that it passed constitutional muster.

2. resolve /rɪˈzɑːlv/ *v.* ~ **sth./itself** to find an acceptable solution to a problem or difficulty 解决（问题或困难）

In its filing, China is requesting consultations with the EU to try to resolve the matter.

3. dispute /dɪˈspjuːt/

① *n.* [C,U] an argument or a disagreement between two people, groups or countries; discussion about a subject where there is disagreement 争论；辩论；争端；纠纷

② *v.* to question whether sth. is true and valid 对……提出质询；对……表示异议（或怀疑）

The dispute over the border is so long-running that India's reaction was no stronger than official outrage and unofficial weariness.

4. negotiation /nɪˌɡoʊʃiˈeɪʃn/ *n.* formal discussion between people who are trying to reach an agreement 谈判；磋商；协商

Negotiation is a course that at least two groups of people trying to reach an agreement with the others for their own benefit.

5. sanction /ˈsæŋkʃn/

① *n.* **~ (against sb.)** an official order that limits trade, contact, etc. with a particular country, in order to make it do sth., such as obeying international law 制裁

② *v.* **(~ sb./sth.)** to punish sb./sth.; to impose a sanction on sth. 惩罚；实施制裁

If Iraq sticks to road map laid out for it by United Nations, sanction could be lifted without more ado.

6. diplomatic /ˌdɪpləˈmætɪk/ *adj.* connected with managing relations between countries 外交的；从事外交的

Obama said the US diplomatic route to curb Iran's nuclear program would be reassessed by year's end.

7. electoral /ɪˈlektərəl/ *adj.* connected with elections 有关选举的

The two opposition parties made an electoral pact.

【词汇点拨】Proper Names

United Nations peacekeeping operations(PKOs) 联合国维和行动

【参考译文】

联合国维和行动

联合国所有行动的基本准则都来自1945年生效的《联合国宪章》。根据该宪章，联合国的基本目标是"维护国际和平与安全"，为此，成员国合作解决国际争端，促进和平，并鼓励其他国家遵守国际法。

如果发生冲突，联合国更倾向于当事方通过谈判和法律途径自行解决冲突。但是，这并不总是可能的。在这种情况下，联合国安理会将就如何最好地维护国际和平与安全提出建议。此类建议可能包括经济制裁、终止外交关系或使用武装部队。从这个意义上说，必须指出缔造和平与维持和平之间的区别：

- 缔造和平：利用外交手段试图在各方之间达成协议。
- 维持和平：一旦达成和平协议，维和特派团将同时使用民事和军事人员来帮助

执行和监督协议。

维持和平行动（PKO）取决于冲突中主要各方的合作。因此，联合国维和部队必须是客观的。它不能偏袒任何一方。

维和行动的组成部分：

维和行动由军事和民事两部分组成。民事部分包括：

- 外交人员
- 人权观察员和教育工作者
- 人道主义部门：协调提供人道主义援助的非政府组织（NGO），例如食品或医疗用品，并协助当地重建和发展
- 有助于组织和监督自由和公正选举的选举部分
- 民政管理和民警部门

军事部分可能有助于：

- 建立缓冲区（分隔敌对势力的中立区）
- 监督停战和停火协议
- 防止国家之间或国家内部的武装冲突；这可能包括帮助解除武装或复员士兵（确保士兵离开作战单位，并恢复正常生活）
- 值守观察哨和检查站
- 提供地面巡逻和车队护送
- 维护法律和秩序
- 支持人道主义工作，例如帮助难民返回本国或清除地雷
- 支持人权监测员或选举部门

维和行动的军事部门通常是轻装上阵，必须以最低限度的武力行事。这意味着武力不是用来完成任务的，尽管它可以用于自卫。军事观察员通常是完全手无寸铁的。军事单位将由一名部队指挥官领导，并组织成营级单位（包括相应的军官、士官和士兵）。

 词汇表

Alpha United Nations peacekeeping missions

peacekeeping 维护和平

headquarters 总部；司令部

Head of Mission 特派团团长

comprise 组成；构成

sector （尤指军事管制的）区域，地带

contingent 代表团；（军队）分遣队

areas of operation 作战区域、行动区
component 组成部分，成分，部件
infantry 步兵

Bravo Peacekeeping then and now
blue berets 蓝盔部队
buffer zone 缓冲区
military observer 军事观察员
truce 休战；停战协定；停战期
demobilize 使退伍；使复员
disarm 缴械；裁军；裁减军备（尤指核武器）
refugee 避难者；难民
monitor 监视；检查；监听
patrol 巡逻；巡查
ceasefire（通常指永久性的）停火，停战
supervise 监督；管理；指导；主管
clear mines 清除地雷
many more 相当多的
authorized 授权的

Charlie We are United Nations peacekeepers
assign 分配（某物）；布置；指定；指派
repatriation 遣送回国
landmine 地雷
duration 持续时间；期间
for good 永久地；一劳永逸地
typical 典型的；有代表性的；一贯的；平常的
punctual 按时的；准时的；守时的
prohibit（尤指以法令）禁止
authorisation 批准；授权
hostile 敌意的；敌对的；敌军的；敌人的

detain 拘留；扣押
set up 发起；开办；创立
middy 正午
daily routine 日常工作
brief on 就某事进行简要汇报

Delta Mines and demining
AT mine 反坦克地雷
AP mine 防步兵地雷
body armour 防弹服；胸甲
deminer 排雷人员
visor（头盔上的）面甲，面罩，护面
tripwires 绊索；绊网
bobby traps 饵雷；诡雷
probe 追问；探究；探测仪
UXO 未爆弹药；未爆炸物
vegetation（地面）植被

Echo Sector headquarters
RPG 火箭助推榴弹
chain of command 指挥系统；指挥链
GPS 全球卫星定位系统
compound 有围栏（或围墙）的场地（内有工厂或其他建筑群）
subordinate 下级；部属；使从属于
superior 级别更高的人；上级；上司
smuggle 走私；私运；偷运
manifest（船或飞机的）货单，旅客名单；表明；清楚显示
custody 拘留；拘押；羁押；监护权；保管权
issue 问题；担忧；（正式）发给；供给
stand by 待命

blast wall 防爆墙

sangar 散兵壕；石砌矮墙

searchlight 探照灯

sentry 哨兵

Alert State 警戒状态

VCP 汽车检查站

as many as 多达

in shifts 轮班作业；分班轮换

anti-sniper 防狙击的

Foxtrot Review

NGOs 非政府组织

humanitarian aid 人道主义救助

come into effect 开始实施；开始生效

resolve 解决（问题或困难）

dispute 争论；争端；纠纷

negotiation 谈判；磋商；协商

sanction 制裁；惩罚

diplomatic 外交的；从事外交的

electoral 有关选举的

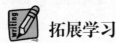 拓展学习

Embarking on Missions for World Peace

UN Peacekeeping, as an instrument developed for peace, has made a significant contribution to world peace. In 1971, China recovered its legitimate seat in the UN and began to play a more active role in international affairs. After reform and opening up began in 1978, China gradually increased its involvement in UN peacekeeping affairs. In April 1990, China's armed forces dispatched five military observers to the United Nations Truce Supervision Organization (UNTSO) and embarked on a new voyage as a participant in the UNPKOs. In the past three decades, China's armed forces have engaged in the UNPKOs with courage and determination, always aspiring to fulfill their missions of meeting the responsibilities of a major country, safeguarding world peace, and contributing to the building of a community with a shared future for mankind. China's Blue Helmets have become a key force in UN peacekeeping.

China's armed forces participate in the UNPKOs, because the pursuit of peace is in the genes of the Chinese nation. The Chinese nation values peace and harmony. Ideas such as "unity of man and nature" "harmony among all nations" "harmony without uniformity" and "kindness towards fellow human beings," voice the mind of the Chinese people on the universe, international relations, society and ethics. The pursuit of peace, amity and harmony has long been the primary aspiration of our nation. The philosophy of upholding peace, harmony, cooperation and common development has been passed down from generation to generation in China. For millennia, peace has been in the veins and the DNA of the Chinese nation. It is a consistent goal of China's armed forces.

China's armed forces participate in the UNPKOs, because the Chinese people care about the wellbeing of humanity. The Chinese people always dream of living in a harmonious world where everyone belongs to one and the same family. They advocate that "a just cause should be pursued for the common good" and that one should put concern for the wellbeing of other people before personal interests. They hope for a better life not only for themselves, but also for other peoples across the world. Chinese service members join the UN efforts to bring hope and promote peace.

China's armed forces participate in the UNPKOs, because serving the people is the fundamental purpose of the people's armed forces. China's armed forces come from the people, have their roots in the people, developed to serve the people, and fight for the people. They serve the people wholeheartedly at all times and under all circumstances, remain close to the people, and always put the people's interests first. With love and humanity, Chinese peacekeeping troops make efforts to bring peace and happiness to people in mission areas.

China's armed forces participate in the UNPKOs, because China honors its responsibilities as a major country. As a founding member of the UN and a responsible member of the international community, China honors its obligations, firmly supports the UN's authority and stature, and actively participates in the UNPKOs. China is a permanent member of the UN Security Council, and therefore, it is incumbent on China as a major country to play an active part in the UNPKOs. World peace is indivisible and humanity shares a common destiny. To participate in the UNPKOs is integral to China's joint efforts with other countries to build a community with a shared future for mankind.

China's armed forces commit themselves to the following policy stances on UN peacekeeping:

Upholding the purposes and principles of the UN Charter. China always abides by the primary principles of the UN such as sovereign equality of all members and settlement of international disputes by peaceful means. It respects the social systems and development paths independently chosen by other countries, and respects and accommodates the legitimate security concerns of all parties.

Following the basic principles of the UNPKOs. China always adheres to the basic principles of UN peacekeeping, including consent of the host nation, impartiality, and non-use of force except in self-defense and defense of the mandate. It respects the territorial integrity and political independence of sovereign states, always remains impartial, and strictly fulfills the mandate of the Security Council.

Championing the vision of global governance based on extensive consultation, joint contribution and shared benefits. China stays committed to building a world of lasting peace through dialogue and consultation, to combining its efforts with others to bring about a world of common security for all, and to creating a world of common prosperity through win-win cooperation, an open and inclusive world through exchanges and mutual learning, and a clean and beautiful world by pursuing green and low-carbon development.

Pursuing common, comprehensive, cooperative and sustainable security. China always respects and ensures the security of each and every country. It upholds security in both traditional and non-traditional fields, promotes the security of both individual countries and broader regions through dialogue and cooperation, and focuses on development and security so that security would be durable.

Staying committed to peaceful means in settling disputes. China advocates that disputes and differences between countries or within a country should be resolved through peaceful means. Countries should increase mutual trust, settle disputes and promote security through dialogue. Willful threat or use of force should be rejected.

Building stronger peacekeeping partnerships. China strives to bring about greater involvement of host nations, TCCs and fund contributing countries (FCCs) through UN peacekeeping reform. It leverages the role of regional and sub-regional organizations, and promotes closer partnerships in peacekeeping operations.

(*Selected from https://baijiahao.baidu.com/s?id=1678148348275563119&wfr=spider& for=pc*)

【Notes】

1. embark 着手；开始
2. UNTSO (United Nations Truce Supervision Organization) 联合国停战监督组织
3. millennia 千年期
4. vein 静脉；血管
5. advocate 拥护；提倡
6. wholeheartedly 全心全意地
7. incumbent 在职的；现任的
8. abide by 遵守；容忍

9. **mandate** 授权；任命

10. **TCCs (Troop Contributing Countries)** 出兵国

11. **FCCs (Fund Contributing Countries)** 出资国

12. **sub-regional** 次区域的

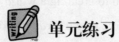 单元练习

I. Finish the following abbreviations

1. UN _____   2. HQ _____   3. AO _____
4. FC _____   5. HOM _____   6. PKO _____
7. UNSG _____   8. UNSC _____

II. Multiple choice

1. An UN peacekeeping mission comprises a Headquarters Peacekeeping Force and areas of operation called _____.

　A. battalions　　　B. sectors　　　C. brigades　　　D. troops

2. The HOM is responsible to the _____ in New York.

　A. UNSG　　　B. HQ PKF　　　C. UNSC　　　D. FC

3. Sides in a conflict don't deploy soldiers in _____.

　A. demilitarized zone　　　　B. weapon exclusion zone
　C. areas of operation　　　　D. buffer zone

4. _____ is an agreement to stop fighting.

　A. Negotiation　　　B. Demobilise　　　C. Truce　　　D. Supervise

5. Which of the following doesn't belong to demining items?

　A. visor　　　B. blue beret　　　C. prodder　　　D. body armour

6. ___ is the activity when people vote to choose someone for an official position.

　A. Election　　　B. Disarm　　　C. Patrol　　　D. Ceasefire

7. To ___ means to respect people's rights and opinions.

　A. monitor　　　B. demobilise　　　C. deploy　　　D. tolerant

8. ___ refers to the room for giving and receiving information or instructions.

　A. Briefing room　　　　B. Operations room
　C. Headquarters　　　　D. Living room

9. The ___ is a person with a lower rank.

A. subordinate   B. sergeant   C. secretary   D. superior

10. ___ means to use diplomacy to try to reach an agreement between the parties.

A. Peacekeeping   B. Peace enforcement

C. Peacebuilding   D. Peacemaking

III. Translation: English to Chinese

1. Traditionally, UN peacekeeping meant the deployment of military personnel. The Blue Berets were normally deployed in international conflicts after a truce was signed.

2. We had three principal tasks, first of all, to monitor the ceasefire, secondly, to demobilize soldiers, and thirdly, to organize elections.

3. My typical day was to get up at 6 am and to check with the soldiers that there was no fighting between the two sides in the buffer zone during the night.

4. Landmines were first used in 1862 in the American Civil War, but the biggest use of mines was in World War II.

5. We carry weapons at all times, inside and outside the compound. Even when we do PT.

IV. Translation: Chinese to English

交战规则

规则一，所有军事行动都要按照战争法进行。

规则二，军官应采取一切必要的行动，使部队随时做好自卫的准备：这意味着步枪应保持清洁，所有车辆应维护以便立即使用等等，以便部队能够随时自卫。

规则三，维和部队只有在遭到射击时才会开火。他们永远不会对任何人开火，除非他们先受到攻击。

规则四，只使用执行任务所需的最低兵力。

最后两条交战规则是关于对待当地人民的。

规则五，所有人都应该受到尊重。

最后，规则六，平民只能因自卫等安全原因被拘留。

V. Oral practice

1. What are the differences between peacemaking and peacekeeping?

2. How much do you know about Chinese troops' active efforts for international peacekeeping cooperation?

附录：Unit 5 练习答案

I. Finish the following abbreviations

1. Unitied Nations   2. Headquarter   3. Area of Operation   4. Force Commander

5. Head of Mission   6. Peacekeeping Operation

7. United Nations Secretary General   8. United Nations Security Council

II. Multiple choice

1-5 BADCB   6-10 ADAAD

III. Translation: English to Chinese

1. 传统意义上讲，联合国维和意味着要部署军事人员。蓝盔部队通常在停战协议签署后部署在国际冲突地区中。

2. 我们有三个主要任务：第一，监督停火；第二，遣散士兵；第三，组织选举。

3. 我日常的一天是早上6点起床，然后和士兵们确认在夜间在缓冲区内双方没有战斗。

4. 地雷在1862年的美国南北战争中首次使用，但使用最多的一次是在第二次世界大战期间。

5. 不管在基地院子里或者基地外面，甚至当我们做体能训练的时候，我们都要随时携带武器。

IV. Translation: Chinese to English

Rules of Engagement

Rule number one, conduct all military operations according to the laws of war. Secondly, all necessary action should be taken by the CO for the unit to be ready for self-defence at all times. This means that rifles should be kept clean, all vehicles should be maintained for immediate use, and so on, so that the unit can defend itself at all times. Thirdly, peacekeeping forces will only open fire if they are fired upon. They will never open fire against anyone unless they are attacked first. Rule number four, only the minimum force required to carry out the mission will be used. The last two rules of engagement refer to treatment of the local people. Rule number five, all people should be treated with respect. Finally, rule number six, civilians can only be detained for security reasons of in self-defence.

V. Oral practice

1. The term peacemaking refers to using diplomacy to reach an agreement between the

parties. It might use diplomacy and mediation to convince hostile parities to stop fighting and negotiate a peaceful settlement. It does not allow the use of direct military force against either of the parties. Peacekeeping uses both civilian and military personnel to help carry out and supervise the agreement after an agreement has been made. All parties involved in a conflict must consent both to a ceasefire, and to outside involvement in the resolution of the conflict. As is commonly said, before a peace can be kept, there must be a peace to keep. Peacekeeping operation forces are expected to maintain strict impartiality. If not, they risk damaging the credibility and objectivity of the mission. They may only use the minimum force necessary to defend the mission. They may not use force to coerce either side to actually cooperate with the mission.

2. World peace is the responsibility of all countries and peacekeeping calls for expanding multilateral cooperation. China's armed forces have cooperated on peacekeeping with over 90 countries and 10 international and regional organizations. They have enhanced mutual understanding, shared experience, extended practical cooperation, strengthened bilateral and multilateral relations, and promoted peacekeeping capability through exchange of visits, expert discussions, joint exercises and training, and personnel training.

Strengthening Strategic Communication to Build Consensus on Peacekeeping

Better strategic communication with the UN leadership is an important means to move the UNPKOs forward. Since 2012, President Xi Jinping has had 11 meetings with UN Secretary-Generals, proposed Chinese ideas and Chinese solutions for world peace and development on multiple international occasions, and reiterated China's support for the UNPKOs. In 2015, President Xi Jinping attended the Leaders' Summit on Peacekeeping at UNHQ and presented proposals that the basic principles of peacekeeping should be strictly followed, the peacekeeping system needs to be improved, rapid response needs to be enhanced, and greater support and help should be given to Africa. Accordingly, China's armed forces are resolved to implement the consensus reached by the leaders. They have strengthened communication with relevant UN agencies, attended several sessions of the UN Peacekeeping Defense Ministerial and the UN Chiefs of Defense Conference, and actively promoted peacekeeping cooperation.

China's armed forces are committed to strengthening bilateral and multilateral communication for better understanding and mutual trust. They have carried out active peacekeeping cooperation with the militaries of countries including Russia, Pakistan, Cambodia, Indonesia, Vietnam, France, Germany, the UK, and the US. Through reciprocal visits, China's armed forces and their foreign counterparts have strengthened communication on policies, made cooperation plans, and advanced friendly state-to-state and military-to-military relations. In May 2010, the first China-US consultation on the UNPKOs was held in Beijing. In April 2015, the defense ministers of China and Vietnam signed a memorandum of understanding (MOU) on peacekeeping cooperation between the two ministries in Beijing. That same year, China conducted the first BRICS consultation on the UNPKOs with Brazil, Russia, India and South Africa. In February 2017, the first China-UK dialogue on peacekeeping operations was held in the UK. In April 2018, military advisers of Russia, France, the UK and the US to the UN Military Staff Committee visited China and exchanged extensive views on the UNPKOs with the Chinese side. In May, the defense ministries of China and Pakistan signed a protocol on policy collaboration with regard to the UNPKOs. In October, the German defense minister visited the Training Base of the Peacekeeping Affairs Center of the Chinese Ministry of National Defense (MND), and a peacekeeping delegation from the Chinese MND visited the German Armed Forces United Nations Training Centre.

Contributing Chinese Wisdom and Sharing Experience

Sharing experience and learning from each other is an effective approach to improving the UNPKOs. China's armed forces have actively conducted international exchanges on peacekeeping. The PLA sent delegations to visit the peacekeeping training facilities of countries including Argentina, Finland and Germany, and received more than 180 visits from other countries and international organizations including the UN and the AU. China has hosted over ten international events on peacekeeping, including the Sino-UK Seminar on Peacekeeping Operations, the International Seminar on Challenges of Peace Operations—Into the 21st Century, the China-ASEAN Seminar on Peacekeeping Operations, and the 2009 Beijing International Symposium on UN Peacekeeping Operations. Meanwhile, Chinese peacekeeping troops in Mali, Sudan, South Sudan, the DRC, Liberia, and Lebanon have exchanged experience with their counterparts from France, Senegal and Spain.

China's armed forces have participated extensively in UN peacekeeping consultations and policy-making, and provided input on the UNPKOs. They have played a dynamic role in the Special Committee on Peacekeeping Operations of the UN General Assembly and the TCC Contingent-Owned Equipment (COE) Working Group, invited officials from the UN High-level Independent Panel on Peace Operations and the UN Security Council to China, and offered suggestions on reforming UN peacekeeping, raising its effectiveness, and ensuring the safety and security of peacekeepers. Expert meetings have been hosted by China to draft and review documents including the United Nations Peacekeeping Missions Military Engineer Unit Manual and the Military Peacekeeping-Intelligence Handbook, and Chinese experts have been sent to participate in updating the manuals of UN peacekeeping infantry, force protection, aviation, transport, medical support units and civil-military cooperation.

Extending Cooperation on Joint Exercises and Training to Build Capability

Joint exercises and training are important as a means of improving the UN's peacekeeping capability and its talent pool. To learn from each other and improve skills, China's armed forces have conducted various peacekeeping exercises and training with the UN, and with relevant countries and regional organizations. In June and July 2009, China and Mongolia held a joint exercise codenamed Peacekeeping Mission-2009 in Beijing. In addition, China's armed forces have sent military personnel to participate in multilateral engagements including the ADMM-Plus Experts' Working Group Table-Top Exercise on Peacekeeping Operations in the Philippines in February 2014, the Khan Quest multinational peacekeeping exercises in Mongolia from 2015 to 2019, the ADMM-Plus Experts' Working Group on Peacekeeping Operations and Humanitarian Mine Action field training exercises in India in March 2016 and in Indonesia in September 2019, peacekeeping table-top exercises in Thailand in May 2016 and May 2018, and the multinational computer-assisted command-post exercise Viking 18 in Brazil in April 2018.

China's armed forces established a specialized peacekeeping training institution in June 2009. Since then, the PLA has run over 20 international training programs for UN peacekeepers, including the UN Military Observers Course, the UN Staff Officers Course, the UN Peacekeeping Training of Trainers Course for Francophone Countries, and the UN Senior National Planners Course. The PLA has also invited UN experts and senior instructors from other countries for pre-deployment training of Chinese peacekeeping troops and military professionals, and sent instructors to assist peacekeeping training in countries including Australia, Germany, the Netherlands, Switzerland, Thailand, and Vietnam. More than 100 PLA officers have attended courses or observed exercises hosted by the UN or other TCCs.

Unit 6　Convoy

单元导学清单

模块	主题	学习目标	任务	军事知识	核心词汇
Alpha	军事护送基本知识	1）了解军事护送基本知识；2）掌握维和护送任务相关的核心军语。	Task 1 Task 2 Task 3 Task 4	■ Humanitarian aid	convoy, escort, humanitarian, rear, release point, reporting point, start point, estimated time of arrival
	军事护送简报	1）掌握下达维和护送任务指令的功能句型。	Task 5 Task 6 Task 7	■ Military briefing	brief, briefing, under sb.'s command, junction, rest area
Bravo	军事护送简报	1）了解护送车辆的编排原则；2）归纳总结护送简报要素。	Task 1 Task 2	■ Convoy formations ■ UN	compose
	军事护送路线图识读	1）掌握表示护送路线的核心词汇；2）复习并掌握地图识读的方法。	Task 3 Task 4	■ Strip map	roundabout, signpost
Charlie	军事检查站沟通协商案例	1）了解通过军事检查站的情况；2）熟练掌握军事检查站协商沟通相关的词汇与句型；3）对比分析不同的协商沟通案例，提炼经验。	Task 1 Task 2 Task 3 Task 4 Task 5 Task 6	■ Ceasefire agreement ■ Military checkpoint ■ UN peacekeepers	checkpoint, en route to, pass, issue, valid, headquarters, ceasefire, authorisation, position, command post, commanding officer, refugee camp, hold, senior non-commissioned officer
	沟通协商原则与策略	1）了解军事检查站协商沟通的原则；2）熟练掌握不同情形下的协商沟通应对策略。	Task 7 Task 8 Task 9 Task10	■ Principles of UN Peacekeeping	negotiation, guard, inspect, rebel, a restricted military zone, identity paper, military ID card, hatch, magazine, loaded, cock, discharges, insurgent, incident, tactful, diplomatic, under the terms of, non-verbal communication, irrespective

续表

模块	主题	学习目标	任务	军事知识	核心词汇
Delta	护送突发情况处置	1）了解护送途中可能出现的突发情况；2）熟练掌握与突发情况处置相关的词汇表达；3）讨论突发情况处置程序和应对方案。	Task 1 Task 2 Task 3 Task 4	■ SITREP	breakdown, repair, maintenance, roadblock, ambush
	车辆故障处理	1）识别维修工具并掌握相关词汇；2）熟练掌握与处理突发情况相关的词汇表达。	Task 5 Task 6 Task 7 Task 8	■ First aid	adjustable, wrench, jack, pliers, screwdriver, tape, spanner, tape measure, flat, tyre, headlight, change, aerial, windscreen, spare, give me a hand, to start with
Echo (Review)	军事护送行动	1）了解军事护送的基本知识、执行护送任务的步骤、护送指挥官的职责以及突发情况处置；2）熟练掌握军事护送相关的术语、词汇和表达；3）综合运用所学知识介绍军事护送中的注意事项。	Task 1 Task 2	■ Peace agreement ■ Start Point ■ Mechanised infantry ■ Debriefing	diplomatic, shipment, escorted, appropriate, in advance, refuelling, close off, in case of, towing vehicle, hostile area, withdraw, reinforcement, debriefing

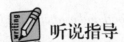 听说指导

Alpha　Convoy task 1

【军事知识】

1. Humanitarian aid 人道主义救援

Humanitarian aid, also known as humanitarian assistance, is a material or logistical support to recipients on a humanitarian basis, with the main purpose of saving lives, alleviating unfortunate situations and preserving human dignity. Provided by government agencies, non-governmental organizations and other non-governmental humanitarian agencies, humanitarian aid must be provided in accordance with the principles of humanity, neutrality and impartiality as listed in United Nations General Assembly resolution 46/182. United Nations Office for the Coordination of Humanitarian Affairs (OCHA) is the part of the United

Nations Secretariat responsible for bringing together humanitarian actors to ensure a coherent response to emergencies.

2. Mission briefing 任务部署（会）；下达（作战）命令

There are four types of briefings that are often used: information, decision, mission, and staff. Mission briefings are informal briefings that occur during operations or training. Briefers may be commanders, staff, or special representatives. Mission briefings serve to convey critical mission information not provided in the plan or order to individuals or small units. Mission briefings also ensure participants to know the mission objective, their contribution to the operation, problems they may confront, and ways to overcome them. The nature and content of the information being provided determines the mission briefing format. Typically a briefer will use the operation plan or order as a format for a mission briefing.

【词汇点拨】Words and Expressions

1. convoy /ˈkɒnvɔɪ/

① *n.* a group of ships or vehicles travelling together, typically one accompanied by armed troops, warships, or other vehicles for protection（尤指有士兵护送的）车队；舰队

② *n.* the act of escorting while in transit 护航；护送

③ *v.* to escort in transit for protection 护航；护送

A UN aid convoy loaded with food and medicine finally got through to the besieged town.

The mission for our forces is to carry out coastal convoy.

Our battleship was assigned to convoy an aircraft carrier to the Caribbean Sea.

2. escort

① /ˈeskɔːt/ *n.* a person, vehicle or aircraft or ship which accompanies an individual or group in order to protect or guard them 护送者；护卫队；护卫舰（车队、飞机）

② /ɪˈskɔːt/ *v.* to act as an escort 护航；护送；护卫

Armed escorts are provided for visiting heads of state.

The shipment was escorted by armed patrol boats.

3. humanitarian /hjuːˌmænɪˈteəriən/ *adj.* marked by humanistic values and devotion to human welfare; intended to prevent or reduce human suffering and hardship 人道主义的；人道的

humanitarian aid/assistance/relief 人道主义救援

humanitarian crisis 人道主义危机

4. armed /ɑːmd/ *adj.* (used of persons or the military) characterized by having or bearing arms 武装的；使用武器的

armed forces 武装力量

5. rear /rɪə(r)/

① *n.* the back of a military formation or position 后部；背部；（队列的）后方

② *adj.* moving toward or located at the back of a formation or position 后部的；后方的

rear area 后方

We engaged the rear platoon.

6. start point (SP) 出发点

Our start point will be the junction of AF34 with BF29.

7. reporting point a point or position on a route where vehicles report 报告点

The commander reported this accident to the control station at the second reporting point.

8. release point (RP) a point on a route where sub-units leave their parent unit and continue independently by different routes; a point where the convoy commander release control of the convoy 分进点；归建点

These trucks loaded with humanitarian relief supplies would head for Trevisham after leaving the release point.

9. estimated time of arrival (ETA) 预计到达时间

Could you please tell me the ETA of flight CA981?

10. brief /briːf/

① *n.* orders or instructions; a detailed explanation or summary 通报；指示

② *v.* to give orders or instructions or to explain a situation in detail 介绍情况；通报；指示

We received a brief on the enemy's organization.

brief sb. on sth. 就某事向某人进行简要介绍

11. briefing /ˈbriːfɪŋ/ *n.* orders or instructions; a meeting where a briefing is given（军事）会议［包括情况介绍（会）、任务部署（会）、交接班（会）、协调（会）等］

mission briefing 任务部署（会）

daily briefing 例会

12. under one's command 由……指挥

The platoon is under Captain Hidas' command from the starting point.

13. junction /ˈdʒʌŋkʃn/ *n.* a place where roads or railway lines meet each other 岔道；交叉口

at the junction of... 在……交叉口

14. rest area 休息区

There's a rest area partway up the mountain; we could make a pit stop there.

【词汇点拨】Proper Names

1. ARZIKI 虚构地名

2. GAMBA 虚构地名

【长句解读】

1. I'm going to brief you on tomorrow's convoy operation. (6-1)

➢ 句子分析：该句中 brief 作动词，brief sb. on sth. 意为"向某人简要报告某事"、"就……进行任务部署"，"convoy operation" 意为"护送行动"。

➢ 翻译：我将向大家简要介绍明天的护送行动。

2. Our mission is to escort a humanitarian aid convoy from ARZIKI to the village of GAMBA. (6-1)

➢ 句子分析：该段对话中有两个词值得注意。第一个是 escort，在该句中作动词，意为"护送、护卫"，第二个是 convoy，在本句中作名词，意为"车队"，humanitarian aid convoy 意为"人道主义救援车队"。

➢ 翻译：我们的任务是护送一支从 ARZIKI 到 GAMBA 村的人道主义援助车队。

3. The convoy is composed of four 5-ton trucks with food and medical supplies. (6-1)

➢ 句子分析：be composed of 意为"由……组成或构成"，5-ton trucks 中 ton 是重量单位"吨"，在这里，5-ton 构成复合形容词修饰 trucks，意为"5 吨重的卡车"，medical supplies 意为"医疗物资"。

➢ 翻译：车队由四辆载重 5 吨的卡车组成，卡车上运有食品和医疗物资。

4. We will provide an armed escort with three APCs. (6-1)

➢ 句子分析：armed 一词意为"全副武装的"，an armed escort 意为"武装护送"。APCs 表示 Armoured Personnel Carriers，装甲运兵车。

➢ 翻译：我们将派出三辆装甲运兵车执行武装护送任务。

【口语输出】

1. What expressions have you learned when it comes to convoy briefing?

【参考词汇】brief...on, convoy operation, humanitarian aid, an armed escort, start point, reporting point, release point, rest area, estimated time of arrival, at the junction of, at the rear

2. Suppose you were the commander, what would you emphasize in the convoy briefing?

【参考词汇】mission, convoy, arms, destination, ETA, formation, total distance

3. What should you pay attention to when reading the strip map of a specific convoy operation?

【参考词汇】start point, route, reporting point, junction, rest area, release point, destination

Bravo More convoy tasks

【军事知识】

1. Convoy formations 车队队形

The convoy must be organized to meet mission requirements and provide organizational control. The convoy commander decides how the convoy is formed for movement. The three basic types of formations are close column, open column, and infiltration. Close column provides the greatest degree of convoy control. It is characterized by vehicle intervals of 25 to 50 meters and speeds under 25 mph. Close column is normally used during limited visibility or on poorly marked or congested roads. Open column is the preferred formation used during movement. It is characterized by vehicle intervals of 100 meters or more and speeds in excess of 25 mph. Open column is normally used on well marked open roads with good visibility. Infiltration has no defined structure. Vehicle intervals and speeds will vary. This type of formation is normally not used during movement. Infiltration should only be used as a last resort in extremely congested areas or when the mission dictates.

2. UN 联合国

UN, the United Nations, is an international organization founded in 1945. Currently made up of 193 Member States, the UN and its work are guided by the purposes and principles contained in its founding Charter. The UN has evolved over the years to keep pace with a rapidly changing world. Peacekeeping is one of the most effective tools available to the United Nations in the promotion and maintenance of international peace and security.

UN Peacekeeping helps countries navigate the difficult path from conflict to peace. The Department of Peace Operations (DPO) is dedicated to assisting the Member States and the Secretary General in their efforts to maintain international peace and security.

3. Strip map 路线草图

The strip map will show a picture of the route over which the convoy will travel. The following eight items must be shown on the strip map: start point(SP), release point (RP), halts, convoy routes, major cities and towns, critical points/checkpoints (CPs), distance between CPs, and north orientation. The strip map will be detailed but not so cluttered with information that it is unreadable.

■ Release point: The RP is the place where convoy elements are released to their owning units.

■ Halts: Scheduled halts provide rest, messing, refueling/refuelling, maintenance, and schedule adjustment, while allowing other traffic to pass. Halt time is included in the road march.

【词汇点拨】Words and Expressions

1. compose /kəmˈpəʊz/ *v.* to combine together to form a whole 组成；构成

be composed of 由……组成；主要由……构成

2. roundabout /ˈraʊndəbaʊt/ *n.* [C] a place where two or more roads meet, forming a circle that all traffic must go around in the same direction（交通）环岛；环形交叉口

At the roundabout, take the second exit.

3. signpost /ˈsaɪnpəʊst/

① *v.* [usually passive] to mark a road, place, etc. 设置路标

The convoy route is well signposted.

② *n.* [C] a sign at the side of a road giving information about the direction and distance of places 路标

Follow the signposts, you will find the petrol station.

4. estimated time of departure (ETD) 预计出发时间

Inform all parties of your ETD.

【词汇点拨】Proper Names

1. Lieutenant Zumikis 祖米克斯中尉

2. Captain Hidas 希道什上尉

3. Ix 虚构地名

4. Kale 虚构地名

5. Dorf 虚构地名

6. Dune 虚构地名

7. Ort 虚构地名

8. River Pil 虚构河流

【长句解读】

1. I'll be in an APC at the front of the convoy, and I'll be followed by the first two trucks. Another APC will follow behind the first two trucks, and the other two APCs will be at the rear of the convoy, behind the other three trucks. (6-3)

> 句子分析：第一句话中 be followed by 意为"……后面跟着……"。第二句话中 follow behind 意为"跟随、紧跟其后"。at the front of 与 at the rear of 意义相反，前者表示"在……的前面"，后者表示"在……的后面"。at the rear of 与 at the back of 同义，但 at the rear of 多用于正式或官方语言。

> 翻译：我乘坐车队前面的一辆装甲运兵车，前两辆卡车跟在我后面。第二辆装甲运兵车跟在这两辆卡车后面，接着是另外三辆卡车，剩下的两辆装甲运兵车紧随其后，跟在整个车队的后面。

2. Our start point is the A5 out of KALE going east. (6-4)

> 句子分析：该句中 out of 表示从某处向该范围以外运动，用于指示方向。going east 现在分词作后置定语，解释说明 A5 的延伸方向是"向东"。

> 翻译：我们的出发点位于从 KALE 出发向东的 A5 公路。

3. Our reporting point is at a bridge about thirteen kilometers down the road. (6-4)

> 句子分析：该句中的 down the road 意为"沿着路走 13 千米处"，down 作介词。

> 翻译：我们的报告点位于这条路 13 千米处的一座桥上。

4. We'll take the A6, signposted Ort, south from Dune airport and stop to report three times during the movement. (6-5)

> 句子分析：该句中的 signposted Ort 相当于 which is signposted Ort，起补充说明作用，signpost 和 A6 之间存在逻辑被动关系，因此 signpost 须用过去分词形式。south 作副词，表示"向南"。

> 翻译：我们将从 Dune 机场向南，走标识为 Ort 的 A6 号公路，在前进过程中将停车报告三次。

【口语输出】

1. What do you think are the factors concerning the formation of vehicles in the convoy?

【参考词汇】mission, danger, personnel, cargo, armoured vehicles, communication, speed

2. What are the basic types of convoy formations?

【参考词汇】mission requirements, close column, open column, infiltration, vehicle intervals

3. What are the elements of convoy commander's briefing?

【参考词汇】mission (cargo, personnel, destination), time (ETA, ETD, total time), route (SP, RP)

Charlie　At the checkpoint

【军事知识】

1. Ceasefire agreement 停火协议

A ceasefire agreement refers to a temporary stoppage of war or any armed conflict for an agreed-upon timeframe or within a limited area. Each party to the agreement agrees with the other to suspend aggressive actions, without necessarily making concessions of any kind. These agreements are military in nature and are basically designed to stop warring parties from continuing military actions while political negotiations are conducted to find a more durable solution.

By themselves, ceasefire agreements are typically short-lived and fragile. They must be quickly followed up with further agreements if the ceasefire is to be maintained.

2. Military checkpoints 军事检查站

A checkpoint is an area where vehicles and/or persons are stopped, identities are verified, possessions searched, and a decision is made whether or not to detain the persons/vehicles or to allow them to pass. Checkpoints aim at controlling an area, to allow a "safe area" to protect from outside influence, to deny hostile intelligence gathering opportunities.

3. UN peacekeepers 联合国维和人员

The United Nations Peacekeeping Forces are employed by the UN to maintain or re-establish peace in an area of armed conflict.

UN peacekeepers come from all walks of life, with diverse cultural backgrounds and from an ever-growing number of Member States. When they serve under the United Nations they are united by a commitment to maintain or restore world peace and security. They share a common purpose to protect the most vulnerable and provide support to countries in transition from conflict to peace.

Peacekeepers are civilian, military and police personnel all working together. The roles and responsibilities of peacekeepers are evolving as peacekeeping mandates become more complex and multidimensional. Peacekeeping operations have developed from simply monitoring ceasefires to protecting civilians, disarming ex-combatants, protecting human rights, promoting the rule of law, supporting free and fair elections, minimizing the risk of land mines and much more. More than one million men and women have served under the UN flag since 1948. Tragically, more than 3,500 have lost their lives in the cause of peace.

4. Principles of UN Peacekeeping 联合国维和原则

There are three basic principles that continue to set UN peacekeeping operations apart as a tool for maintaining international peace and security. These three principles are inter-related and mutually reinforcing:

■ **Consent of the parties.** This requires a commitment by the parties to a political process. Their acceptance of a peacekeeping operation provides the UN with the necessary freedom of action, both political and physical, to carry out its mandated tasks.

■ **Impartiality.** United Nations peacekeepers should be impartial in their dealings with the parties to the conflict, but not neutral in the execution of their mandate.

■ **Non-use of force except in self-defence and defence of the mandate.** In certain volatile situations, the Security Council has given UN peacekeeping operations "robust" mandates authorizing them to "use all necessary means" to deter forceful attempts to disrupt the political process, protect civilians under imminent threat of physical attack, and/or assist the national authorities in maintaining law and order.

【词汇点拨】Words and Expressions

1. checkpoint /ˈtʃekpɔɪnt/ *n.*
① a place (usually on a border) where people or vehicles are stopped, inspected 检查站
② a place or feature on the ground which is used as a navigational reference point 检查点
The convoy failed to pass through the military checkpoint.
Our next checkpoint is the track junction at grid 339648.

2. en route to 在前往……途中
They have arrived in London en route to the United States.

3. pass /pɑːs/
① *n.* an official document authorizing the holders to do something 通行证
② *v.* to move past or to the other side of sb./sth. 通过；走过
You are not allowed to enter the barrack without a pass.
to pass a barrier/checkpoint 通过障碍；通过检查站

4. issue /ˈɪʃuː/
① *n.* [C] an important topic that people are discussing or arguing about 议题；争议问题

② *n.* [C] a problem or worry that sb. has with sth.（有关某事的）问题；担忧

③ *v.* **(~sth. to sb./~sb. with sth.)** to give sth. to sb., especially officially（正式）发给；供给

a key/sensitive/controversial issue 关键的 / 敏感的 / 有争议的问题

Money is not an issue.

to issue passports/visas/tickets 签发护照 / 签证 / 机票

5. valid /ˈvælɪd/ *adj.* something that is legally or officially acceptable 法律上有效；正式认可的

a valid passport 有效护照

This passport is invalid.

6. headquarters /ˌhedˈkwɔːtəz/ *n.* [pl.] (*abbr.* HQ)

① an administrative and command centre of a tactical grouping 司令部；指挥部

② the staff of a headquarters 指挥部人员；总部人员

report to the headquarters immediately 立刻向指挥部报告

general headquarters 陆军总司令部

7. ceasefire /ˈsiːsfaɪə(r)/ *n.* [U] a ceasefire is an arrangement in which countries or groups of people that are fighting each other agree to stop fighting 停火；休战

They have agreed to a ceasefire after three years of conflict.

8. authorisation /ˌɔːθəraɪˈzeɪʃən/ *n.* [C]

① official permission or power to do sth., the act of giving permission 批准；授权

② a document that gives sb. official permission to do sth. 批准书；授权书

You may not enter the security area without authorisation.

Can I see your authorisation?

9. position /pəˈzɪʃn/

① *n.* a place occupied by troops or equipment for tactical purposes 阵地

② *n.* [C, usually sing.] **(in a ~)** a situation that sb. is in, especially when it affects what they can and cannot do 处境；状况

The enemy positions were clearly visible in the satellite photograph.

This put him and his comrades in a difficult position.

10. command post 指挥所；战地指挥部

The command post was set up in a forward position.

11. commanding officer 指挥官

He got permission from his commanding officer to join me.

12. refugee camp shelter for persons displaced by war or political oppression or for religious beliefs 难民营；难民收容所

Troops tried to set up a lookout post inside a refugee camp.

13. hold /həʊld/ *v.* if someone holds you in a place, they keep you there as a prisoner and do not allow you to leave 扣留

The enemy is holding the commanding officer hostage.

14. senior non-commissioned officer (SNCO) 高级军士

Our candidates are exclusively Officers (including Warrant Officers) and Senior Non-Commissioned Officers.

15. negotiation /nɪˌɡəʊʃɪˈeɪʃn/ *n.* formal discussions between people who have different aims or intentions, especially in business or politics, during which they try to reach an agreement（尤指商业或政治上的）谈判

A ceasefire is an essential precondition for negotiation.

16. guard /ɡɑːd/

① *n.* [C] a guard is someone such as a soldier, police officer, or prison officer who is guarding a particular place or person 卫兵

② *n.* [U] the act or duty of protecting places or people from attack or danger 警戒；保卫

③ *v.* to protect property, places or people from attack or danger 警戒；保卫

a security guard 安全警卫

to do guard duty 担任警戒任务

Political leaders are guarded by the police.

17. inspect /ɪnˈspekt/ *v.* (**~sth./sb. for sth.**) to look closely at sth./sb., especially to check that everything is as it should be 检查；查看

The guard inspected the convoy at the checkpoint.

18. rebel

① /ˈrebl/ *n.* a peoson who fights against the government of their country 叛乱者；造反者

② /rɪˈbel/ *v.* (**~against sb./sth.**) to fight against or refuse to obey an authority 造反；反叛

rebel forces 叛乱武装

19. a restricted military zone 军事禁区

20. identity paper 身份证件

21. military ID card 军官证

22. hatch /hætʃ/ *n.* a small door or opening (as in an airplane or other vehicles) 舱门；舱口

loading hatch 装货舱口

escape hatch 安全门；逃生舱

23. magazine /ˌmæɡəˈziːn/ *n.* [U] in an automatic gun, the magazine is the part that contains the bullets 弹匣

This machine gun had an empty magazine.

24. loaded /ˈləʊdɪd/ *adj.* (of weapons) charged with ammunition 装满弹药的

He carried a loaded gun.

25. cock /kɒk/ *v.* to set the firing pin, hammer, or breech block of (a firearm) so that a pull on the trigger will release it and thus fire the weapon 扣扳机准备射击；做好战斗准备

cock a gun/pistol/rifle 扣上扳机准备射击

26. discharge /ˈdɪstʃɑːdʒ/ *n.* the act of discharging a gun 射击；开火

accidental discharges 擦枪走火

27. insurgent /ɪnˈsɜːrdʒənt/ *n.* people who are fighting against the government or army of their own country 叛乱者；反叛分子

Pakistan denies sponsoring any insurgent groups.

28. incident /ˈɪnsɪdənt/ *n.* [C] a disagreement between two countries, often involving military forces （两国间的）摩擦；冲突；常指军事冲突

a border/diplomatic incident 边境/外交冲突

29. tactful /ˈtæktfl/ *adj.* careful not to say or do anything that will annoy or upset other people 得体的；妥善的

I tried to find a tactful way of telling her the truth.

30. diplomatic /ˌdɪpləˈmætɪk/ *adj.* having or showing skill in dealing with people in difficult situations 有手腕的；灵活变通的；有策略的

a diplomatic answer 巧妙的回答

31. under the terms of 根据；根据……条款

The ship was impounded under the terms of the UN trade embargo.

32. non-verbal communication 非言语交际

People have an instinct for interpreting non-verbal communication.

33. irrespective /ˌɪrɪˈspektɪv/ *adj.* (~ **of**) without considering sth. or being influenced by it 不考虑；不管；不受……影响

Everyone is treated equally, irrespective of race.

【词汇点拨】Proper Names

1. San Carlo 虚构地名
2. UN Force Headquarters 联合国维和部队司令部
3. Divisional Headquarters 分区司令部
4. Manzanares 虚构地名
5. San Marcos 虚构地名
6. Corporal Jones 上等兵琼斯
7. Lola /ˈləulə/（人名）罗拉
8. Trevisham 虚构地名
9. Morak 虚构地名
10. Denom 虚构地名
11. Trevnik 虚构地名
12. Bouvet 虚构地名

【长句解读】

1. Your Divisional Headquarters has been informed of this convoy and they are aware that under the terms of the ceasefire agreement we have authorisation to travel freely. (6-6)

➢ 句子分析：该句中有两个并列分句，由并列连词and连接。第一句中的Divisional Headquarters 意为"分区司令部"，has been informed of 意为"已知悉"。aware that... 引导宾语从句，意为"了解；知道"，ceasefire agreement 指两国间达成的"停火协议"。

➢ 翻译：你所在的分区司令部已获悉这支车队的情况，他们也知道根据停火协议的条款，我们已获授权，可自由出行。

2. Could you radio through to your command post and get authorisation from them? (6-6)

➢ 句子分析：radio through to 中的 radio 作动词，意为"用无线电通信（或发送消息）"，相当于 contact your command post through radio；command post 表示"指挥所"。

➢ 翻译：你能用无线电联系你的指挥所并获得他们的授权吗？

3. Remember, stay calm and let me do the talking. (6-8)

➢ 句子分析：句中的 do the talking 是一个用于非正式语体的短语，通常用于表示处境较为困难情况下的沟通、谈判和解释行为。

➢ 翻译：记住，一定要保持冷静，让我来给他解释。

4. He wants the soldiers to line up at the side of the road with their identity papers. (6-9)

➢ **句子分析**：句中的动词不定式 to line up... 作宾语补足语，意为"按顺序列队"，with their identity papers 中的 with 意为"拿着……"。

➢ **翻译**：他希望士兵们拿着身份证件在路边列队。

5. Have the men open the rear hatches of the vehicles and be ready to present their military ID cards. (6-9)

➢ **句子分析**：该句为祈使句，句首的 have 为实义动词，句子结构即 have sb. do sth.。rear hatches 表示车辆的后门（后舱门）。

➢ **翻译**：让士兵们打开车辆的后门，并准备好出示他们的军官证。

6. Magazine on the weapon but not cocked. I don't want any accidental discharges. (6-9)

➢ **句子分析**：第一句并不是完整的句子，而是由两个并列的独立主格构成，由表示转折的并列连词 but 连接。Magazine on the weapon 是由"名词＋介词短语"构成的独立主格，意为"装好弹匣"，后一部分相当于 but (weapons) not cocked，可理解为是由"名词＋过去分词"构成的独立主格，都强调了战斗准备的状态。accidental discharges 意为"擦枪走火"。

➢ **翻译**：装好弹匣，但不要扣动扳机，我不希望出现任何擦枪走火的情况。

7. How long do you think it will take via Denom? (6-10)

➢ **句子分析**：via 作介词，表示"经由（某地）"，相当于 go through。

➢ **翻译**：你认为经由 Denom 前往 Trevisham 需要多长时间？

【口语输出】

1. What should you take into consideration when negotiating with a guard at the checkpoint?

【参考词汇】situation, position, pass, valid, identification, authorisation, orders, terms of ceasefire agreement, respect, polite, tactful, diplomatic, stay calm, attitude, Divisional Headquarters

2. What circumstances might you encounter when passing the checkpoint?

【参考词汇】guard, commanding officer, command post, proposal, refuse, in exchange for, supplies, permission, open fire, threaten, accidental discharges, fail, insurgent, rebel

3. What kind of strategies can be employed in negotiating?

【参考词汇】polite, professional, tactful, diplomatic, official evidence, convincing, clarify your position, bottom line, understanding, avoid problems

4. What are the dos and don'ts when negotiating at the checkpoint?

【参考词汇】Dos: respect local customs, address by rank, patient, understand, interests

Don'ts: interrupt, lose temper, humiliate

Delta Vehicle security and maintenance

【军事知识】

1. SITREP (situation report) 情况报告

A situation report (SITREP) is exactly what the name implies: a report on a situation containing verified, factual information that gives a clear picture of the "who, what, where, when, why and how" of an incident or situation. SITREPs are not intended to replace the normal detailed communications between Missions and Headquarters on specific matters. The daily SITREP should cover the period from midnight to midnight local time. The topics mentioned under the individual headlines will be reported to the extent necessary in accordance with the mandate of the mission and the situation on the ground. If a single topic takes more space than one page, the topic is commonly reported on as an annex to the SITREP for practical reasons.

2. First aid 急救

First aid is the emergency care given to the sick, injured, or wounded before being treated by medical personnel. There are several principles of first aid:

(1) Safety first. In case of an accident, bring the patient to a safe place (shade, away from a vehicle line on a highway, away from fuel leaks or minefields).

(2) Stay calm, act with care, but decisively.

(3) Assess carefully:

- Does he/she breathe and is the air passage clear?
- How does he/she react? (conscious and alert, drowsy, unconscious)
- How is the pulse?
- Is the victim bleeding?

These four situations may require immediate action. Gently assess the nature of injuries/illness and administer the necessary care. Position the patient appropriately and reassure the patient. You should remain calm. Never leave an unconscious or severely injured patient without supervision. Call for help or get somebody to call for CASEVAC at closest Medical Clinic or Headquarters.

【词汇点拨】Words and Expressions

1. **breakdown** /ˈbreɪkdaʊn/ n.

① [C] an occasion when a vehicle or machine stops working（车辆或机器的）故障；损坏

② [C,U] a failure of a relationship, discussion or system（关系）破裂；（讨论、系统）失败

a breakdown on the motorway 在高速公路上出的故障

a breakdown recovery service 车辆抢修服务

a breakdown in communications 通信中断

The breakdown of the negotiations was not unexpected.

2. repair /rɪˈpeə(r)/

① *v.* to restore sth. that is broken, damaged or torn to good condition 修理；修补；修缮

② *v.* to say or do sth. in order to improve a bad or unpleasant situation 补救；纠正；弥补

③ *n.* [C,U] an act of repairing sth. 修理；修补；修缮

to repair a car/roof/road/television 修理汽车 / 屋顶 / 道路 / 电视

repair the damage done to their relationship 弥补他们的关系遭受的创伤

repair and maintenance 检修；修理和维护

beyond repair 无法修复

under repair 正在修复

3. maintenance /ˈmeɪntənəns/ *n.*

① ~ **(of sth.)** the act of keeping sth. in good condition by checking or repairing it regularly 维护；保养

② ~ **(of sth.)** the act of making a state or situation continue 维持；保持

vehicles maintenance 车辆保养

the maintenance of international peace 维护世界和平

4. roadblock /ˈrəʊdblɒk/ *n.*

① a barrier put across the road by the authorities to stop and search vehicles 路障

② something that stops a plan from going ahead 障碍

place roadblock 设置路障

overcome the roadblock 克服障碍

5. ambush /ˈæmbʊʃ/

① *n.* [C,U] the act of hiding and waiting for sb. and then making a surprise attack on them 伏击；埋伏

② *v.* to make a surprise attack on sb./sth. from a hidden position 伏击

Two soldiers were killed in a terrorist ambush.

They were ambushed and taken prisoner by the enemy.

6. adjustable /əˈdʒʌstəbəl/ *adj.* that can be moved to different positions or changed in shape or size 可调整的；可调节的

adjustable seat belts 可调节的座位安全带

7. wrench /rentʃ/ *n.* [C] a metal tool with a specially shaped end for holding and turning things, including one which can be adjusted to fit objects of different sizes, also called a monkey wrench or an adjustable spanner 扳钳；扳手

adjustable wrench for turning off gas and water 用于关闭煤气和水的活动扳手

8. jack /dʒæk/

① *n.* [C] a device for raising heavy objects off the ground, especially vehicles so that a wheel can be changed 千斤顶，起重器（换车轮时常用）

② *v.* (~ **sth. up**) to lift sth., especially a vehicle, off the ground using a jack 用千斤顶托起

You need to jack up the car before you try to remove the wheel.

a jack of all trades 博而不精的人；万金油；三脚猫

9. pliers /ˈplaɪəz/ *n.* [pl.] a metal tool with handles, used for holding things firmly and twisting and cutting wire 钳子；夹钳

a pair of pliers 一把钳子

10. screwdriver /ˈskruːdraɪvə(r)/ *n.* a tool with a narrow blade that is specially shaped at the end, used for turning screws 螺丝刀；改锥

Dean tried to assemble the toy gun with a screwdriver.

11. tape /teɪp/

① *n.* [U] a long narrow strip of material with a sticky substance on one side that is used for sticking things together 胶带；胶条

② *v.* to fasten sth. by sticking or tying it with tape 用胶带粘住；用带子系紧

Put it in a box and tape it up securely.

12. spanner /ˈspænə(r)/ *n.* a metal tool with a specially shaped end for holding and turning nuts and bolts (= small metal rings and pins that hold things together) 扳手；扳子；扳钳

I need a spanner to change the back wheel.

13. tape measure 卷尺

Make sure the tape measure is level all the way around.

14. flat /flæt/ *adj.*

① not containing enough air, usually because of a hole 瘪了的；撒了气的

② having a level surface, not curved or sloping 水平的；平坦的

a flat tyre 爆胎

People used to think the earth was flat.

15. tyre /ˈtaɪə(r)/ *n.* a thick rubber ring that fits around the edge of a wheel of a car, bicycle 轮胎

That back tyre needs changing.

16. headlight /ˈhedlaɪt/ *n.* a large light, usually one of two, at the front of a vehicle; the beam from this light（车辆的）前灯；头灯；前灯的光束

a broken headlight 车头灯坏了

17. charge /tʃɑːdʒ/ *v.*

① **~ (sb./sth.) for sth./~ (sb.) sth. (for sth.)** to ask an amount of money for goods or a service 收（费）；（向……）要价

② **~ sb. (with sth./with doing sth.)** to accuse sb. formally of a crime so that there can be a trial in court 控告；起诉

③ **~ sb. (with sth./with doing sth.)** (formal) to accuse sb. publicly of doing sth. wrong or bad 指责；谴责

④ **~ (sth.) (up)** to pass electricity through sth. so that it is stored there 给……充电

We won't charge you for delivery.

He was charged with treason.

Opposition MPs charged the minister with neglecting her duty.

charge the battery 给电池充电

18. aerial /ˈeəriəl/ a piece of equipment made of wire or long straight pieces of metal for receiving or sending radio and television signals 天线

The aerial doesn't look very secure to me.

19. windscreen /ˈwɪndskriːn/ *n.* the window across the front of a vehicle（机动车前面的）挡风玻璃；风挡

Rioters hurled a brick through the car's windscreen.

20. spare /speə(r)/

① *adj.* that is not being used or is not needed at the present time 不用的；闲置的

② *adj.* kept in case you need to replace the one you usually use; extra 备用的；外加的

③ *adj.* available to do what you want rather than work 空闲的；空余的

④ *v.* **~ sth./sb. (for sb./sth.), ~ (sb.) sth.** to make sth. such as time or money available to sb. or for sth., especially when it requires an effort for you to do this 抽出；拨出；留出；匀出

spare cash 多余的现金

spare tyre 备用胎

in spare time 在空余时间

I'd love to have a break, but I can't spare the time just now.

21. give me a hand 帮我一下

Could you give me a hand moving this cupboard, please?

22. to start with 一开始；首先

She wasn't keen on the idea to start with.

【词汇点拨】Proper Names

1. Control station 调度站；控制站

2. Emergency procedure 应急操作程序

【长句解读】

1. If you have a breakdown, move to the side of the road and report to the control station. (6-12)

➢ 句子分析：该句为由 if 引导的条件状语从句，主句是两个由 and 连接的并列祈使句，所以根据原则，if 从句要用一般现在时态。

➢ 翻译：如果你的车遇到了故障，将车移到路边并向控制站报告。

2. If you stop at a checkpoint, don't leave your vehicle and send a SITREP, that's a situation report, to the control station. (6-12)

➢ 句子分析：由 if 引导的条件状语从句，主句是一个祈使句，句中的 that 指代 SITREP。

➢ 翻译：如果你在检查站被拦下，不要下车，立即给控制站发送 SITREP，也就是情况报告。

3. The battery's flat to start with. (6-13)

➢ 句子分析：句中 flat 表示"（电池）电力不足的，没电的"。

➢ 翻译：电池没电了，无法启动。

4. You have to clean it—you can't see a thing through it. (6-13)

➢ 句子分析：句中 can't see a thing 是常用表达，意思为"什么也看不见"。

➢ 翻译：你得把车窗擦干净，透过这个窗户什么也看不见。

【口语输出】

1. What emergencies might happen to the vehicles during the convoy?

【参考词汇】traffic accidents, control station, breakdown, checkpoint, roadblock, report, help

2. What can you do if you have a breakdown?

【参考词汇】report to the control station, move to the side of the road, check, damage, call for

3. What are the guidelines for facing an unexpected attack?

【参考词汇】seek appropriate protection, security, fire back if necessary, stop other vehicles, hostile areas, report, ask for help, stay calm

4. What tools should you prepare for vehicle maintenance?

【参考词汇】adjustable wrench, jack, pliers, screwdriver, spanner, tape, measure

5. What aspects can you check when you are trying to repair a vehicle and what can you do?

【参考词汇】battery, flat, charge, spare tyre, headlight, petrol tank, fill the tank, aerial, windscreen, clean

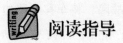

阅读指导

【文章导读】

This passage mainly focuses on convoy operations. After explaining what convoy is and the purpose of convoy, it elaborates the basic operating procedures of convoys as well as the commanders' responsibilities. Finally, the planned stops and unexpected events during convoy operations are introduced.

【军事知识】

1. Peace agreement 和平协定

Peace agreements are contracts intended to end a violent conflict, or to significantly transform a conflict, so that it can be more constructively addressed. There are various types of agreements that can be reached during a peace process. Each type of agreement has a distinct purpose and serves a value in itself towards building positive momentum for a final settlement. These agreements, however, are not easily distinguished, as the content may sometimes overlap. Not all types of agreements are needed for each conflict. Some processes may have step-by-step agreements that lead towards a comprehensive settlement. Other peace processes may seek to negotiate one agreement comprehensively.

While categorizing each document that is negotiated during a peace process is often difficult, the following are common classifications used by the United Nations to differentiate the various types of peace agreements:

- Cessation of Hostilities or Ceasefire Agreements
- Pre-Negotiation Agreements
- Interim or Preliminary Agreements
- Comprehensive and Framework Agreements
- Implementation Agreements

2. Start Point 出发点

In military operations, reconnaissance or scouting is the exploration of an area by military forces to obtain information about enemy forces, terrain, and other activities. Examples of reconnaissance include patrolling by troops, ships or submarines, crewed or uncrewed reconnaissance aircraft, satellites, or by setting up observation posts. Espionage is usually considered to be different from reconnaissance, as it is performed by non-uniformed personnel operating behind enemy lines.

3. Mechanised infantry 机械化步兵

Mechanised infantry is infantry equipped with armoured personnel carriers (APCs) or infantry fighting vehicles (IFVs) to move soldiers during heavy fire fights. They often work together with tanks in a symbolic setting, providing shelter and fire support for other troops.

Mechanised infantry is distinguished from motorized infantry in that its vehicles provide a degree of protection from hostile fire, as opposed to "soft-skinned" wheeled vehicles (trucks or jeeps) for motorized infantry. Most APCs and IFVs are fully tracked, or are all-wheel drive vehicles (6×6 or 8×8), for mobility across rough ground.

4. Debriefing 总结汇报（会）；讲评（会）

Military debriefing is used to receive information from a soldier or a pilot after a mission. It aims to analyze if mission objectives were met, the encountered challenges, and any valuable information. Military debriefing is also designed to instruct which information can be made public and censored. It is also a session which can help assess if an individual is ready to go back to work.

During debriefing sessions, unit leaders gathered information from troops returning from operations. This information concerned events that occurred on the battlefield, and so each soldier was encouraged to add to the discussion to ensure a full and accurate account of the operation.

Before long, the additional psychological benefits of debriefing sessions became apparent. By giving the soldiers a voice and respecting their experiences, unit leaders were reinforcing group coherence and increasing morale. Sessions also offered soldiers a chance to purge themselves of emotional weight as they recounted events and acknowledged grief.

【词汇点拨】Words and Expressions

1. diplomatic /ˌdɪpləˈmætɪk/ *adj.* connected with managing relations between countries 外交的；从事外交的

to break off/establish/restore diplomatic relations with a country 断绝 / 建立 / 恢复外交关系

2. shipment /ˈʃɪpmənt/ *n.*

① [U] the process of sending goods from one place to another 运输；运送；装运

② [C] a load of goods that are sent from one place to another 运输的货物

the illegal shipment of arms 非法军火运输

arms shipments 运送的几批军火

3. appropriate /əˈprəʊpriət/ *adj.* ~ **(for/to sth.)** suitable, acceptable or correct for the particular circumstances 合适的；恰当的

as appropriate 酌情；视情况而定

4. rank /ræŋk/ *n.*

① [C,U] the position, especially a high position, that sb. has in a particular organization, society, etc.（尤指较高的）地位；级别

② [C,U] the position that sb. has in the army, navy, police, etc. 军衔；军阶；警衔

officers of junior/senior rank 有低级 / 高级军衔的军官

5. in advance 预先；提前；事先

The subject of the talk is announced a week in advance.

6. planned /plænd/ *adj.* planned in advance 有计划的；根据计划的

The exercises went exactly as planned.

7. unexpected /ˌʌnɪkˈspektɪd/ *adj.* if sth. is unexpected, it surprises you because you were not expecting it 出乎意料的；始料不及的

an unexpected result 意想不到的结果

8. refuel /ˌriːˈfjuːəl/ *v.* to fill sth., especially a plane, with fuel in order to continue a journey; to be filled with fuel（尤指给飞机）补充燃料，加燃料；加油

The planes needed to refuel before the next mission.

9. close off 封锁；隔绝；使隔离

They have to close off the stores.

10. in case of 假如；万一；若发生某事

In case of an accident, bicycle riders should know how to give first aid.

11. towing vehicle 牵引车；拖拽车

The towing vehicle is towed on other vehicles travelling on the high way according to the electromagnetic principle.

12. hostile /ˈhɒstaɪl/ *adj.*

① **~ (to/towards sb./sth.)** very unfriendly or aggressive and ready to argue or fight 敌意的；敌对的

② belonging to a military enemy 敌军的；敌人的

The politician was openly hostile towards this reform.

hostile territory 敌方领土

hostile area 敌对区域

13. withdraw /wɪðˈdrɔː/ *v.* **~ (sb./sth.) (from sth.)** to move back or away from a place or situation; to make sb./sth. do this（使）撤回，撤离

The troops were forced to withdraw.

14. reinforcement /ˌriːɪnˈfɔːsmənt/ *n.*[pl.] extra soldiers or police officers who are sent to a place because more are needed 援军；增援警力

to send in reinforcements 派出增援部队

15. debriefing /ˌdiːˈbriːfɪŋ/ *n.* A debriefing is a meeting where someone such as a soldier, diplomat, or astronaut is asked to give a report on an operation or task that they have just completed. 总结汇报（会）；讲评（会）

A debriefing would follow this operation, to determine where it went wrong.

【词汇点拨】**Proper Names**

Peace agreement 和平协定

【参考译文】

军事护送

即便签订了和平协定，出入冲突地区或在其周边活动也是很危险的。为保护军事和外交要员的人身安全，保护人道主义援助物资安全抵达，必须进行护送。护送车队一般由两台或多台车辆编队，以维护秩序。护送车队可以有也可以没有押运（押运就是由多台车、多名人员组成小队，为需要护送的车辆提供保护）。每个有押运的护送车队均有明确的行动指令，该指令依据具体任务及整个局势制定。不过，几乎所有的护送车队都遵循以下基本程序：

- 护送/押送的一般规程
  - 每辆车都要始终使后车在其视线范围内
  - 尽可能白天行进
  - 尽可能沿预定路线行驶
  - 每台车辆都配一幅地图（有时也称路线图或路线专用地图）

- 指挥
  - 车队队长负责组织车队和保护车队，规划行进路线，提供地图，向有关人员说明情况。
  - 从出发点开始，车队队长全权负责车队的所有车辆；出发点即护送车辆与被护送车辆汇合，护送正式开始的地点。
  - 车队队长在整个车队到达分进点（交接点）之前，对车队负全责；所谓分进点，即护送车队把被护送车队送至目的地或预先商定的脱离点（由此地点开始，车辆按各自方向继续行进）。（注：车队队长对所有车辆及车辆里的人员负责。不论被护送的人员军衔高低；比如，即便是一名上尉护送一位将军，在达到交接点之前，也是上尉说了算。）
  - 车队队长一般乘坐第一辆车行进，并用无线话机同其他车辆保持联络（或至少与尾车保持联络）。

- 计划停靠
  - 应提前规划好中途必要的停靠点，并且尽可能设在主路线上。计划停靠包括休息、加油和吃饭。

- 意外事件

  在遭遇紧急情况时，应采取以下步骤以减少风险：
  - 交通事故：封闭事故区，警示其他车辆，向下一个指挥部报告事故情况；（如果指挥部未联系到军事警察）联系军事警察。

□ 车辆抛锚：如车辆抛锚（车辆故障），（尽可能）将车辆移至路边；向下一个指挥部打报告；车辆能修则修，必要时可以呼叫拖车。

□ 遭遇伏击（突遭袭击）：寻找恰当的保护；必要时还击；阻止其他车辆进入敌对区域；决断是退还是进；把交火情况向下一个指挥部汇报，下一个指挥部应立刻调集人员给予援助或增援。

在到达目的地后，车队队长可视具体情况下达行动命令，比如就餐、住宿或是返程等。待返回营部后，车队队长要作简要汇报（说明护送车队遭遇的情况）。

词汇表

Alpha Convoy task 1

convoy 护航；护送；车队

escort 护送者；护卫队；护卫

blanket 毯子；毛毯

medical supplies 医疗物资；医疗用品

humanitarian aid 人道主义救援

commander 指挥员；指挥官

briefing（军事）会议［包括情况介绍（会）、任务部署（会）、交接班（会）、协调（会）等］

compose 组成；构成

armed 武装的；使用武器的

rear 后面（的）；后部（的）

start point 出发点

reporting point 报告点

release point 分进点；归建点

intersection 十字路口；交叉路口；交点

estimated time of arrival 预计到达时间

rest area 休息区

route 路线；路径

petrol station 加油站

junction 交叉路口；汇合处

service station 服务区；服务站；加油站

total distance 全程

Bravo More convoy tasks

APC 装甲运兵车

truck 卡车

estimated time of departure 预计出发时间

exit 出口

roundabout （交通）环岛；环形交叉口

signpost 路标；指示牌；设置路标

rear area 后方（地域）

insignia 标识；徽章；军装服饰

Charlie At the checkpoint

checkpoint 检查站

negotiation 协商；沟通

sergeant 中士

a restricted military zone 军事禁区

come through 安然度过；经过
turn around 掉头
sector 战区；防区；分区
en route to 在前往……途中
permission 同意；许可；权限
pass through 经过……；通过……
go through 通过
pass 通行证
United Nations 联合国
UN Force Headquaters 联合国部队总部
valid 有效的；正当的；合法的
Divisional Headquarters 分区总部；分区司令部
inform 通知；告知
aware 知道的；明白
under the terms of 根据……条款
ceasefire agreement 停火协议
authorisation 授权；批准
let...through 允许通行；允许通过
position 处境；地位；状况
command post 指挥所
commanding officer 指挥官
in the town 在城镇上
refugee camp 难民营
get into trouble 使自己或他人陷入困境
hold 扣留；控制
avoid 避免；防止
corporal 下士
barrier 障碍；关卡
in luck 运气好
get used to 习惯于
guard 卫兵
inspect 检查；查看

rebel 叛乱者；反叛分子
the armoured vehicles 装甲车
line up 依次排开；按顺序排好队
identity papers 身份证件
interrupt 打断；打岔；擦嘴
UN Peacekeepers 联合国维和人员
be taken prisoner 被俘虏
hatch 舱门；舱口
military ID cards 军官证
acceptable 可接受的；允许的
magazine 弹匣
cock 扣上扳机准备射击
accidental 意外的
discharge 射击；开火
identification 身份证明
via 经过；经由
due to 由于
carry on 继续
take care 注意；小心
insurgent 叛乱者；叛乱的
order 命令；指示

Delta Vehicle security and maintenance
breakdown 故障；损坏；关系破裂；失败
maintenance 维护；保养
repair 修理；修缮；弥补
emergency 突发情况；紧急情况
roadblock 路障；障碍
traffic accidents 交通事故
first aid 急救
control station 控制站；调度站
Situation Report (SITREP) 情况报告
flat 瘪了的；撒了气的

tyre 轮胎
ambush 伏击；埋伏
adjustable 可调整的；可调节的
wrench （英式英语）扳手
pliers 老虎钳；钳子
screwdriver 螺丝刀
tape 胶带
spanner （美式英语）扳手
tape measure 卷尺
broken 破损的；破坏的
headlight 汽车前灯；头灯
to start with 一开始；首先
charge 充电
spare 备用的
petrol tank 油箱
battery 电池
aerial 天线
windscreen 汽车挡风玻璃
give a hand 帮助某人

Echo Review
conflict 冲突
peace agreement 和平协定
diplomatic personnel 外交人员

shipment 物资；货物
keep...in sight 保持在视线范围内
established 确定的；既定的
force operation 部队军事行动
emergency personnel 应急人员
brief 介绍情况；通报；指示
officially 正式地；官方地
destination 目的地
separation 分离；分别
in charge of 负责；主管
maintain 保持；维持
at the very least 至少；起码
planned 有计划的；根据计划的
in advance 提前；预先
refuel 加油
unexpected 意外的
in case of 假如；万一
towing vehicle 拖车
hostile 敌对的
withdraw 撤退
reinforcement 增援；增兵
as appropriate 酌情；视情况而定
battalion headquarters 营部
debriefing 总结汇报（会）；讲评（会）

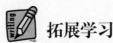 拓展学习

Convoy Organization

The organization of a convoy consists of the following:

Organizational elements

A convoy commander can better control a convoy if it is broken into smaller, more manageable groups. Whenever possible, convoys are organized along organizational lines, such as platoon, company, and battalion. The three organizational elements of a convoy are a march column, a serial, and a march unit. They are described as follows:

• A march column is a group of two to five serials. It represents approximately a battalion-to-brigade size element. Each column has a column commander.

• A serial is a subdivision of the march column. It consists of elements of a march column (convoy) moving from one area over the same route at the same time. All the elements move to the same area and are grouped under a serial commander. The serial commander is directly responsible to the convoy commander. A serial may be divided into two or more march units.

• A march unit is a subdivision of the serial. It comes under the direct control of the march unit commander. It is the smallest organized subgroup of the convoy and usually will not exceed 20 vehicles.

Functional elements

All convoys, regardless of size, are made up of three functional elements. These elements consist of a head, a main body, and a tail. These elements are explained as follows:

The head is the first vehicle of each column, serial, or march unit. It carries the pacesetter, who sets the pace to maintain the prescribed schedules and rates of march. The pacesetter leads the convoy on the proper route. With the head performing these duties, the convoy commander is free to move up and down the convoy to enforce march discipline.

The main body follows right behind the head (pacesetter) and consists of the majority of vehicles in the convoy. It is the largest part of the convoy. It can be subdivided into serials and march units for easier control and management.

The tail is the last section of a march element. The tail consists of recovery, maintenance, and medical support. The tail officer is responsible for march discipline, breakdowns, straggling vehicles, and control at the scene of any accident involving his march unit until the arrival of civilian authorities.

Types of formation

The convoy must be organized to meet mission requirements and provide organizational control. The convoy commander decides how the convoy is formed for movement. The three basic types of formations are close column, open column, and infiltration. They are described as follows:

• Close column provides the greatest degree of convoy control. It is characterized by vehicle intervals of 25 to 50 meters and speeds under 25 mph. Close column is normally used during limited visibility or on poorly marked or congested roads.

- Open column is the preferred formation used during movement. It is characterized by vehicle intervals of 100 meters or more and speeds in excess of 25 mph. Open column is normally used on well-marked open roads with good visibility.

- Infiltration has no defined structure. Vehicle intervals and speeds will vary. This type of formation is normally not used during movement. Infiltration should only be used as a last resort in extremely congested areas or when the mission dictates.

NOTE: Dimension and weight limitations on vehicles vary greatly. Check local rules and restrictions before any military motor movement. However, for gross planning purposes, vehicles are normally considered over dimensional or overweight if they exceed the following:

Width	102 inches
Height	162 inches (13 feet, 6 inches)
Weight	20,000 pounds for single axles
	34,000 pounds for tandem axles
	80,000 pounds for gross weight
Length	48 to 60 feet for semitrailers

Vehicle placement

The placement of the vehicles in an organizational element of a convoy is determined by many factors. One of the major factors is the danger of rear-end collisions. To reduce the possibility of injury to personnel, place vehicles transporting troops in the first march unit of the main body of the convoy. When empty trucks or trucks loaded with general cargo are available, use them as buffer vehicles between those transporting personnel and those loaded with hazardous cargo. Other factors to consider include the following:

- Position those vehicles that require the longest unloading time near the front of the main body of the convoy. This will shorten the turnaround time.

- If the convoy consists of vehicle-trailer combinations, have one prime mover without trailer (bobtail) per 10 vehicle-trailer combinations to support the recovery operations.

- Place vehicles transporting hazardous cargo in the last serial of the convoy but not in the trail party. (767 words)

(*Selected from https://www.globalsecurity.org/military/library/policy/army/fm/55-65/ch5.htm*)

【Notes】

1. serial 连续的
2. subdivision 分支；分部
3. organizational 组织的；结构上的
4. functional 功能的；职能的
5. pacesetter 引导的；步调调整者
6. tail 尾部
7. close column 密集纵队
8. open column 开放纵队
9. infiltration 渗透式
10. bobtail 截短的尾巴；短尾

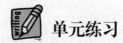单元练习

I. Match the items with the descriptions

Crossroads — 1. a sign at the side of a road giving information about the direction and distance of places

Exit — 2. a place where two roads meet and cross each other

Roundabout — 3. a place where vehicles can leave a road to join another road

Signpost — 4. a place where three or more roads meet, forming a circle that all traffic must go around in the same direction

1._____ 2._____ 3._____ 4._____

II. Multiple choice

1. At this point during a convoy, the sub-units leave their parent unit and continue independently by different routes. This location refers to a (an) _____.

A. start point B. reporting point
C. release point D. ending point

2. _____ refers to the time when a plane, ship or vehicle is expected to depart.

A. Estimated time of arrival B. Estimated time of departure

Unit 6　Convoy

C. Start point　　　　　　　　D. Departure time

3. Convoys and escorts are often used to protect important military and diplomatic personnel and shipments of _____.

　　A. big companies　　　　　　B. humanitarian aid
　　C. non-governmental organizations　　D. UN forces

4. When negotiating, it is important to address the person you are dealing with by _____.

　　A. rank　　B. age　　C. title　　D. name

5. Which of the following doesn't belong to the basic operating procedures of convoy?

　　A. Travel during daylight hours.　　B. Keep the vehicle behind in sight.
　　C. Travel along established routes.　　D. Travel along the shortest route.

6. The _____ is responsible for organising the convoy and its protection.

　　A. personnel with the highest rank in the convoy

　　B. commanding officer

　　C. convoy commander

　　D. guide

7. If you stop at a checkpoint, don't leave your vehicle and send a _____ to the control station.

　　A. SITREP　　B. FIREREP　　C. message　　D. signal

8. The convoy commander usually travels in the _____ vehicle and maintains radio contact with other vehicles.

　　A. first　　B. second　　C. third　　D. last

9. Necessary stops during the convoy include rest stops, _____, meals and reporting.

　　A. briefing　　B. refuelling　　C. preparation　　D. reinforcement

10. Which one of the following is not mentioned as the common unexpected events during a convoy?

　　A. Traffic accidents　　B. Breakdowns　　C. Ambush　　D. Cyber attack

III. Translation: English to Chinese

1. This is a UN convoy en route to San Carlos. We have permission to pass through this sector.

2. Your Divisional Headquarters has been informed of this convoy and they are aware that under the terms of the ceasefire agreement we have authorisation to travel freely.

3. Do make sure you know what your non-verbal communication means in the local culture.

4. The second situation is less successful because the escort commander is confrontational and the situation is potentially explosive.

5. I promise you that if you do not let us through, I will report this incident to your headquarters and you will be in trouble.

IV. Translation: Chinese to English

中国维和步兵营根据联合国驻南苏丹特派团授权和朱巴战区命令，需将载有联合国物资的 21 辆大型运输车护送至 300 千米外的姆沃洛地区，返程时再护送肯尼亚部队车辆物资返回朱巴。这次长途武装护卫任务由维和步兵营步兵三连承担，共出动人员 85 人，车辆 17 台。刚刚出发不久，护卫车队经过一处武装检查站，武装人员对车辆上的所有物资进行开箱检查，并索要大量过路费。双方经过两个多小时的协商，车队才被放行。一路上，像这样的武装盘查多达 9 次。在行程 600 多千米的武装护卫中，车队多次遇到突发状况，在车队指挥员的带领下，官兵们 7 次建立临时行动基地，原计划 7 天完成的任务，最终耗时 11 天。

V. Oral practice

1. What are the responsibilities of the convoy commander before the convoy, during the convoy and after the convoy?

2. What are the general principles of negotiation at the checkpoint?

附录：Unit 6 练习答案

I. Match the items with the descriptions

1. signpost   2. crossroads   3. exit   4. roundabout

II. Multiple choice

1-5 CBBAD   6-10 CAABD

III. Translation: English to Chinese

1. 我们是前往圣卡洛斯的联合国车队，我们获准可以通过这个防区。

2. 你所属的分区司令部已获悉这支联合国车队的情况，他们知道根据停火协议的条款，我们已获授权可以自由行驶。

3. 你一定要清楚你的非语言交流在当地文化中的含义。

4. 第二种情况就不那么成功了，因为护送指挥官的语言充满挑衅，双方间有可能会爆发冲突。

5. 我向你保证，如果你不让我们通过，我就向你的总部报告，那你就有麻烦了。

IV. Translation: Chinese to English

According to the mandate of the United Nations Mission in South Sudan (UNMISS) and the command of the Juba Sector, the Chinese peacekeeping infantry battalion was required to escort 21 large transport vehicles carrying UN supplies to the Mwolo area, 300 kilometers away, and then escort the Kenyan troops' vehicles and supplies back to Juba when they returned. This long-distance armed escort mission was undertaken by the Third Infantry Company of the Peacekeeping Infantry Battalion, involving 85 personnel and 17 vehicles. Shortly after setting off, the convoy passed an armed checkpoint. Armed men unpacked the vehicles for all their supplies to be inspected and demanded a large toll. The convoy was finally released after more than two hours of negotiations between the two sides. Along the way, there were as many as nine armed stops like this. During the armed escort covering more than 600 kilometers, the convoy encountered unexpected situations many times. Under the leadership of the convoy commander, the soldiers established the temporary operation base for 7 times. The task originally planned to be completed in 7 days took 11 days in the end.

V. Oral practice

1. Before the convoy starts, the convoy commander is responsible for giving the convoy briefing. In the briefing, the personnel should be informed of the general information of the mission, including date, distance, route, ETA, ETD, start point, release point and so on. The commander is also in charge of organising both the convoy and its protection. During the convoy, the commander should supervise the movement of the convoy and have contact, communications with all subordinate commanders and vehicle drivers during the movement. Besides, it is important to provide a movement report to the next higher HQ. Stop as scheduled for rest and meals, if it is a long-distance escort mission. Stay alert to unexpected attack or ambush during the mission and get prepared for any emergency. When the mission is completed, the commander should give orders for further activity, meals, accommodation and the return journey. Last but not the least, the commander will give a debriefing to the personnel involved.

2. The principles concerning negotiation at the checkpoint are as follows: first, the politeness principle. It is of great importance to make a special effort to be polite. Second, the cooperative principle, which means to offer evidence with official language to make yourself convincing. Third, the stand principle, that is to clarify your position and stick to your bottom line. Last, the diplomatic principle. During the negotiation, it is suggested that you can give a warning if necessary.

Unit 7 Patrol

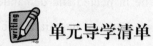 单元导学清单

模块	主题	学习目标	任务	军事知识	核心词汇
Alpha	巡逻	1）了解巡逻基本常识，包括巡逻类型、巡逻方式、巡逻装备等；2）掌握巡逻装备词汇和表达。	Task 1 Task 2 Task 3 Task 4	■ Flak jacket ■ Kevlar helmet ■ Blue UN beret	patrol, camouflage, pack, kit, bayonet, webbing, waterproof, torch
Bravo	巡逻路线与任务	1）了解巡逻基本任务；2）制订巡逻路线注意事项；3）学习在地图上标注巡逻路线。	Task 1 Task 2 Task 3 Task 4	■ Observation Post ■ Reconnaissance	check, conduct, observe, prevent, set up, establish, stay on, forecast, mission, mine
Charlie	巡逻与报告	1）描述巡逻途中观察到的情况；2）了解交战规则；3）熟练使用巡逻词汇与术语。	Task 1 Task 2 Task 3 Task 4	■ AK-47 ■ Sniper ■ ROE (Rules of Engagement) ■ Militia	on foot, uniform, suspicious, impatient, vicinity, authorize, disarm, engage, civilian, in danger
Delta	无线电通信	1）了解无线电通信的特点；2）熟练掌握无线电通信用语。	Task 1 Task 2 Task 3 Task 4 Task 5	■ Radio communications ■ Call sign	close down, destination, mobile from, radio check, roger, acknowledge, affirmative, negative, wilco, read
Echo	报告与请求	1）运用无线电通信知识报告情况并请求支援；2）熟练掌握相关的词汇和表达方式。	Task 1 Task 2 Task 3	■ Evacuation ■ Ambulance	under control, hostile, on the spot, site, extract, evacuation
Foxtrot (Review)	巡逻	1）巡逻类型及方式；2）巡逻任务。	Task 1 Task 2	■ United Nations peacekeeping missions ■ UN peacekeeping missions after the Cold War	reconnaissance, confirm, supervise, terrain, halt, debrief

Unit 7 Patrol

 听说指导

Alpha Patrol kit

【军事知识】

1. Flak jacket 防弹背心

Flak jacket is a special jacket that is worn by soldiers, police officers, etc., for protection against flak or bullets—called also flak vest, originally used by air force pilots.

2. Kevlar helmet 卡夫拉头盔

Kevlar (trademark) is a strong, lightweight synthetic fiber developed to substitute for steel in automobile tires, cables, and many other products. It is lighter than nylon and five times as strong as steel.

3. Blue UN beret 蓝色贝雷帽

Blue UN beret is an informal name for a soldier of a United Nations peacekeeping force.

【词汇点拨】Words and Expressions

1. patrol /pəˈtrəʊl/ *v.* to walk or go around or through an area, building, etc., especially in order to make sure that is safe 巡逻

The squad had orders to patrol the area.

2. camouflage /ˈkæməˌflɑːʒ/ *n.* a way of hiding something (such as military equipment) by painting it or covering it with leaves or branches to make it harder to see; the green and brown clothing that soldiers and hunters wear to make them harder to see 伪装；迷彩服

The rabbit's white fur acts as a camouflage in the snow.

3. pack /pæk/

① *v.* to put something into a bag, suitcase, etc., so that you can take it with you 打包；捆扎

② *n.* a container made to be carried on the back of a person or animal 背包

We have only one week to pack and move out of our apartment.

They loaded the packs onto the horses.

4. kit /kɪt/ *n.* a set of parts ready to be made into sth.; a set of tools or equipment that you use for a particular purpose; a set of clothes and equipment that you use for a particular activity 成套设备；全套服装

Even the little things like the food at the training ground or the training kit, everything is different and you have to get used to it.

5. bayonet /ˈbeɪənɪt/ *n.* a long, sharp blade that can be fixed to the end of a rifle and used as a weapon 刺刀；枪刺

A soldier plunged a bayonet into his body.

6. webbing /ˈwebɪŋ/ *n.* strong strips of cloth that are used to make belts, etc., and to support the seats of chairs, etc.（用以制作带子等的）带状结实织物；作战装具挂带（携行具）

Nylon webbing 尼龙带

7. waterproof /ˈwɔːtəpruːf/ *adj.* that does not let water through or that cannot be damaged by water 不透水的；防水的；耐水的

waterproof notebook 防水笔记本

waterproof clothing 防水衣

a waterproof camera 防水照相机

8. torch /tɔːtʃ/ *n.*

① a small electric lamp that uses batteries and that you can hold in your hand 手电筒

② a long piece of wood that has material at one end that is set on fire and that people carry to give light 火炬；火把

Shine the torch on the lock while I try to get the key in.

a flaming torch 燃烧着的火炬

the Olympic torch 奥林匹克火炬

【词汇点拨】Proper Names

1. Camel Bak water bladder 驼峰储水袋

2. Firesteel lighter 火石打火机

3. Leatherman Supertool 莱瑟曼多用途工具

【长句解读】

1. It's very popular in the United States and many US marines carry a Camel Bak on operations and exercises. (7-2)

➢ **句子分析**：句子中的 popular 意为"颇受欢迎"；operation 指作战行动，exercise

指平时训练。

➢ **翻译**：驼峰储水袋在美国颇受欢迎，许多美国海军陆战队士兵在作战和训练中携带这种储水袋。

2. The Firesteel was invented by a captain in the Swedish Army. It's made of aluminium. You can light about 12,000 fires with the Firesteel. (7-3)

➢ **句子分析**：句中 Firesteel 是 Firesteel lighter 的简称，为简洁起见，省掉了 lighter 一词，Firesteel 是品牌名称；词组 light fire 意为"点火"。

➢ **翻译**：一名瑞典陆军上尉发明了火石打火机。这种铝质打火机可以点火 12000 次。

3. The Leatherman has 17 different tools, including pliers, different knives and four different screwdrivers. (7-4)

➢ **句子分析**：Leatherman 是品牌名称，即"莱瑟曼"牌多用途工具。

➢ **翻译**：莱瑟曼集 17 种不同的工具为一体，包括各种钳子、刀具以及四种不同类型的螺丝刀。

【口语输出】

1. If you are to conduct a patrol, what equipment would you like to pack?

【参考词汇】first aid kit, flak jacket, Kevlar helmet, sleeping bag, water bottle, personal weapon, torch

2. Use adjectives to describe your favorite piece of kit.

【参考词汇】useful, convenient, portable, comfortable, camouflage, light, multipurpose, excellent, durable

3. What's the purpose of a UN patrol?

【参考词汇】to confirm..., to supervise..., to find out..., to check..., to observe..., to show UN presence

Bravo Patrol route and tasks

【军事知识】

1. Observation Post 观察哨

An observation post (commonly abbreviated OP), temporary or fixed, is a position from which soldiers can watch enemy movements, to warn of approaching soldiers (such as in trench warfare), or to direct artillery fire. In strict military terminology, an observation post is any preselected position from which observations are to be made-this may include very temporary installations such as a vehicle parked as a roadside checkpoint, or even an airborne aircraft.

2. Reconnaissance 侦查

A mission undertaken to obtain, by visual observation or other detection methods, information about the activities and resources of an enemy or potential enemy, or to secure data concerning the meteorological, hydrographic, or geographic characteristics of a particular area. It is also called RECON.

【词汇点拨】Words and Expressions

1. check /tʃek/ *v.* the act or an instance of inspecting or testing something, as for accuracy or quality 查证；核查

We should check the equipment to make sure that it's working properly.

2. conduct /kənˈdʌkt/ *v.* to direct the course of; manage or control 指导；管理；控制

a police officer who conducts traffic; a scientist who conducts experiments.

3. observe /əbˈzɜːv/ *v.* to be or become aware of, especially through careful and directed attention; notice; to watch attentively 观察

Our primary mission is to observe and report all activities in the area.

4. prevent /prɪˈvent/ *v.* to keep from happening; to keep (someone) from doing something 阻止

prevented us from winning

prevented the disease from spreading.

5. set up to prepare for something by putting things where they need to be 设立

The church set up a school for the city's homeless children.

The commander ordered to set up a checkpoint on the crossroad.

6. establish /ɪˈstæblɪʃ/ *v.* to start or create an organization, a system, etc. that is meant to last for a long time 设立

The patrol is tasked to establish an OP position on the hill for three hours.

7. forecast /ˈfɔːkɑːst/

① *v.* to estimate or predict in advance, especially to predict (weather conditions) by analysis of meteorological data 预测；预报

② *n.* a statement about what will happen in the future, based on information that is available now 预测；预报

weather forecast 天气预报

They're forecasting rain for this weekend.

8. mission /ˈmɪʃən/ *n.* a task or job that somebody is given to do 任务

go on a rescue mission (= go somewhere to rescue someone) 执行营救任务

9. mine /ˈmaɪn/

① *n.* a bomb that is placed in the ground or in water and that explodes when it is touched, also landmine 地雷

② *v.* to place mines below the surface of an area of land or water; to destroy a vehicle with mines 埋雷于；布雷于；用雷炸毁（车辆）

The enemy had mined the harbor.

The road was mined.

【词汇点拨】Proper Names

1. SILA /ˈsaɪlə/ 村庄名

2. Hill 602 山名

3. ABA 河流名称

【长句解读】

1. Reports from other patrols indicate the area is calm, although we will probably pass a police checkpoint on the route from base to SILA. (7-6)

➤ 句子分析：indicate 意为"显示"；calm 表示"平静，无事，安全"；although 引导让步状语；SILA 表示巡逻途经的地点；police checkpoint 的存在预示着巡逻区域可能不太平。

➤ 翻译：从基地到 SILA 要经过一个警察检查站，但来自其他巡逻队的报告显示，该区域目前平安无事。

2. Walk on the road. Don't walk on the side of the road. (7-6)

➤ 句子分析：这是祈使句的典型用法，军事语境下多以命令、指示的形式出现。on the road 意思是"沿着主路前进"，不是"在路上"；on the side of the road 指的是"偏离主路"。根据上文提示，这个地区仍有很多地雷，所以走大路比较安全。

➤ 翻译：沿着主路行进，不要偏离大路。

【口语输出】

1. What's the main task of a patrol?

【参考词汇】set up an observation post, patrol the border, conduct a reconnaisance, set up a roadblock

2. What should we do when patrolling in an area that has been mined?

【参考词汇】watch carefully, stay on the main road, do not walk on the side of the road, call for help if necessary

3. How could we carry out a patrol task?

【参考词汇】it depends on the circumstances and the need, on foot, in vehicles, by air, by boat, on skis, etc.

Charlie Patrol and report

【军事知识】

1. AK-47 卡拉什尼科夫突击步枪

The AK-47 is a selective-fire, gas-operated 7.62mm×39mm assault rifle, first developed in the USSR by Mikhail Kalashnikov. It is officially known as Avtomat Kalashnikova. It is also known as a Kalashnikov, an AK or in Russian slang, Kalash.

Design work on the AK-47 began in the last year of World War II (1945). After the war in 1946, the AK-46 was presented for official military trials. In 1948 the fixed-stock version was introduced into active service with selected units of the Soviet Army. In 1949, the AK-47 was officially accepted by the Soviet Armed Forces and used by the majority of the member states of the Warsaw Pact. The AK-47 remains the most widely used and popular assault rifles in the world because of their durability, low production cost, and ease of use. It has been manufactured in many countries and has seen service with armed forces as well as irregular forces worldwide.

2. Sniper 狙击手

A sniper is a military/paramilitary marksman who engages targets from positions of concealment or at distances exceeding the target's detection capabilities. Snipers generally have specialized training and are equipped with high-precision rifles and high-magnification optics, and often also serve as scouts/observers feeding tactical information back to their units or command headquarters.

In addition to long-range and high-grade marksmanship, military snipers are trained in a variety of special operation techniques: detection, stalking, target range estimation methods, camouflage, field craft, infiltration, special reconnaissance and observation, surveillance and target acquisition.

3. ROE (Rules of Engagement) 交战规则

A military directive that delineates the limitations and circumstances under which forces will initiate and prosecute combat engagement with other forces encountered. For example, UN peacekeepers are not allowed to use deadly force or open fire unless their lives are in immediate danger.

4. Militia 民兵

Militia may refer to 1) an army composed of ordinary citizens rather than professional soldiers; 2) a military force that is not part of a regular army and is subject to call for service in an emergency; 3) the whole body of physically fit civilians eligible by law for military service.

【词汇点拨】Words and Expressions

1. on foot 徒步；步行

I knew the rest of the expedition would be on foot, and I'd hope to be leaving the bike at this place.

The station was not very far from his office, and Mr White always went on foot, and he always went along the same street.

2. uniform /ˈjuːnɪfɔːm/ *n.* the special set of clothes worn by all members of an organization or a group at work, or by children at school 制服

soldiers in uniform; militany uniform

3. suspicious /səsˈpɪʃəs/ *adj.* feeling that sb. has done sth. wrong, illegal or dishonest, without having any proof 怀疑的，可疑的

Didn't you notice anything suspicious in his behaviour?

I was suspicious of his motives.

4. impatient /ɪmˈpeɪʃ(ə)nt/ *adj.* ~ (with sb./sth.)/~ (at sth.) annoyed or irritated by sb./sth., especially because you have to wait for a long time 不耐烦的；没有耐心的

I'd been waiting for twenty minutes and I was getting impatient.

Sarah was becoming increasingly impatient at their lack of interest.

5. vicinity /vəˈsɪnəti/ *n.* the area around a particular place 周围地区；邻近地区；附近

Crowds gathered in the vicinity of Trafalgar Square.

There is no hospital in the immediate vicinity.

6. authorize /ˈɔːθəraɪz/ *v.* to give official permission for sth., or for sb. to do sth. 批准；授权

I have authorized him to act for me while I am away.

The soldiers were authorized to shoot at will.

7. disarm /dɪsˈɑːm/ *v.* to reduce the size of an army or to give up some or all weapons, especially nuclear weapons 裁军；解除武装；裁减军备（尤指核武器）

Most of the rebels were captured and disarmed.

We will agree to disarming troops and leaving their weapons at military positions.

8. engage /ɪnˈɡeɪdʒ/ *v.* to participate or become involved in 参与；从事；卷入

Even in prison, he continued to engage in criminal activities.

9. civilian /səˈvɪliən/

① *n.* a person who is not a member of the armed forces or the police 平民百姓

② *adj.* in military situation, civilian is used to describe people or things that are not military 平民的；民事的

They bombed military and civilian targets.

He is in charge of the civilian side of the UN mission.

10. in danger 处在危险当中

It could also get people out of cities in danger of being bombed.

Student will be in danger of getting expelled from school if he breaks school rules.

【长句解读】

1. It looks like he's waiting for a sign or an order... (7-8)

➢ 句子分析：look like 表示"看起来像是要……"，后面可以接从句，或者动词的 -ing 形式；sign 表示"某种信号"。

➢ 翻译：看起来他好像在等待信号或者命令。

2. We have a stove, two mess tins, a torch, two sleeping bags and a small, brown, empty ammo pouch. (7-9)

➢ 句子分析：句中 have 意为 find out，即"发现"的意思；多个名词并列使用时，倒数第二个词与最后一个词之间应该用 and 连接起来；多个形容词修饰同一个名词时，注意形容词的排列顺序，通常应遵循大小、形状、年龄、新旧、颜色、材质等顺序排列。

➢ 翻译：我们发现了一个炉子、两个饭盒、一把手电筒、两个睡袋，还有一个空的小号棕色弹药袋。

3. Your mission is to patrol the area in the immediate vicinity of the bridge during daylight hours and to keep the area clear of snipers. (7-10)

➢ 句子分析：be to do 表示"按照命令或者预定计划做某事"；keep clear of 表示"清

除，清理干净"; and 连接两个并列谓语"to patrol..."和"to keep..."。

> 翻译：你们的任务是白天巡逻大桥附近的区域，并确保该区域没有狙击手。

4. You are not authorized to use deadly force to stop or disarm persons crossing the bridge unless they are engaged in hostile acts. (7-10)

> 句子分析：not...unless 表示"除非……否则不……"，引导条件状语。

> 翻译：你们无权使用致命武力阻止人员过桥或解除他们的武装，除非他们从事了敌对活动。

5. Civilian vehicles may not normally be attacked — unless they attempt to drive straight over the bridge and then you are only authorized to fire warning shots and then to shoot at the tyres if these vehicles do not stop at the checkpoint, unless of course you are fired on. (7-10)

> 句子分析：civilian vehicle 指"民用车辆"，drive straight over 表示"直冲……而去"，to fire warning shots 意为"鸣枪警告"。

> 翻译：通常情况下不能攻击民用车辆，除非他们试图冲桥而过。即使在那种情况下，你们也只能鸣枪警告；如果车辆在检查点还不停下来，你们可以射击车辆轮胎。当然，如果遭到火力攻击（你们可以开火还击）。

【口语输出】

1. What do patrols generally do when they conduct a patrol task?

【参考词汇】keep a written record, record any changes, observe the conditions of roads, terrain and obstacles, report anything unusual, maintain radio contact

2. What's the duty of a patrol leader?

【参考词汇】full responsibility, know well about his patrol area, brief, debrief

3. How do you understand rules of engagement (ROE)?

【参考词汇】be careful of using deadly force, fire back, authorized, unauthorized, warning shots, in immediate danger, for the purpose of, shoot at the tyres

Delta Radio communications

【军事知识】

1. Radio communication 无线电通信

Radio communication refers to telecommunication by means of radio waves. Radio communication requires the use of both transmitting and receiving equipment. The transmitting equipment, which includes a radio transmitter and a transmitting antenna, is installed at the point from which messages are transmitted. The receiving equipment, which includes a radio receiver and a receiving antenna, is installed at the point at which messages are received.

2. Call sign 呼号

Call sign is a set of characters (code symbols, letters, or numbers) or an audible signal (word, musical phrase, bird call) that serves as the distinctive sign of a radio station—usually for the purpose of identification of the station during reception. As a rule, it is transmitted at the beginning of each period of the station's operation.

The set of symbols in a call sign identifies the station's national affiliation. The initial characters are established by the International Radio Regulations. The complete structure of a call sign depends on the class (purpose) of a station and differs for broadcasting, official, and amateur radio stations. Call signs for amateur radio stations have a complicated makeup and contain the most information. They often indicate through a code the station's operating wave band, the station's group (a collective or individual station) and site, and the individual letter symbol or registration number of the station.

【词汇点拨】Words and Expressions

1. close down proword, not in radio contact 关闭；停止通话

A: Hello L1. This is L3. Closing down for 30 minutes at delta 2. Over.

B: L1. Roger. Out.

2. destination /ˌdestɪˈneɪʃn/ *n.* a place to which sb./sth. is going or being sent 目的地；终点

A: Hello L1. This is L3. Leaving delta 1 in convoy. Destination delta 6. Over.

B: L1. Roger. Out.

3. radio check proword, to check to see if the radio is in good state 无线电通信检查

Radio check, how do you read this transmitter?

A: Hello L1. This is Radio check. Over

B: L1. OK. Over.

A: L3. OK. Out.

4. mobile from proword, moving from one place to another 移动；转移

A: Hello D1. This is R3. Leaving Z1 in convoy. Now mobile from Z1. Over.

B: D1. Roger. Out.

5. acknowledge /əkˈnɒlɪdʒ/ *v.* proword, to confirm that you have the message and say you will comply 确认

A: *L1. We will send a medical team as soon as possible. Stay at the site. Acknowledge. Over.*

B: *L3. Wilco. Over.*

6. affirmative /əˈfɜːmətɪv/ *adj.* proword, an affirmative word or reply means 'yes' or expresses agreement 肯定的；同意的

A: *L1. Question. Is this area safe? Over.*

B: *L3. Affirmative. Situation is under control. There are no hostile elements in the vicinity.*

7. negative /ˈneɡətɪv/ *adj.* proword, containing a word such as 'no', 'not', 'never', etc. 没有；否定

A: *L1. Question. Is there a doctor on the spot? Over.*

B: *L3. Negative. We don't have a doctor. Over.*

8. wilco /ˈwɪlkəʊ/ *int.* people say Wilco in communication by radio to show that they agree to do sth.（无线电用语）照办；遵办

A: *L1. We will send a medical team as soon as possible. Stay at the site. Acknowledge. Over.*

B: *L3. Wilco. Over.*

9. read /riːd/ *v.* proword, to hear and understand sb. speaking on a radio set 听到；听明白 *Calling DR 40. I say again. Calling DR 40. Do you read me? Over.*

10. roger /ˈrɒdʒə(r)/ *int.* people say Roger in communication by radio to show that they have understood a message（用于无线电通信，表示已听懂信息）信息收到；明白

A: *Hello L1. This is L3. Mobile from delta 2. Destination delta 3. Over.*

B: *L1. Roger. Out.*

【长句解读】

1. Hello L1. This is L3. Leaving delta 1 in convoy. Destination delta 6. Over. (7-13)

➤ 句子分析：通信用语要求简洁明了，避免拖泥带水。此句中 leaving 表示"我们正准备离开……"，in convoy 表示"与车队在一起"；destination 表示"将要到达的目的地"。

➤ 翻译：一号，我是三号。我们正准备随车队离开1号区域，前往6号区域。完毕！

2. Hello L1. This is L3. Closing down for 30 minutes at delta 2. Over. (7-13)

➢ 句子分析：closing down 相当于"I am going to close down my radio"，意为"关闭电台"。

➢ 翻译：一号，我是三号。我的电台将在 2 号区域关闭 30 分钟。完毕！

3. A: Hello L1. This is L3. Mobile from delta 2. Destination delta 3. Over.

B: L1. Roger. Out. (7-13)

➢ 句子分析：此句为祈使句，其中 mobile from 表示"从……离开"，destination 为目的地。

➢ 翻译：A：一号，我是三号。现在命令你离开 2 号区域，前往 3 号区域。完毕！

B：一号收到，完毕！

【口语输出】

1. What prowords do you know?

【参考词汇】roger, out, over, send, message, negative, affirmative, acknowledge, question, radio check, closing down, destination, wilco, etc.

2. Read the following conversation. Can you guess what SALUTE means?

White Horse: Brown Horse. This is White Horse. Stand by for SALUTE Report 2-1. Over.

Brown Horse: White Horse. This is Brown Horse. Stand by to copy.

White Horse: Line SIERRA-20 Personnel.

　　　　　Line ALPHA-Patrolling what appears to be a headquarters. Break.

　　　　　Line LIMA-38SPV324429.

　　　　　Line UNIFORM-Republic guard. Break.

　　　　　Line TANGO-23 MARCH 2003 1730 ZULU.

　　　　　Line ECHO-AK47s and RPKs.

　　　　　How copy so far. Over.

【参考词汇】Each letter in the word "SALUTE" stands for a specific situation on the battle field. S stands for the enemy SIZE; A stands for enemy ACTIVITY; L stands for enemy LOCATION; U stands for enemy UNIT; T stands for TIME and E stands for enemy EQUIPMENT.

Echo Report and request

【军事知识】

1. Evacuation（救援）撤离

Evacuation may refer to (1) the removal of wounded or sick personnel and prisoners of war as well as inoperative or superfluous equipment or captured material from combat areas to the rear; (2) the withdrawal of troops from areas of military occupation, for military or political reasons or as the result of the conclusion of an agreement or a treaty; (3) the removal of civilians, businesses, institutions, and property as well as valuables, such as works of art, from an area threatened by enemy attack or stricken by a natural disaster, such as a flood.

2. Ambulance 救护车

An ambulance is a medically equipped vehicle which transports patients to treatment facilities, such as hospitals. Typically, out-of-hospital medical care is provided to the patient. Ambulances are used to respond to medical emergencies by emergency medical services. For this purpose, they are generally equipped with flashing warning lights and sirens. They can rapidly transport paramedics and other first responders to the scene, carry equipment for administering emergency care and transport patients to hospital or other definitive care. Most ambulances use a design based on vans or pick-up trucks. Others take the form of motorcycles, cars, buses, aircraft and boats. Ambulances can be grouped into types depending on whether or not they transport patients, and under what conditions. In some cases, ambulances may fulfill more than one function (such as combining emergency ambulance care with patient transport)

Emergency ambulance—The most common type of ambulance, which provides care to patients with an acute illness or injury. These can be road-going vans, boats, helicopters, fixed-wing aircraft (known as air ambulances), or even converted vehicles such as golf carts.

Patient transport ambulance—A vehicle, which has the job of transporting patients to, from or between places of medical treatment, such as hospital or dialysis center, for non-urgent care. These can be vans, buses, or other vehicles.

Ambulance bus—A large ambulance, usually based upon a bus chassis, that can evacuate and transport a large number of patients.

Charity ambulance—A special type of patient transport ambulance is provided by a charity for the purpose of taking sick children or adults on trips or vacations away from hospitals, hospices, or care homes where they are in long-term care. Examples include the

United Kingdom's "Jumbulance" project. These are usually based on a bus.

Bariatric ambulance—A special type of patient transport ambulance designed for extremely obese patients equipped with the appropriate tools to move and manage these patients.

Rapid organ recovery ambulance—A kind of transport vehicle designed to collect the bodies of people who have died suddenly from heart attacks, accidents and other emergencies and try to preserve their organs.

【词汇点拨】Words and Expressions

1. hostile /ˈhɒstaɪl/ *adj.* **~ (to/towards sb./sth.)** very unfriendly or aggressive and ready to argue or fight 敌意的；敌对的

The speaker got a hostile reception from the audience.

hostile elements 敌对因素

hostile area 敌对区域

2. under control 在掌控之中

Army officers said the situation was under control.

He believes that from a military standpoint, the situation is under control.

3. on the spot 现场；当场

If mistakes occurred, they were assumed to be the fault of the commander on the spot.

An ambulance was on the spot within minutes.

4. site /saɪt/ *n.* the place, scene, or point of an occurrence or event 场地；场景

He was able to pinpoint the site of the medieval village on the map.

Stay at the site. 留在现场

5. extract /ɪkˈstrækt/ *v.* to remove or obtain a substance from sth., for example by using an industrial or a chemical process 提取（此处可理解为救援或者撤离）

Stay at your current position and we will send two APCs to extract you.

They need to extract as much information as possible from the first or second reading.

6. evacuation /ɪˌvækjuˈeɪʃn/ *n.* the act or process of evacuating 撤离；疏散

air evacuation 空中（救援）撤离

The evacuation is being organized at the request of the United Nations Secretary General.

【长句解读】

1. C21. Shooting incident. One soldier wounded in the head at grid 987624. Near small

house. We request urgent medical evacuation. Over. (7-16)

> 句子分析：通信用语力求简洁明了，往往以短语或断句的形式出现，但是并不影响通信的准确性。本句中 shooting incident, near small house 都属于这类用法。

> 翻译：C21，（这里）发生枪击事件，一名战士头部受伤。坐标987624，靠近一间小房子。请求紧急医疗后送。

2. A: C21. Affirmative. Situation is under control. There are no hostile elements in the vicinity. Over.

B: C20. We are sending an air evacuation team. Over. (7-16)

> 句子分析：affirmative 是对前面问题"Is the area safe?"的肯定回答，表示"安全"；no hostile elements 是对 affirmative 的补充说明，强调安全性。

> 翻译：A：C21。安全。已控制形势，附近地域无敌对情况。完毕！
B：C20。正在派遣空中救援队。完毕！

3. B: F20. Question. Can you move the casualty? Over.

A: F24. Wait. Over. F24. Affirmative. We can move the casualty a short distance. Over. (7-19)

> 句子分析：question 表示有问题要询问；wait 表示等待，需要进一步确定信息。这两个词都属于常用通信用语。

> 翻译：B：F20。询问。可否移动伤员？完毕！
A：F24。稍候。完毕。F24。可以短距离移动伤员。完毕！

【口语输出】

1. If you are reporting a situation, what elements should be included?

【参考词汇】time, incident, grid, distance, safety, request, casualty, etc.

2. What orders should be given in a certain evacuation situation?

【参考词汇】stay at the site, stay at the current position, move to grid..., wait for further orders, send air evacuation/ambulance/APC to extract you

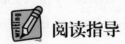 阅读指导

【文章导读】

This passage briefly answers four questions about patrol. Firstly, it explains what a patrol is, including various means of patrolling. Secondly, it lists the common patrol aims, such as to confirm or supervise a ceasefire, to find out information, to observe and report activities in a given area, etc. Thirdly, it tells us what a patrol should do as their routine activities. And at last, it emphasizes the responsibilities of a patrol leader.

【军事知识】

1. United Nations peacekeeping missions 联合国维和任务

Peacekeeping, as defined by the United Nations, is a way to help countries torn by conflict create conditions for sustainable peace. UN peacekeepers—soldiers and military officers, police officers and civilian personnel from many countries—monitor and observe peace processes that emerge in post-conflict situations and assist ex-combatants in implementing the peace agreements they have signed. Such assistance comes in many forms, including confidence-building measures, power-sharing arrangements, electoral support, strengthening the rule of law, and economic and social development.

The Charter of the United Nations gives the Security Council the power and responsibility to take collective action to maintain international peace and security. For this reason, the international community usually looks to the Security Council to authorize peacekeeping operations. Most of these operations are established and implemented by the United Nations itself with troops serving under UN operational command. In other cases, where direct UN involvement is not considered appropriate or feasible, the Council authorises regional organisations such as the North Atlantic Treaty Organisation, the Economic Community of West African States or coalitions of willing countries to implement certain peacekeeping or peace enforcement functions. In modern times, peacekeeping operations have evolved into many different functions, including diplomatic relations with other countries, international bodies of justice (such as the International Criminal Court), and eliminating problems such as landmines that can lead to new incidents of fighting.

2. UN peacekeeping missions after the Cold War 冷战后联合国维和任务

In 1991, the political situation created by the collapse of the USSR allowed the first explicitly-authorized operation of collective self-defense since the Korean War: expelling Iraq from Kuwait in the Gulf War. Following the cessation of hostilities, the UN authorized United Nations Iraq–Kuwait Observation Mission (UNIKOM) to monitor the DMZ between the two countries. Two other inter-state conflicts have been the cause for UN peacekeeping since. In 1994, the UN Aouzou Strip Observer Group (UNASOG) oversaw the withdrawal of Libya from a strip of contested territory in accordance with the decision of the International Court of Justice. In 2000, UN Mission in Ethiopia and Eritrea (UNMEE) was established to monitor the cessation of hostilities after the Eritrean–Ethiopian War.

The 1990s also saw the UN refocus its attention on genocide and ethnic cleansing. The Civil War in Rwanda and the breakup of Yugoslavia both were occasions of widespread atrocities and ethnic violence. Eight UN peacekeeping missions have been sent to the former Yugoslavia, UNPROFOR, UNCRO, UNPREDEP, UNMIBH, UNTAES, UNMOP, UNPSG, and UNMIK as well as two to Rwanda, UNAMIR and UNOMUR.

Despite the cessation of international, Cold-War inspired aid, civil wars continued in many regions and the UN attempted to bring peace. Several conflicts were the cause of multiple peace-keeping missions.

The collapse of Somalia into the Somali Civil War in 1991 saw UNOSOM I, UNITAF, and UNOSOM II fail to bring peace and stability, though they did mitigate the effects of the famine.

The First Liberian Civil War resulted in the authorization of UNOMIL in September 1993 to assist and supervise the troops of the Economic Community of West African States (ECOWAS), which had intervened militarily at the request of the Liberian government, and oversee the maintenance of the peace agreement in the nation. However, two rebel groups instigated the Second Liberian Civil War in 2003, and UNMIL was dispatched to oversee the implementation of the ceasefire agreement and continues to assist in national security reform.

A coup in Haiti in 1991, followed by internal violence, was the impetus for the UN Mission in Haiti (UNMIH). In 1996 and 1997 three missions, UNSMIH, UNTMIH, and MIPONUH, were organized with the goal of reforming, training, and assisting the police through a period of political turmoil. A coup d'état in 2004 saw the ouster of the president and the UN authorized MINUSTAH to stabilize the country.

In Sudan, the UN initially sponsored UNMIS to enforce a ceasefire between the Sudan People's Liberation Army/Movement and the Sudanese government. Since then, rebel groups in Darfur have clashed with government-sponsored forces, resulting in UNAMID, the AU/UN Hybrid Operation in Darfur. Violence in Darfur, spilled over the border into Chad and the Central African Republic. In 2007, MINURCAT was deployed to minimize violence to civilians and prevent interference of aid distribution related to violence in Darfur.

The UN has also organized single peacekeeping missions aimed at ending civil wars in a number of countries. In Central African Republic, MINURCA (1998) was created to oversee the disarmament of several mutinous groups of former CAR military personnel and militias as well as to assist with the training of a new national police and the running of elections. The mission was extended after successful elections to help ensure further stability. In Sierra Leone, UNOMSIL/UNAMSIL) in 1999, followed the ECOMOG-led restoration of the government after a coup. In 1999, in the Democratic Republic of the Congo MONUC was designed to monitor the ceasefire after the Second Congo War—it continues to operate due to continuing violence in parts of the DRC. In Cote d'Ivoire, UNOCI was dispatched to enforce a 2004 peace agreement ending the Ivorian Civil War, though the country remains divided. Following ceasefire agreements ending the Burundi Civil War, ONUB was authorized in 2004 to oversee the implementation of the Arusha Peace Accords.

【词汇点拨】Words and Expressions

1. reconnaissance /rɪˈkɒnɪsns/ *n.* the activity of getting information about an area for military purposes, using soldiers, planes, etc. 侦查

a reconnaissance aircraft/mission/satellite 侦查飞机 / 任务 / 卫星

2. carry out to act according to a certain instruction 执行

She needed a clear head to carry out her instructions.

3. confirm /kənˈfɜːm/ *v.* to state or show that sth. is definitely true or correct, especially by providing evidence（尤指提供证据来）证实；证明；确认

We will always confirm the revised amount to you in writing before debiting your account.

4. supervise /ˈsuːpəvaɪz/ *v.* to be in charge of sb./sth. and make sure that everything is done correctly, safely, etc. 监督；管理；指导；主管

A team was sent to supervise the elections in Nicaragua.

5. terrain /təˈreɪn/ *n.* used to refer to an area of land when you are mentioning its natural features, for example, if it is rough, flat, etc. 地形；地势；地带

difficult/rough/mountainous, terrain etc. 难以通过的地带、崎岖不平的地形、山地等

6. halt /hɔːlt/ *v.* to stop; to make sb./sth. stop（使）停止；停下

The peace process has ground to a halt while Israel struggles to form a new government.

7. debrief /ˌdiːˈbriːf/ *v.* ~ **sb. (on sth.)** to ask sb. questions officially, in order to get information about the task that they have just completed 正式询问；了解（某人执行任务的情况）

He was taken to a US airbase to be debriefed on the mission.

8. appropriate /əˈprəʊpriət/ *adj.* **~ (for/to sth.)** suitable, acceptable or correct for the particular circumstances 合适的；恰当的

Now that the problem has been identified, appropriate action can be taken.
In the circumstances, Paisley's plans looked highly appropriate.

【参考译文】

巡 逻

巡逻队是指被派出执行某种特殊任务的士兵小队，如侦查、安保或观察。根据环境和需求的不同，巡逻可分为徒步巡逻、乘车巡逻、驾机巡逻、舰艇巡逻，甚至滑雪巡逻。

巡逻任务因巡逻目的或目标的不同而不同。巡逻目标一般包括：
- 确认或监督停火；
- 搜集信息，包括区域侦查；
- 观察并报告特定区域内活动情况；
- 实地检查从观察哨无法观测到的区域；
- 显示联合国的存在（告知他人此区域驻有联合国部队）。

所有巡逻队都有一个巡逻计划，详细说明他们巡逻的具体目标。除此之外，巡逻队通常还有以下任务：
- 书面记录观察到的所有情况，必要时可以绘图表示；
- 记录武装部队或者平民活动的任何变动情况；
- 观察道路、地貌和所设障碍的变动情况；
- 报告一切非正常情况；
- 保持常规无线电联络；
- 如果遭到挑衅要立即停止巡逻，并通过电台迅速报告情况。

巡逻开始前，巡逻队长需要确保巡逻队员清楚此次巡逻的目的。巡逻队员同时还需要了解任务区的整体态势和当地具体情况，清楚任务区内其他巡逻队和联合国的活动状态。巡逻结束后，巡逻队长需要提交一份巡逻报告，并向有关人员汇报巡逻情况。

 词汇表

Alpha Patrol kit
patrol 巡逻

camouflage 伪装；迷彩
pack 背包；打背包

kit 全套衣服及装备；成套设备
bayonet 刺刀；枪刺
webbing 带子；织带
waterproof 防水的
torch 手电筒；火把；火炬

Bravo Patrol route and tasks

check 检查；核对
conduct 执行；实施
observe 观察；遵守
prevent 阻止；阻拦
set up 设立；建立；树立
establish 创立；建立
forecast 预测；预告
mission 任务；使命
mine 地雷；布雷

Charlie Patrol and report

on foot 徒步；步行
uniform 制服；军服
suspicious 可疑的；令人怀疑的
impatient 躁动不安的；不耐烦的
vicinity 附近；邻近地区
authorize 授权；批准
disarm 缴械；解除武装
engage 交火；卷入
civilian 平民；民间的，民事的
in danger 处于危险中，受到威胁

Delta Radio communication

close down 关闭；关门
destination 目的地；终点
mobile from 移动；转移
radio check 通信检查
acknowledge 确认
affirmative 正确；肯定
negative 否定；错误
wilco 遵命
read 听到；听明白
roger 收到；明白

Echo Report and request

under control 受控；控制
hostile 敌对的；不友好的
on the spot 在现场；即刻
site 现场；场地；位置
extract 撤离，提取
evacuation 撤离；撤退；疏散

Foxtrot Patrols

reconnaissance 侦查
carry out 执行
confirm 确认
supervise 监督
terrain 地形
halt 停止
debrief 正式询问；盘问
appropriate 合适的；恰当的

 拓展学习

Three Basic Principles of UN Peacekeeping Operations

There are three basic principles that continue to set UN peacekeeping operations apart as a tool for maintaining international peace and security. These three principles are inter-related and mutually reinforcing: Consent of the parties; Impartiality; Non-use of force except in self-defense and defense of the mandate.

1. Consent of the parties

UN peacekeeping operations are deployed with the consent of the main parties to the conflict. This requires a commitment by the parties to a political process. Their acceptance of a peacekeeping operation provides the UN with the necessary freedom of action, both political and physical, to carry out its mandated tasks.

In the absence of such consent, a peacekeeping operation risks becoming a party to the conflict; and being drawn towards enforcement action, and away from its fundamental role of keeping the peace.

The fact that the main parties have given their consent to the deployment of a United Nations peacekeeping operation does not necessarily imply or guarantee that there will also be consent at the local level, particularly if the main parties are internally divided or have weak command and control systems.

Universality of consent becomes even less probable in volatile settings, characterized by the presence of armed groups not under the control of any of the parties, or by the presence of other spoilers.

2. Impartiality

Impartiality is crucial to maintaining the consent and cooperation of the main parties, but should not be confused with neutrality or inactivity. United Nations peacekeepers should be impartial in their dealings with the parties to the conflict, but not neutral in the execution of their mandate.

Just as a good referee is impartial, but will penalize infractions, so a peacekeeping operation should not condone actions by the parties that violate the undertakings of the peace process or the international norms and principles that a United Nations peacekeeping operation upholds.

Notwithstanding the need to establish and maintain good relations with the parties, a peacekeeping operation must scrupulously avoid activities that might compromise its image of impartiality. A mission should not shy away from a rigorous application of the principle of impartiality for fear of misinterpretation or retaliation.

Failure to do so may undermine the peacekeeping operation's credibility and legitimacy, and may lead to a withdrawal of consent for its presence by one or more of the parties.

3. Non-use of force except in self-defense and defense of the mandate

UN peacekeeping operations are not an enforcement tool. However, they may use force at the tactical level, with the authorization of the Security Council, if acting in self-defense and defense of the mandate.

In certain volatile situations, the Security Council has given UN peacekeeping operations "robust" mandates authorizing them to "use all necessary means" to deter forceful attempts to disrupt the political process, protect civilians under imminent threat of physical attack, and/or assist the national authorities in maintaining law and order.

Although on the ground they may sometimes appear similar, robust peacekeeping should not be confused with peace enforcement, as envisaged under Chapter VII of the United Nations Charter.

Robust peacekeeping involves the use of force at the tactical level with the authorization of the Security Council and consent of the host nation and/or the main parties to the conflict.

By contrast, peace enforcement does not require the consent of the main parties and may involve the use of military force at the strategic or international level, which is normally prohibited for Member States under Article 2(4) of the Charter, unless authorized by the Security Council.

A UN peacekeeping operation should only use force as a measure of last resort. It should always be calibrated in a precise, proportional and appropriate manner, within the principle of the minimum force necessary to achieve the desired effect, while sustaining consent for the mission and its mandate. The use of force by a UN peacekeeping operation always has political implications and can often give rise to unforeseen circumstances.

Judgments concerning its use need to be made at the appropriate level within a mission, based on a combination of factors including mission capability; public perceptions; humanitarian impact; force protection; safety and security of personnel; and, most importantly, the effect that such action will have on national and local consent for the mission.(712 words)

(Selected from https://peacekeeping.un.org/en/principles-of-peacekeeping)

Unit 7　Patrol

【Notes】

1. UNPKOs (United Nations Peacekeeping Operations) the activities of keeping peace by military forces (especially when international military forces enforce a truce between hostile groups or nations) 联合国维和行动

2. **consent** 同意；允许；准许

3. **impartiality** 公平；公正

4. **mandate** 授权；委托；委任

5. **enforcement** 执行；实施；强制

6. **guarantee** 保证；担保

7. **volatile** 反复无常的；无定性的

8. **retaliation** 报复

9. **envisage** 想象；设想；展望

10. **authorization** 授权；批准

单元练习

I. Match the prowords with the definitions

1. wilco	a. Yes/Correct.
2. radio check	b. Confirm you have the message and say you will comply.
3. roger	c. This is the end of my transmission. I don't expect a reply.
4. negative	d. I have a message for you.
5. question	e. I am moving in a vehicle and I am going to a place.
6. destination	f. I have a question.
7. message	g. No/Incorrect.
8. out	h. I received your message.
9. acknowledge	i. Is my signal strong and clear?
10. affirmative	j. I have your message and I will comply.

II. Multiple choice

1. What does "over" mean in a radio transmission?

A. The end of a transmission and expect a reply.

B. The end of a transmission and doesn't expect a reply.

C. Confirm the message.

D. Deny the message.

2. You are authorized to use deadly force when _____.

A. you are patrolling a hostile area

B. you are conducting a patrol task

C. your life is in immediate danger

D. you feel your life is in danger

3. Depending on the circumstance and the need, patrol can be carried out except _____.

A. by air   B. by train   C. by car   D. by boat

4. Before the patrol, the patrol leaders need to make sure that they understand the _____ of the patrol clearly.

A. destination   B. local situation

C. other patrols   D. aims

5. After the patrol, the patrol leaders need to hand in _____ and debrief the appropriate people.

A. personal weapon   B. patrol report

C. patrol kit   D. patrol pack

6. UN peacekeepers should stick to the following principles when they are on operations except _____.

A. consent of all parties in the conflict

B. use of force at will

C. non-use of force except in self-defense

D. impartiality

7. If a civilian vehicle refuses to stop at a checkpoint, what should the soldiers on duty do?

A. Fire warning shots.

B. Give a challenge first in English and then in local language.

C. Shoot at the tyres.

D. All of the above.

8. Which of the the following is not the task of a patrol team?

A. Supervise a ceasefire.   B. Conduct a reconnaissance.

C. Set up an observation post.   D. Help refugees return home.

9. There are three principles in using prowords except _____.

A. complexity   B. conciseness   C. accuracy   D. completeness

10. If you send a message and wait for a reply on radio transmission, which proword should you use?

A. Out.   B. Wait, out.   C. Over.   D. Wilco.

III. Translation: English to Chinese

1. They are not wearing a uniform or insignia. I think they are probably irregular forces.

2. There is a man on the right in the front of the crowd—he looks very suspicious. I think he's carrying a gun under his jacket.

3. Remember you are UN peacekeepers, not an attacking force, so no force is authorized unless it is to protect yourself or other people.

4. Hostile acts include threatening to use a weapon. This means, if a person threatens to shoot you with a rifle or attack you with a knife, etc. then you can use your weapon to protect yourself.

5. But remember: Always give a challenge in English first and then in the local language before you open fire.

IV. Translation: Chinese to English

维持和平仍然是联合国协助东道国从冲突走向和平最有效的工具之一。我们特派团的工作环境在不断变化，并受到部署地的社会、政治和经济变化的影响。为确保联合国行动在未来五到十年继续做出积极贡献，我们必须研究可能影响到联合国和平行动方向的因素，为未来做出调整和准备。秘书长安东尼奥·古特雷斯（António Guterres）2018年发起的"维持和平行动"（A4P）倡议，重点是强化今天的维和行动。"维持和平行动"仍然是维和政策和改革的核心框架。秘书处目前正在为2021年及下一阶段A4P的实施制定优先事项。

V. Oral practice

1. What's the primary task of a patrol?

2. How much do you know about UN peacekeeping operation principles?

附录：Unit 7 练习答案

I. Match the prowords with the definitions

1. j   2. i   3. h   4. g   5. f   6. e   7. d   8. c   9. b   10. a

II. Multiple choice

1-5 ACBDB　　　6-10 BDDAC

III. Translation: English to Chinese

1. 他们没穿军装，也没有佩戴标识。我估计他们是非正规军。

2. 站在人群前面右边的那个男人看起来十分可疑，我觉得他外套里面可能藏有枪。

3. 记住：你们是联合国维和人员，不是攻击部队。所以，除非为了保护自己或他人，否则不得使用武力。

4. 敌对行动包括威胁使用武器。这意味着，如果有人用枪支或刀具威胁你，你才可以使用武器保护自己。

5. 但是记住：要遵循开火前先用英语，然后用当地语言进行警告的惯例。

IV. Translation: Chinese to English

Peacekeeping remains one of the most effective tools available to the UN to assist host countries navigate the difficult path from conflict to peace. Our missions operate in a constantly changing environment, impacted by the social, political and economic changes in the places where we are deployed. To ensure that UN operations continue to make a positive contribution in the next five to ten years, it is critical that we examine the factors that may influence the direction of UN peace operations, to adjust and prepare for the future. The Action for Peacekeeping (A4P) initiative, which was launched in 2018 by Secretary-General António Guterres, is focused on strengthening the peacekeeping operations of today and remains the central framework for peacekeeping policy and reform. The Secretariat is currently developing its priorities for a next phase of A4P implementation in 2021 and beyond.

V. Oral practice

1. The task of a patrol may vary according to different patrol aims. The primary task of a patrol may include observing and reporting any activities in a given area, either military or civilian. The patrol should observe the conditions of roads, terrain, obstacles and keep a written record of all observations. After the patrol, the patrol leader should hand in a patrol report and debrief the appropriate people.

2. There are three basic principles that UN peacekeeping force should abide by for maintaining international peace and security. These three principles are inter-related and mutually reinforcing. Namely, Consent of the parties, Impartiality and Non-use of force except in self-defense and defense of the mandate.

First, UN peacekeeping operations are deployed with the consent of the main parties to the conflict. This requires a commitment by the parties to a political process. Second, United Nations peacekeepers should be impartial in their dealings with the parties to the conflict, but not neutral in the execution of their mandate. Third, UN peacekeeping operations are not an enforcement tool. However, they may use force at the tactical level, with the authorization of the Security Council, if acting in self-defense and defense of the mandate.

Unit 8　The Battalion

单元导学清单

模块	主题	学习目标	任务	军事知识	核心词汇
Alpha	营级编制	1）了解陆军营级单位基本知识；2）掌握常用词汇和表达；3）能够通俗讲解英美陆军营级单位常见构成。	Task 1 Task 2 Task 3 Task 4	■ battlegroup	tactical, strength, mortar, ration, remit, division, adventure training exercise, sentry guard, prisoner of war
Alpha	职能分配	1）了解营房排长及军士常见职务；2）能够运用常用句型描述个人职能。	Task 5 Task 6 Task 7 Task 8	■ Quartermaster in the British Army	operational effectiveness, ammunition, supervise, travel claim, discharge documentation, maintenance, dispatch
Bravo	上下级称谓	1）了解英国陆军上下级称谓常识；2）掌握各级军衔相关的词汇和表达。	Task 1 Task 2	■ Regimental Sergeant Master ■ Color Sergeant	colonel, corporal, sergeant, staff sergeant, colour sergeant, captain
Bravo	上下级命令传达及请求表达	1）了解上下级称谓风格和措辞；2）掌握如何用中立或礼貌口吻表达请求。	Task 3 Task 4 Task 5 Task 6	■ Warrant officer	drop off, range, station, transfer request
Charlie	靶场装备	1）了解靶场常见武器装备；2）熟练掌握与装备相关的词汇与术语；3）介绍武器操作考核一天行程安排等。	Task 1 Task 2 Task 3	■ SA80 ■ kevlar helmet	armoury, cleaning kit, ear defenders, mess tin, respirator, waterproof poncho, webbing
Charlie	靶场射击	1）了解靶场射击相关常识；2）熟练掌握与靶场射击相关的词汇和句型；3）讨论靶场射击场景下的明令禁止用语。	Task 4 Task 5 Task 6 Task 7	■ range flag ■ standing order ■ live round ■ firing line ■ down range	concurrent, ammunition point, backstop, down range, empty case, firing line, live round, muzzle, range flag

续表

模块	主题	学习目标	任务	军事知识	核心词汇
Delta	"山痕"演习	1）复习不同的地形特点；2）熟练掌握与装备相关的词汇和表达；3）能够运用地图知识熟练标记军事部署地点。	Task 1 Task 2 Task 3 Task 4 Task 5	■ Estimated Time of Arrival (ETA) ■ Estimated Time of Departure (ETD) ■ grid ■ last light	base camp, campsite- compass bearing, garrison, last light, grid, ETA, ETD, terrain, leg, elevation
Echo	营级以上编制	1）了解英国陆军营级以上单位编制大小；2）熟练掌握相关的词汇和表达；3）了解英国军队海外部署情况。	Task 1 Task 2 Task 3 Task 4 Task 5	■ 1st Armoured Brigade (the United Kingdom) ■ 2nd Cavalry Regiment (the United States) ■ 18th Airborne Corps (the United States) ■ 3rd Signal Regiment (the United Kingdom)	section, squad, platoon, company, battalion, regiment, brigade, division, corps, deployment
Foxtrot (Review)	陆军编制	1）了解英国陆军基本编制构成；2）熟练掌握相关的词汇和表达；3）掌握常见军队单位的缩写形式。	Task 1	■ British 1st Armoured Division ■ British Army Organisation	hierarchy, colonel, major, lieutenant, corporal, organigram, variation, generic, mechanized, major general, brigadier

 听说指导

Alpha Battalion organisation

【军事知识】

1. Quartermaster in the British Army 英国军队军需官

In the British Army, the Quartermaster (QM) is the officer in a battalion or regiment responsible for supply. By longstanding tradition, he or she is always commissioned from the ranks (and is usually a former Regimental Sergeant Major) and holds the rank of captain or major. Some units also have a Technical Quartermaster, who is in charge of technical stores. The Quartermaster is assisted by the Regimental Quartermaster Sergeant (RQMS) and a staff of storemen. The QM, RQMS and storemen are drawn from the regiment or corps in which they work, not from the Royal Logistic Corps, which is responsible for issuing and transporting supplies to them. Units which specialize in supply are known as "supply" units, not "quartermaster" units, and their personnel as "suppliers".

2. battlegroup 作战群

A battlegroup (British/Commonwealth term), or task force (U.S. term) in modern military theory, is the basic building block of an army's fighting force. A battlegroup is formed around an infantry battalion or armoured regiment, which is usually commanded by a Lieutenant Colonel. The battalion or regiment also provides the command and staff element of a battlegroup, which is complemented with an appropriate mix of armour, infantry and support personnel and weaponry, relevant to the task it is expected to perform.

【词汇点拨】Words and Expressions

1. division /dɪˈvɪʒn/ *n.* a division is a group of military units which fight as a single unit 师（军队的作战单位）

Several armoured divisions are being moved from Germany.

2. tactical /ˈtæktɪkəl/ *adj.* an action or plan which is intended to help someone achieve what they want in a particular situation 战术上的

The battalion is the basic tactical unit in the infantry and is commanded by a lieutenant colonel.

3. strength /streŋθ/ *n.* capability in terms of personnel and material that affect the capacity to fight a war（军事方面的）实力

Each section has a strength of eight men and is divided into two fire teams known as Charlie and Delta.

4. brigade /brɪˈɡeɪd/ *n.* an army unit smaller than a division 旅

The brigade forms part of a division, which is made up of three or four brigades with a total strength of about 15,000 soldiers.

5. mortar /ˈmɔːrtər/ *n.* a big gun that fires missiles high into the air over a short distance 迫击炮

The two sides exchanged fire with mortars and small arms.

6. comprise /kəmˈpraɪz/ *v.* include or contain; have as a component 包括；构成

The Support Company comprises the battalion's fire support assets and is equipped with mortars, anti-tank weapons and machine guns.

7. ration /ˈræʃn/ *v. & n.* the food allowance for one day (especially for service personnel)（单兵）口粮；供给（兵士）伙食，限定（粮食等）

We're on short rations (= allowed less than usual) until fresh supplies arrive.

8. remit /rɪˈmɪt/ *n.* the area of autbority or responsibility of an individual or a group（官

方）责任；职权范围

By taking that action, the committee has exceeded its remit.

9. sentry guard a soldier who guards a camp or a building 哨兵

The sentry guard would not let her enter.

10. operational effectiveness the ability to operation during the wartime 作战效能

The commander is responsible for training, fitness, operational effectiveness and welfare for 30 men.

11. ammunition /ˌæmjuˈnɪʃn/ *n.* any nuclear or chemical or biological material that can be used as a weapon of mass destruction 弹药；军火

We only have three rounds of ammunition left.

12. supervise /ˈsuːpəvaɪz/ *v.* keep under surveillance 监督（任务，计划，活动）

I supervise the work of a section of clerks.

13. travel claim *claim for the travelling expenses* 差旅报销

The accounting department will reimburse you for all travel claims.

14. discharge documentation 退任（撤职文件）

The officer is processing the discharge documentation.

15. casualty /ˈkæʒuəlti/ *n.* someone injured or killed or captured or missing in a military engagement 伤亡人员

Our remit includes casualty reporting and prisoner of war registration.

16. maintenance /ˈmeɪntənəns/ *n.* activity involved in maintaining something in good working order 维护；保养

Careful maintenance can extend the life of the aircrafts.

17. dispatch /dɪˈspætʃ/ *v. & n.* send away towards a designated goal 派遣

We have to dispatch several troops to the area.

【词汇点拨】**Proper Names**

1. **Officer Commanding** 连长
2. **Commanding Officer** 营长
3. **Headquarters (HQ)** /ˈhedˌkwɔrtər/ 总部
4. **Regimental Quartermaster Sergeant Major** 团级军需官一级军士长
5. **Cpl (Corporal) Crawley** 克劳利下士
6. **Regimental Administration Office** 团行政办公室
7. **Platoon Sergeant** 副排长

8. Motor Transport Platoon 机动车辆运输排

9. Lieutenant Freeman 弗里曼中尉

10. Platoon Commander 排长

【长句解读】

1. With a total strength of 625 officers and men, the battalion is the basic tactical unit in the infantry and is commanded by a lieutenant colonel, known as the Commanding Officer or CO. (8-1)

> ➢ 句子分析：该句中 With a total strength of 625 officers and men 作伴随状语，known as the Commanding Officer or CO 作后置定语。the basic tactical unit 意为"基本战术单位"，be commanded by 被动语态，意为"被……指挥"。lieutenant colonel 中校。

> ➢ 翻译：该营总兵力为625名官兵，是步兵中的基本战术单位，由一名中校指挥，被称为营长。

2. Several times a year we have inspections and I'm personally responsible for every piece of equipment that the battalion is issued with. (8-2)

> ➢ 句子分析：be responsible for 意为"对……负责"。be issued with 意为"被配给"。后半句中"that the battalion is issued with"为定语从句。

> ➢ 翻译：我们每年都要进行几次检查，我亲自负责营里的每一件装备。

3. We carry out a wide range of duties, including processing travel claims, applications for leave and transfer and discharge documentation. (8-2)

> ➢ 句子分析：a wide range of 意为"大量的"，travel claims 意为"差旅报销"，applications for leave and transfer 意为"请休假申请和调任"，discharge documentation 意为"退任或撤职文件"。

> ➢ 翻译：我们承担广泛的职责，包括处理差旅报销、请休假申请和调任、离职文件等。

4. It's not the same thing to transport a tank, for instance, as it is to carry troops, or to take stores out to the men on exercise—so each mission requires its own type of transport and we have to train our drivers so they can handle all of these. (8-4)

> ➢ 句子分析：not the same thing to 意为"和……不一样"，on exercise 意为"训练中的"，take stores out to 意为"给……运送物资"。

> ➢ 翻译：这和运输一辆坦克、运送士兵或向训练中的士兵运送物资是不一样的——因此，每个任务都需要自己的运输工具，我们必须训练司机，以便他们能够处理所有这些。

5. We have to group loads going to the same destination and decide on which transport to use and also to avoid our trucks travelling empty. (8-4)

➤ 句子分析：该对话中的 group 用作动词，意为"分组"；going to the same destination 作后置定语。

➤ 翻译：我们必须将前往同一目的地的货物分组，并决定使用哪种运输工具，同时也要避免我们的卡车空车行驶。

【口语输出】

1. What is the general strength of an infantry battalion in China?

【参考词汇】tactical, command, compose, comprise, lieutenant colonel

2. What is your understanding of a battlegroup?

【参考词汇】task-organised, specific mission, be assigned to, be made up of, strength

3. What is your platoon commander's job?

【参考词汇】be responsible for, routine, operational effectiveness, adventure training exercises

4. What are your duties?

【参考词汇】clerk, be comprised of, supervise, carry out, stressful, remit, sentry guard

Bravo　Battalion HQ

【军事知识】

1. Regimental Sergeant Major　团级军士长；一级准尉

Regimental Sergeant Major (RSM) is an appointment held by warrant officers class 1 (WO 1) in the British Army, the British Royal Marines and in the armies of many Commonwealth and former Commonwealth nations, including Ireland, Australia and New Zealand; and by chief warrant officers (CWO) in the Canadian Forces. Only one WO 1/CWO holds the appointment of RSM in a regiment or battalion, making him the senior warrant officer; in a unit with more than one WO 1, the RSM is considered to be "first amongst equals". The RSM is primarily responsible for maintaining standards and discipline and acts as a parental figure to his or her subordinates.

2. Warrant officer　准尉军官

A warrant officer (WO) is an officer in a military organization who is designated an officer by a warrant, as distinguished from a commissioned officer who is designated an officer by a commission, and a non-commissioned officer who is designated an officer, often by virtue of seniority. The rank was first used in the (then) English Royal Navy and is today used in most services in many countries, including the Commonwealth nations and the United States.

【词汇点拨】Words and Expressions

1. colonel /ˈkɜːrnl/ *n.* a commissioned military officer in the United States Army or Air Force or Marines who ranks above a lieutenant colonel and below a brigadier general（陆军、空军或海军陆战队）上校

Mills was regimental colonel.

2. corporal /ˈkɔːrpərəl/ *n.* a non-commissioned officer in the army or airforce or marines（陆军、海军陆战队或英国空军的）下士

The corporal shouted an order at the men.

3. sergeant /ˈsɑːrdʒənt/ *n.* any of several non-commissioned officer ranks in the army or air force or marines ranking above a corporal（陆军、海军陆战队或空军的）中士

The sergeant snapped out an order.

4. staff sergeant a non-commissioned officer ranking above sergeant and below warrant officer class 2 in the British Army or ranking above sergeant and below sergeant first class in the U.S. Army or ranking above senior airman and below technical sergeant in the U.S. Air Force（英陆军）上士；（美陆军）上士；（美空军）中士

Staff Sergeant Robert Bales, accused of killing 16 Afghan civilians, is set to appear in a military court today.

5. colour sergeant a rank of non-commissioned officer found in several militaries, such as a non-commissioned title in the Royal Marines and infantry regiments of the British Army, ranking above sergeant and below warrant officer class 2（英陆军步兵、海军陆战队等）上士

His father is a color sergeant in the army.

6. captain /ˈkæptɪn/ *n.* an officer holding a rank below a major but above a lieutenant 上尉

He served as a captain in the army.

7. drop off leave or unload, especially of passengers or cargo 让……下车

I need the soliders to pick up and drop off.

8. range /reɪndʒ/ *n.*

① a place for shooting (firing or driving) projectiles of various kinds 靶场

They are departing the range.

② **the limit of the shooting capability** 射程

The range of the rifle is stunning.

9. transfer request a formal message requesting for a change from one position to another 调任请求

Can you send this transfer request to Division HQ?

【词汇点拨】Proper Names

1. **RSM(Regimental Sergeant Master)** 团级军士长
2. **NCO(Non-commissioned Officer)** 军士
3. **Sergeant Berkeley** 伯克利中士
4. **Sergeant Willis** 威利斯中士
5. **Corporal Jones** 琼斯下士
6. **Captain Marks** 马克斯上尉
7. **Motor Transport** 机动车运输
8. **Sergeant Cooper** 库珀中士
9. **Churchill Field** 丘吉尔菲尔德（英国地名）
10. **Catterick** /ˈkætrik/ 卡特里克

【长句解读】

1. You can't ask for a transfer every time you have a problem with an NCO. (8-5)

➤ 句子分析：该句子中 every time you have a problem with an NCO 作时间状语。ask for a transfer 意为"申请调职"。

➤ 翻译：你不能每次跟军士长闹矛盾就要求调职。

2. Could you tell me the departure and destination points, please? (8-6)

➤ 句子分析：本句属于委婉语表达。为更礼貌地表达个人请求，可以用句型：Could you...please? 稍微逊色一些的表达：Could you...? 比较中性的礼貌表达：Can you...?

➤ 翻译：请告诉我出发站和目的站好吗？

3. You need two trucks, for pick up and drop off. (8-6)

➤ 句子分析：该句中的 pick up and drop off 属于口语式表达，严格意义上应表述为 for picking up and dropping off。

➤ 翻译：你需要两辆卡车，用于接送任务。

【口语输出】

1. How do you usually address a junior NCO? and How does an officer address an NCO?

【参考词汇】rank, sir/ma'am, rank+name or rank only

2. Suppose you are a corporal, and your partner is a colour/staff sergeant. Your partner wants his or her photo copied, and you are the clerk in the printing room. Please make a conversation.

【参考词汇】photocopy, complete, form, corporal, be responsible

3. You are a corporal, and you are the clerk in Motor Transport. Your partner, a lieutenant, wants to arrange transport for next Tuesday. Please make a conversation.

【参考词汇】lieutenant, troop, departure point, destination point, pick up, drop off

Charlie   Range day

【军事知识】

1. SA80 突击步枪

The SA80 (Small Arms for the 1980s) is a British family of weapons that were designed by RSAF Enfield. The family comprises the L85 assault rifle, the L86 light support weapon, and the L22 carbine. The L85 is currently the standard service rifle of the British Army, and has been produced by Enfield and Heckler & Koch.

SA80 is of a bullpup configuration. It has a simple fire selector with two settings: single and auto. The SA80 is compatible with any STANAG standard magazine. It is fitted with a bayonet lug. Full-sized versions of SA80 can be fitted with the AG36 underbarrel grenade launcher (which reportedly enhances the somewhat poor balance of the weapon).

2. Kevlar helmet 凯夫拉尔头盔

The kevlar helmet was a piece of armor that provided high protection from bullets, therefore making it the preferred choice of armor for PvP situations. Kevlar will reduce damage taken by weapons at a certain level, depending on the damage it would've dealt, except if damage was taken to the legs.

【词汇点拨】Words and Expressions

1. armoury /'ɑːrməri/ *n.* a military structure where arms and ammunition and other military equipment are stored and training is given in the use of arms 军械库

Nuclear weapons will play a less prominent part in NATO's armoury in the future.

2. cleaning kit gear consisting of a set of tools to clean the accessories of the guns 擦枪套件；清洁套件

Then, carefully follow the directions of a compatible sensor cleaning kit.

3. ear defenders the protectors used to protect eyes from being hurt 护耳塞

For tinnitus, protection like ear plugs or ear defenders help.

4. mess tin A mess tin is an item of mess kit, designed to be used over portable cooking apparatus（军用）饭盒

The "healthy nanny" is an intelligent mess tin with USB port, that is designed specifically for office worker.

5. respirator /ˈrespəreɪtər/ *n.* a protective mask with a filter; protects the face and lungs against poisonous gases 防毒面具

It's smart to use a respirator during the NBC training process.

6. waterproof poncho a water-resistant blanket-like cloak with a hole in the center for the head 防水雨披

If the weather looks iffy, you'll want to add rain gear: a waterproof poncho or nylon jacket.

7. webbing /ˈwebɪŋ/ *n.* a narrow closely woven tape; used in upholstery or for seat belts（作战装具）背带

Soldiers have to wear webbings while they are shooting.

8. concurrent /kənˈkɜːrənt/ *adj.* occurring or operating at the same time 同时发生的

The Range Officer will be Lieutenant Deacon and I will supervise the concurrent activity.

9. ammunition point the place where projectiles are placed and supplied 弹药补给点

Soldiers get their ammunitions at the ammunition point.

10. backstop /ˈbækstɑːp/ *n.* a fence or screen to prevent the projectiles from traveling out of the range 挡弹体

The backstop is used to prevent the bullets from traveling out of the range.

11. down range Down range is the horizontal distance traveled by a spacecraft, or the spacecraft's horizontal distance from the launch site. In this section, it refers to the safe direction that the bullets travel（射击枪支指向）安全方向

Always point the muzzle of the weapon down range whenever on the firing point.

12. empty case the case in which no ammunitions are left 空弹夹

You must not leave the range before being inspected for live rounds and empty cases.

13. firing line the line from which soldiers deliver fire 射击线

Any hostages in the firing line would have been sacrificed.

14. live round live ammunition 实弹

Soldiers use live round on the range.

15. muzzle /ˈmʌzl/ *n.* the open circular discharging end of a gun 枪口

Mickey felt the muzzle of a rifle press hard against his neck.

16. range flag 边旗

Soldiers have to point weapons or fire inside the range flags.

【词汇点拨】**Proper Names**

1. Sergeant Peters 中士彼得斯

2. Lieutenant Deacon 中尉迪肯

3. NBC (Nuclear, Biological, Chemical) Training 核生化训练

【长句解读】

1. You must never touch your weapon while personnel are down range or in front of the firing line. (8-8)

➢ 句子分析：该句中 must never 相当于 mustn't。语气较为强烈。在靶场上要严格按照命令执行。

➢ 翻译：当靶场执行人员在射击范围内或在射击线前时，禁止触碰武器。

2. The two red range flags, located on the far left and right, are the range's left and right limits. (8-8)

➢ 句子分析：本句中 located on the far left and right 为非谓语动词，起定语作用。

➢ 翻译：位于最左和最右的两个红色边旗是靶场的左边界和右边界。

3. You must not leave the range before you are inspected for live rounds and empty cases. (8-8)

➢ 句子分析：该句中 before 引导时间状语从句，must not 语气较为强烈。

➢ 翻译：在你被检查是否有实弹和空弹匣之前，你不能离开靶场。

【口语输出】

1. What kinds of rules must you obey while you are on the ranges?

【参考词汇】instruction, range personnel, down range, firing line, muzzle, firing point, trigger area, hearing protection, helmet, live round, empty case, ammunition point

2. Please make a briefing on a shooting training on the weekend.

【参考词汇】weapons test, on the ranges, helmets, ear defenders, webbing, waterproof ponchos, mess tin, programme, reveille, draw individual weapons, armoury, parade, detail, supervise, concurrent, award, barrack

Delta   Exercise Mountain Trail

【军事知识】

1. Estimated Time of Arrival (ETA) 预计抵达时间

Estimated time of arrival (ETA) is the time interval at which a certain vehicle will arrive its destination. It is a transportation term that defines the time remaining for certain aircraft, automobile, ship or emergency service to reach the place it is directed to.

This term is frequently employed to inform passengers about the remaining period of time when certain transport is supposed to arrive or it is supposed to reach certain destination. It is also employed to inform recipients about the estimated date when they will receive certain cargo or mail delivery. Additionally, emergency services also provide the patient or the receiving health care facility with an estimated time of arrival for them to be properly prepared to attend the situation.

2. Estimated Time of Departure (ETD) 预计出发时间

Estimated Time of Departure is the projection of time that is expected for a transport system to depart its point of origin/location. Also, it is an indicated time to start a particular trip/journey. This term is the same as Expected Time of Departure.

【词汇点拨】Words and Expressions

1. base camp a camp where soldiers start their journey 大本营

Every 100 kilometers there will be a base camp you can have a rest at.

2. campsite /ˈkæmpˌsaɪt/ *n.* a site where soldiers can pitch a tent 宿营地

We'll start back towards the campsite after lunch.

3. compass bearing the directions that the compass points at 罗盘指示方向

Description of the general location. Include the site's distance (in a straight line) and compass bearing from the nearest significant town or city.

4. garrison /ˈɡærɪsn/ *n.* troops who occupy a fortress or town in order to defend it 驻军；卫戍区；卫戍部队；守备部队

a garrison of 5000 troops 有 5000 士兵驻守的防地

The garrison held out for three weeks.

5. last light the twilight 暮色，天黑时间

The last light of day was fading in the window.

6. grid /ɡrɪd/ *n.* a network of horizontal and vertical lines that provide coordinates for locating points on an image 网格

The grid reference is C8.

7. ETA (Estimated Time of Arrival) 预计到达时间

8. ETD (Estimated Time of Departure) 预计出发时间

9. terrain /təˈreɪn/ *n.* a piece of ground having specific characteristics or military potential 地形；地势

The terrain changed quickly from arable land to desert.

10. leg /leɡ/ *n.* a section or portion of a journey or course（旅程或赛程的）一段，一程

On the return leg, you leave the summit of Mt Annan at 1200 hours.

11. elevation /ˌelɪˈveɪʃ(ə)n/ *n.* a raised or elevated geological formation 海拔，高地

We're probably at an elevation of about 13,000 feet above sea level.

【词汇点拨】Proper Names

1. Lieutenant Phillips 菲利普斯中尉

2. Exercise Mountain Trail "山痕"演习

3. Corporal Hetman 盖特曼下士

4. Sergeant Major Thompson 汤普森准尉

5. Glen hotel 格伦酒店（虚构旅店）

6. Mt Annan 安南山（虚构山地）

7. Mt Moffat 莫法特山（虚构山地）

8. Sergeant Peters 彼得斯中士

9. Corporal Smith 史密斯下士

【长句解读】

1. I estimate one hour to the camp, so we should make it by last light with no problems. (8-10)

➢ 句子分析：本句中 make it 为口语式表达，意为"成功抵达"。last light 意为"天黑时间"。

> 翻译：我估计到营地要一个小时，所以我们应该能在天黑之前到达。

2. Why don't we split the group in two? You take one group and do Annan, and I'll take the other group and climb Mt Moffat. (8-11)

> 句子分析：本句中 split 意为"把……分成"。do Annan 为口语式表达，意为"前往安南山"。

> 翻译：我们为什么不把小组分成两组呢？你带一组去爬安南山，我带另一组去爬莫法特山。

3. B: Let's go through the route card for your team. We'll call it team A. Team leader: Sergeant Peters. You leave camp at 0800 hours, travel nine km on a bearing northeast and your ETA at the summit of Mt Annan is 1130. Terrain is rough.

A: Have that.

> 句子分析：该句主语中 bearing 意为"……方向"，Have that. 口语式表达，意为"明白"。

> 翻译：

B：让我们为你的队伍过一遍路线卡。我们叫它 A 队，队长彼得斯中士。你们 8 点离开营地，向东北方向行驶 9 千米，预计 11 点 30 分到达安南山峰顶。地形是粗糙的。

A：明白。

4. On the return leg, you leave the summit of Mt Annan at 1200 hours, travel nine km on a bearing southwest and reach camp at 1515. (8-11)

> 句子分析：句中 leg 意为"一程"，bearing 意为"……方向"。

> 翻译：返程时，你们在 12 点离开安南山峰顶，向西南方向行驶 9 千米，在 15 点 15 分到达营地。

【口语输出】

1. Please talk about "Exercise Moutain Trail" based on the audio clip.

【参考词汇】adventure training exercise, survival exercise, the minimum equipment

2. Please plan your own exercise based on the map of Task Four.

【参考词汇】depart, garrison, drop off point, set off, terrain, last light, base camp, ETA, ETD, bearing, return leg, summit, elevation

Echo Larger formations

【军事知识】

1. 1st Armoured Brigade (the United Kingdom) 第一装甲旅

The 1st Armoured Brigade was a regular British Army unit formed on 3 September 1939, by the redesignation of the 1st Light Armoured Brigade.

At the start of the war, the brigade was based in the United Kingdom, initially as part of the 1st Armoured Division and then as part of the newly formed 2nd Armoured Division. In November 1940, it was shipped to Egypt, arriving on 1 January 1941. In March 1941, the brigade was dispatched to Greece as part of General Maitland Wilson's unsuccessful attempt at stopping the German invasion. On 29 April 1941, the brigade was evacuated to Egypt.

The 1st Armoured Brigade served in the Western Desert Campaign with the 7th Armoured Division at the Battle of El Alamein. The brigade was disbanded on 21 November 1942.

2. 2nd Cavalry Regiment (the United States) 第二骑兵团

The 2nd Cavalry Regiment (2 CR), also known as the Second Dragoons, is an active "Strykerized" infantry and cavalry regiment of the United States Army. The Second Dragoons is a component of V Corps and United States Army Europe, with its garrison at the Rose Barracks in Vilseck, Germany. It can trace its lineage back to the early part of the 19th century where the 2nd Cavalry has the distinction of being the longest continuously serving regiment in the Army.

3. 18th Airborne Corps (the United States) 第十八空降军

The 18th Airborne Corps is the corps of the United States Army designed for rapid deployment anywhere in the world. It is referred to as "America's Contingency Corps". Its headquarters are at Fort Bragg, North Carolina. The corps was first activated on 17 January 1936 as the 2nd Armored Corps at Camp Polk in Louisiana. When the concept of Armored Corps proved unnecessary, the 2nd Armored Corps was redesignated as 18th Corps on 9 October 1937 at the Presidio of Monterey, California.

4. 3rd Signal Regiment (the United Kingdom) 第三通信团

The 3rd Signals Regiment is a regiment of the Royal Corps of Signals within the British Army. The regiment has recently converted from a divisional signal regiment to a support signals regiment part of the 11th Signal Brigade, but has remained at Bulford.

The 3rd Signal Regiment can trace its history back to "The Telegraph Battalion, Royal

Engineers". In 1903 it became the telegraph battalion for the 3rd Division. Just a few months later it joined the 2nd Infantry Division, and finally in 1905 it joined the 3rd Division. Since 1905 the regiment has been part of the division and became the Headquarters Signal regiment in 1962. In 1945 after the end of World War II the regiment was retitled to the "3rd Infantry Division Signal Regiment". In 1947 upon returning from British Palestine the regiment was disbanded, but re-formed in 1951 as part of the new Army Strategic Command.

【词汇点拨】Words and Expressions

1. section /ˈsekʃn/ *n.* a small army unit usually having a special function 班（英）

A British infantry section has between eight to ten men.

2. squad /skwɑːd/ *n.* a smallest army unit 班（美）

Here is a great sense of team spirit in the squad.

3. platoon /pləˈtuːn/ *n.* a military unit that is a subdivision of a company; usually has a headquarters and two or more squads; usually commanded by a lieutenant 排

The lieutenant was put in charge of an infantry platoon.

4. company /ˈkʌmpəni/ *n.* small military unit; usually two or three platoons 连

An infantry company has between 100 and 200 men.

5. battalion /bəˈtæliən/ *n.* an army unit usually consisting of a headquarters and three or more companies 营

A U.S. infantry battalion was marching down the street.

6. regiment /ˈredʒɪmənt/ *n.* an army unit smaller than a division 团

We were part of an entire regiment that had nothing else to do but to keep that motorway open.

7. brigade /brɪˈɡeɪd/ *n.* an army unit smaller than a division 旅

The brigade forms part of a division, which is made up of three or four brigades with a total strength of about 15,000 soldiers.

8. division /dɪˈvɪʃn/ *n.* a unit of an army, consisting of several brigades or regiments 师

Ford was attached to the 101st Airborne Division.

9. corps /kɔːr/ *n.* an army unit usually consisting of two or more divisions and their support 军

A corps has between 40,000 and 60,000 men.

10. deployment /dɪˈplɔɪmənt/ *n.* the distribution of forces in preparation for battle or work 部署

Missile deployment did much to further polarize opinion.

【词汇点拨】Proper Names

1. **1st Armoured Brigade** 第一装甲旅
2. **2nd Cavalry Regiment** 第二骑兵团
3. **18th Airborne Corps** 第十八空降军
4. **5th Infantry Division** 第五步兵师
5. **3rd Signal Regiment** 第三通信团
6. **14th Artillery Regiment** 第十四炮兵团
7. **Major Pettifer** 佩蒂弗少校
8. **Cyprus** /ˈsaɪprəs/ 塞浦路斯（地中海东部一岛国）
9. **Bosnia** /ˈbɑːznɪə/ 波斯尼亚（位于原南斯拉夫中西部，现为波斯尼亚和黑塞哥维那北部地区）
10. **Kosovo** /ˈkɒsəvəʊ/ 科索沃（塞尔维亚自治省）
11. **Osnabruck** /ˈɒznəbrʊk/ 奥斯纳布吕克（位于德国下萨克森州）
12. **Bergen** /ˈbɜːɡən/ 卑尔根（位于德国汉诺威州）
13. **Paderborn** /ˌpɑːdərˈbɔːrn/ 帕德博恩（位于德国莱茵—威斯特法伦州）

【长句解读】

1. With combat support and combat service support elements, the total strength of British forces in Germany is about 18,000 personnel. (8-13)

　➢ 句子分析：本句中 With combat support and combat service support elements 为伴随状语，真正的主句为后半句话。

　➢ 翻译：加上作战支援和作战勤务保障单元，英国驻德部队的总兵力约为1.8万人。

2. The UK contributes a battlegroup and support personnel to the NATO-led operations in Bosnia and Kosovo. (8-13)

　➢ 句子分析：本句中 contribute...to... 意为"为……提供……"。NATO-led 为复合词，意为"北约领导下的"。

　➢ 翻译：英国为北约领导下的在波斯尼亚和科索沃的行动提供了一个战斗群和支援人员。

【口语输出】

1. Please talk about the formation of China's Army.

【参考词汇】section, platoon, company, battalion, regiment, brigade, division, corps, comprise, current, contribute...to..., deploy

2. Please talk about your understanding of British Army.

【参考词汇】deploy, station, division, brigade, combat support, combat service support, personnel, joint military command, civilian personnel, servicemen

 阅读指导

【文章导读】

This passage mainly focuses on the general idea of the hierarchy of military formations at the division level and below. The organization of specific units will vary depending on their history, needs, mission, and resources. Brigades and battalions are often named according to their branch classification, or primary function, such as: armoured, infantry, engineer, mechanized, signal, and artillery. This section mainly focused on the 1st (UK) Armoured Division.

【军事知识】

1. British 1st Armoured Division 英国第一装甲师

The 1st (UK) Armoured Divsion is the only British Army division stationed in Germany, being part of the British Forces in Germany. Based at Herford Germany, the majority of British forces in Nordrhein-Westfalen and Niedersachsen, Germany, belong to the 1st (UK) Armoured Division. Although the Division is under direct command of Headquarters, Land Forces in the United Kingdom, it relies on the United Kingdom Support Command (Germany) for infrastructure, administrative and welfare support within Germany. 1st (UK) Armoured Division is also an important element assigned to NATO's Allied Rapid Reaction Corps and is capable of undertaking national or NATO operations ranging from peace-keeping duties to high intensity conflict.

2. British Army Organisation 英国军队编制

The British Army is the land warfare branch of the British Armed Forces. The English Army, founded in 1660, was succeeded in 1707 by the new British Army, incorporating the existing Scottish regiments. It was administered by the War Office from London, which was subsumed into the Ministry of Defence in 1964. The professional head of the British Army is the Chief of the General Staff.

Section: A section usually consists of 7 to 12 men, and is part of a platoon. Sections are usually under the command of a non-commissioned officer, often a corporal or sergeant.

Platoon: A platoon is a part of an infantry company and is further divided into three or four sections. A British platoon usually consists of 25 to 30 men. Platoons are commanded by a lieutenant or second lieutenant.

Troop: Part of a squadron of cavalry or a battery of artillery, a troop is equivalent to an infantry platoon. Troops are normally commanded by a lieutenant or second lieutenant.

Company: A company is part of a battalion and usually consists of between 100 and 150 men. They are usually lettered A through to D, and made up of at least two platoons. But sometimes they have names such as 'Grenadier Company' or 'Fire Support Company'. Companies are commanded by a major or a captain.

Squadron: A squadron is a sub-unit of a cavalry, engineer or armoured regiment. It is equivalent in status and size to an infantry company and normally consists of two or more troops. Squadrons are commanded by a captain or major and usually named by letter.

Battery: A battery is an artillery unit equivalent to an infantry company. Sub-units of batteries are called troops. An administrative collection of artillery batteries was called a battalion in the 18th century, a brigade until 1938, and since then a regiment. Tactical artillery regiments all belong to the Royal Regiment of Artillery. Batteries are commanded by a captain or major.

Battalion: A battalion is a regimental sub-unit of infantry amounting to between 500 and 1,000 soldiers. It normally consists of a headquarter and three or more companies. Traditionally, most British regiments have had more than one battalion. But different battalions of the same regiment have seldom fought together.

A tactical grouping of battalions is called a brigade. Battalions are normally commanded by a lieutenant colonel. At present, the British Army has 47 regular and reserve infantry battalions.

Brigade: This is a formation consisting of three infantry battalions or three cavalry or armoured regiments. During the world wars a brigade numbered between 3,500 and 4,000 men.

When forming part of a division, a brigade has no internal support. But when operating independently (usually called a brigade group), it includes supporting reconnaissance, artillery, engineers, supply and transport. A brigade is commanded by a major-general or brigadier.

Division: A division is made up of three infantry, cavalry or armoured brigades. Divisions are usually equipped to operate independently in the field, and have a full complement of supporting reconnaissance, artillery, engineers, medical, supply and transport troops.

During the World Wars, the average British division numbered around 16,000 men. Divisions are commanded by a lieutenant-general or major-general. The British Army currently has two deployable divisions.

Coprs: This is a tactical formation made up of two or three divisions and commanded by a lieutenant-general. Corps are normally identified by Roman numerals. During the First World War the British Army grew to encompass 22 army corps.

Army: An army is a formation consisting of two or more corps. They are commanded by a general or a field marshal. An army in the Second World War numbered about 150,000 men. Eleven British armies were formed during the First World War. More than one army operating together is known as an army group.

【词汇点拨】Words and Expressions

1. hierarchy /ˈhaɪərɑːrki/ *n.* the organization of people at different ranks in an administrative body 等级体系

Like most other American companies with a rigid hierarchy, workers and managers had strictly defined duties.

2. generic /dʒəˈnerɪk/ *adj.* applicable to an entire class or group 一般的，通用的

Doctors sometimes prescribe cheaper generic drugs instead of more expensive brand names.

3. mechanised /ˈmekənaɪzd/ *adj.* equipped with machinery 机械化的，用机械装置的

Mechanised farming schemes that grow staples have often ended with abandoned machinery rusting in the returning bush.

4. major general a general officer ranking above a brigadier general and below a lieutenant general 少将

On July 3, Lee ordered an assault on that center, spearheaded by Major General George Pickett's fresh division.

5. brigadier /ˌbrɪɡəˈdɪə(r)/ *n.* a general officer ranking below a major general 准将；陆军准将

Brigadier Udaya Nanayakkara said the wreckage and the body of the pilot were recovered.

6. brigade /brɪˈɡeɪd/ *n.* a subdivision of an army, typically consisting of a small number of infantry battalions and/or other units and typically forming part of a division 旅

A third brigade is at sea, ready for an amphibious assault.

7. colonel /ˈkɜːn(ə)l/ *n.* a commissioned military officer in the United States Army or Air Force or Marines who ranks above a lieutenant colonel and below a brigadier general（陆军、空军或海军陆战队）上校

The colonel paraded his men before the Queen.

8. major /ˈmeɪdʒər/ *n.* a commissioned military officer in the United States Army or Air Force or Marines; below lieutenant colonel and above captain 少校

The Officer Commanding is usually a major.

9. lieutenant /lefˈtenənt/ *n.* a commissioned military officer 中尉

The lieutenant stopped and stood stock-still.

10. corporal /ˈkɔːpərəl/ *n.* a non-commissioned officer in the army or airforce or marines 下士

The corporal shouted an order at the men.

11. organigram /ɔːˈɡænɪɡræm/ *n.* a drawing or plan that gives the names and job titles of all the staff in an organization or department, showing how they are connected to each other 组织结构图

The Office organigram is regularly included for the Committee's information within the annual budget proposal.

12. variation /ˌveəriˈeɪʃn/ *n.* an instance of change; the rate or magnitude of change 变化；变动

This variation shows up in the trees growth rings.

【词汇点拨】**Proper Names**

Long-Range Anti-Tank Guided Weapon 长射程反坦克制导武器
Provost (Pro) /ˈprɒvəst/ 宪兵

【参考译文】

陆军建制
师、旅和营级单位

军事编制的一般等级

本节提供了师级及以下军事编制等级的一般性看法。但是这里的描述只是呈现的

对于一些"典型"单位的一般性概述。同时也意在清楚地了解单位与单位之间的关系。具体单位的建制因历史、需求、任务和资源而有所不同。想要更进一步了解某个特定划分，请参考下一节——具体示例：英国第一装甲师。

应当指出，旅和营往往是根据某兵种分类或主要职能命名的。例如：装甲兵、步兵、工兵、机械化兵、通信兵和炮兵。

因此，我们谈到实际的某个单位时，我们可能会提到一个装甲旅或一个机械化营。这些分类不会出现在我们的通用图表中，在下一节将有所呈现——具体示例：英国第一装甲师。

军事编制：从最大到最小排列

师（通常有 3 或 4 个旅）

总兵力：约 1 万~2 万人

指挥官：少将

旅（通常有 5 个营或团）

总兵力：约 3000~5000 人

指挥官：准将或上校（在美国陆军中是准将）

团 / 营（通常有 3 个连）

总兵力：约 700 人

指挥官：中校

连（通常有 1 个总部和 3 个排）

总兵力：约 125 人

连长：少校

排（通常有 1 个总部和 3 个班）

总兵力：通常 36 人

排长：中尉

班（美国陆军称 Squad）

总兵力：8~10 人

班长：下士

具体示例：英国第一装甲师

下方组织架构提供了上述一般性概述如何应用到具体某个师及其旅和营的示例。在这个示例中，我们可以看到英国第一装甲师从较大架构到较小架构的层级。

第 1 级：**师**：这代表了英国第一装甲师的实际师级建制。

第 2 级：**旅**：这代表了典型的装甲旅建制，尽管具体旅会有不一样的地方。

第 3 级：**营**：这代表了典型的装甲步兵营的建制，尽管具体营会有不一样的地方。

下面是常见的缩写形式。

单位大小	单位特点
Bde=Brigade 旅 Bn=Battalion 营 Bty=Battery（炮兵）连 Coy=Company（步兵）连 Div=Division 师 Pl=Platoon 排 Regt=Regiment 团 Sect=Section 班 Sqn=Squadron（装甲兵、工兵）连	Armd=Armoured 装甲兵 Arty=Artillery 炮兵 HQ=Headquarters 总部 Inf=Infantry 步兵 LRTGW=Long Range Anti-tank Guided Weapon 长射程反坦克制导武器 Mech=Mechanised 机械化兵 Pro=Provost(Law enforcement) 宪兵

 词汇表

Alpha Battalion organisation
division 师
tactical 战术上的
strength （军事方面的）实力
brigade 旅
mortar 迫击炮
comprise 包括；构成
ration （单兵）口粮
remit 职权范围
sentry guard 哨兵
operational effectiveness 作战效能
ammunition 弹药；军火
supervise 监督（任务，计划，活动）
travel claim 差旅报销
discharge documentation 退任（撤职文件）
casualty 伤亡人员
maintenance 维护；保养
dispatch 派遣

Bravo Battalion HQ
colonel （陆军、空军或海军陆战队）上校

corporal（陆军、海军陆战队或英国空军的）下士
sergeant（陆军、海军陆战队或空军的）中士
staff sergeant（英陆军）上士；（美陆军）上士；（美空军）中士
colour sergeant（英陆军步兵、海军陆战队等）上士
captain 上尉
drop off 让……下车
range 靶场；射程
station 驻扎
transfer request 调任请求

Charlie Range day
armoury 军械库
cleaning kit 擦枪套件；清洁套件
ear defenders 护耳塞
mess tin （军用）饭盒
respirator 防毒面具
waterproof poncho 防水雨披

webbing（作战装具）背带
concurrent 同时发生的
ammunition point 弹药补给点
backstop 挡弹体
down range（射击枪支指向）安全方向
empty case 空弹夹
firing line 射击线
live round 实弹
muzzle 枪口
range flag 边旗

Delta　Exercise Mountain Trail
base camp 大本营
campsite 宿营地
compass bearing 罗盘指示方向
garrison 驻军；卫戍区；卫戍部队；守备部队
last light 暮色，天黑时间
grid 网格
ETA (Estimated Time of Arrival) 预计到达时间
ETD (Estimated Time of Departure) 预计出发时间
terrain 地形；地势
leg（旅程或赛程的）一段，一程

elevation 海拔，高地
Echo　Larger formations
section 班（英）
squad 班（美）
platoon 排
company 连
battalion 营
regiment 团
brigade 旅
division 师
corps 军
deployment 部署
generic 一般的，通用的
mechanized 机械化的，用机械装置的

Foxtrot　Review
hierarchy 等级体系
colonel（陆军、空军或海军陆战队）上校
major 少校
lieutenant 中尉
corporal 下士
major general 少将
brigadier 准将；陆军准将
organigram 组织结构图
variation 变化；变动

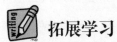 拓展学习

British Army

　　The British Army is purely a professional force since National service came to an end. The full-time element of the British Army is referred to as the Regular Army since the creation of the reservist Territorial Force in 1908. The size and structure of the British Army is continually evolving, but as of 1 December 2013, the British Army employs; 95,800 Regulars, 3,130 Gurkhas and 26,500 Army Reservists for a combined component strength of 125,430 personnel.

The future transformation of the British Army is referred to as "Army 2020", Army 2020 is the result of the October 2010 Strategic Defence and Security Review and the number of following reviews/modifications thereafter. Army 2020 will "ensure that the British Army remains the most capable Army in its class" and enable "it to better meet the security challenges of the 2020s and beyond". Initially, the October 2010 SDSR outlined a reduction of the Regular British Army by 7,000 to a trained strength of 95,000 personnel by 2015. However, following a further independent review on the future structure of the British Army—18 July 2011 "Future Reserves 2020—The Independent Commission to review the United Kingdom's Reserve Forces"—it was announced that the Regular Army will be reduced to a trained strength of 82,000 while the Army Reserve will be increased to a trained strength of around 30,000 personnel. There will of-course be an added margin for soldiers in training. This reform will bring the ratio of regular and part-time personnel of the British Army in-line with US and Canadian allies. Perhaps the most important aspect of Army 2020 is that the Army Reserve will become "fully integrated" with the Regular Army and "better prepared" for overseas deployments and operations. However budget cuts have made Britain a "hostile recruiting environment", with barely half the required number of new reservists actually signing up.

In addition to the active elements of the British Army (Regular and Army Reserve), all ex-Regular Army personnel remain liable to be recalled for duty in a time of need, this is known as the Regular Reserve. The Regular Reserve is separated into two categories: A and D. Category A is mandatory, with the length of time serving in category A depending on time spent in Regular service. Category D is voluntary and consists of personnel who are no-longer required to serve in category A. Regular Reserves in both category A and D serve under a fixed-term reserve contract and are liable to report for training or service overseas and at home. These contracts are similar in nature to those of the Army Reserve. The Long Term Reserve is also part of the Regular Reserve but excludes personnel serving in categories A and D. Unlike the other reserves the Long Term Reserve do not serve under a contract of any sort, instead they retain a "statutory liability for service" and may be recalled to service under Section 52 of the Reserve Forces Act (RFA) 1996 (until the age of 55). In 2007 there were 121,800 Regular Reserves of the British Army, of which, 33,760 served in categories A and D. Publications since April 2013 no-longer report the entire strength of the Regular Reserve, instead they only give a figure for the Regular Reserves serving in categories A and D only. They had a reported strength of 31,300 personnel in 2013. (560 words)

(*Selected from https://military-history.fandom.com/wiki/British_Army*)

【Notes】

1. the Regular Army 正规军
2. the reservist Territorial Force 国土防御预备役部队
3. Gurkha 英国军队中的尼泊尔族士兵
4. Army Reservists 陆军预备役
5. Strategic Defence and Security Review（SDSR）《战略防御与安全审查》
6. The Independent Commission 独立委员会
7. in line with 符合；与……一致
8. hostile 敌对的，怀敌意的
9. mandatory 义务的，强制性的
10. voluntary 志愿的
11. fixed-term 固定期限的
12. The Long Term Reserve 长期预备役
13. statutory 依照法令的，法定的
14. the Reserve Forces Act (RFA) 预备役法案
15. Regular Reserves 常规预备役

 单元练习

I. Label the picture

1._____ 2._____ 3._____
4._____ 5._____ 6._____

II. Multiple choice

1. A _____ is the basic tactical unit in the infantry and is commanded by a lieutenant colonel.

    A. corps   B. division   C. brigade   D. battalion

2. Each section has a strength of _____ men and is divided into two fire teams.

    A. five   B. six   C. seven   D. eight

3. In the US a section is a unit between a _____ and a _____.

    A. platoon, company   B. company, battallion

    C. squad, platoon   D. squad, company

4. As is often the case, NCOs usually address enlisted personnel by_____.

    A. last name   B. sir/ma'am   C. first name   D. rank+name

5. Which of the following is the most polite requirement?

    A. Could you complete this for me please, sir?

    B. Could you complete this for me please, corporal?

    C. Can you complete this?

    D. Complete this for me.

6. Which of the following is seriously prohibited on the range?

    A. Always point the muzzle of the weapon down range whenever on the firing point.

    B. Fire or point weapons outside the limit of poles.

    C. Wear hearing protection and kevlar helmets on range at all times.

    D. Obey all instructions given by the range personnel.

7. _____ refers to the network of squares on a map, numbered for reference.

    A. Terrain   B. Compass bearing

    C. Grid   D. Garrison

8. _____ refer to the food allowance for one day (especially for service personnel).

    A. Remits   B. Rations   C. Ammunitions   D. Ranges

9. Which of the following shows the most difficult terrain?

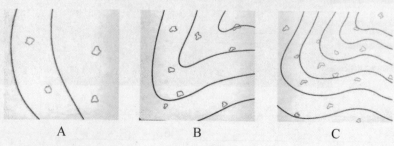

    A   B   C

10. On the range, soliders must not leave the range befor being inspected for live round and_____.

    A. empty cases   B. waterproof poncho

    C. webbing   D. backstop

III. Translation: English to Chinese

1. The organization of specific units will vary depending on their history, needs, mission, and resources.

2. It should be noted that brigades and battalions are often named according to their branch classification, or primary function, such as: armoured, infantry, engineer, mechanized, signal, and artillery.

3. So when we speak of actual units, we might take reference to an armoured brigade or a mechanized battalion.

4. In barracks, they carry out a wide range of duties, including processing travel claims, applications for leave and so on.

5. On operations, he is responsible for providing ammunition, rations and clothing.

IV. Translation: Chinese to English

英国陆军实际兵力为 103350 人；预备役 121800 人；本土防御部队 35020 人。正规军由 11 个装甲团、15 个炮兵团、12 个工兵团、36 个步兵营、5 个航空团、12 个通信团、7 个装备保障营、17 个后勤团以及 8 支卫生部队组成。通常，这些营/团由 500 人/800 人组成。本土军由 4 个装甲团、7 个炮兵团、6 个工程团、15 个步兵营、1 个航空团、9 个通信团、4 个装备保障营、18 个后勤团、2 个情报营以及 8 所医院和 3 个医疗卫生团组成。

V. Oral practice

1. Please talk about some safety rules for operating a weapon, machine or vehicle. Try to use expressions like "must always", "must never" and "must only".

2. Work in pairs and talk about your duties to your partner.

附录：Unit 8 练习答案

I. Label the picture

1. backstop   2. range flag   3. down range

4. target   5. firing line   6. ammunition point

II. Multiple choice

1-5 DDCDA   6-10 BCBCA

III. Translation: English to Chinese

1. 具体单位的建制因历史、需求、任务和资源而有所不同。

2. 应当指出，旅和营往往是根据某兵种分类或主要职能命名的。例如：装甲兵、步兵、工兵、机械化兵、通信兵和炮兵。

3. 因此，我们谈到实际的某个单位时，我们可能会提到一个装甲旅或一个机械化营。

4. 在营房里他们负责一系列职务，包括差旅报销、请假申请等。

5. 作战方面，他负责提供弹药、口粮和衣物。

IV. Translation: Chinese to English

The British Army has the actual strength of 103,350 men, with 121,800 in the reserve and 35,020 men in the Home Defence Force. The regular army consists of 11 armoured regiments, 15 artillery regiments, 12 engineering regiments, 36 infantry battalions, 5 aviation regiments, 12 communications regiments, 7 equipment support battalions, 17 logistics regiments and 8 health units. Typically, these battalions/regiments consist of 500 men /800 men. The Home Army consists of 4 armoured regiments, 7 artillery regiments, 6 engineering regiments, 15 infantry battalions, 1 aviation regiment, 9 communications regiments, 4 equipment support battalions, 18 logistics regiments, 2 intelligence battalions, and 8 hospitals and 3 medical and health regiments.

V. Oral practice

1. In order to make sure of our safeties, there are some rules we need to obey. We must obey all instructions given by the range personnel; never touch weapons while personnel are down range or in front of the firing line. We must always point the muzzle of the weapon down range whenever on the firing point. The firing finger must not be on the trigger area. We must only aim for the targets assigned to you. We must never fire or point weapons outside the limit of poles. We must wear hearing protection and kevlar helmets on range at all times. When leaving the firing line, give all live rounds and empty cases to the ammunition point. We must not leave the range before being inspected for live rounds and empty cases.

2. I am the regimental quartermaster sergeant major. I'm in charge of the battalion's equipment and supplies. In barracks, I spend a lot of time controlling the issue of equipment from the battalion stores. I also handle orders for spare parts for vehicles. On operations, I'm responsible for providing the battalion with ammunition, rations and clothing. I have inspections several times a year and I have to be responsible for every piece of equipment that the battalion is issued with.

Unit 9 Parachute Regiment

单元导学清单

模块	主题	学习目标	任务	军事知识	核心词汇
Alpha	空降前选拔	1）了解伞兵团和空降前选拔基本知识；2）掌握空降前选拔相关的常用词汇和表达；3）讨论不同的训练主题。	Task 1 Task 2 Task 3 Task 4	■ Parachute Regiment ■ RAF Brize Norton	airborne force, camouflage, cross-country run, march in battle order, pass out parade, Milan anti-tank weapon, SA80, military parachutist wings (emblem)
Bravo	伞兵装备	1）了解伞兵基本装备；2）掌握介绍伞兵装备的相关词汇和表达。	Task 1 Task 2 Task 3	■ fighting order ■ marching order	beret, radio headset, webbing, grenade, extra ammunition, water bottle, individual weapon, protective clothing, combat rations, rucksack
Bravo	SA80	1）了解英军配发的制式单兵武器SA80；2）掌握与SA80相关的词汇与表达；3）介绍一种武器或装备。	Task 4 Task 5 Task 6 Task 7 Task 8 Task 9	■ SA80 ■ G-36	calibre, barrel, strip, reassemble, cocked, live round, chamber, eject, cocking handle, change lever, magazine catch, magazine, sights, safety catch, muzzle, rounds, trigger
Charlie	排训练计划	1）了解排训练计划的拟定：时间、人员、场地的确认等；2）熟练掌握与排训练计划相关的词汇与术语；3）讨论排训练计划。	Task 1 Task 2 Task 3 Task 4 Task 5	■ marksmanship ■ obstacle course ■ NBC test	marksmanship, obstacle course, range, instructor, installation, range officer
Charlie	排训练科目	1）了解排训练科目；2）熟练掌握与排训练科目相关的词汇和句型；3）介绍自己的训练科目。	Task 6 Task 7 Task 8 Task 9	■ colour sergeant (CSgt) ■ practice for parade	fitness, tunnel, cargo net, ditch, log, ramp, fence, low wall

续表

模块	主题	学习目标	任务	军事知识	核心词汇
Delta	从军经历	1）熟练掌握与从军经历相关的词汇和表达；2）介绍自己的从军经历。	Task 1 Task 2 Task 3 Task 4	■ Cyprus ■ tour of duty	CV (curriculum vitae), posting, lance corporal, company sergeant major, regimental sergeant major, depot, late entry commission, line infantry regiment, quartermaster
Echo (Review)	马岛战争	1）了解马岛战争的历史背景、过程、影响等相关知识；2）熟练掌握相关词汇和表达；3）运用所学知识简要介绍马岛战争。	Task 1 Task 2	■ The Malvinas War ■ surprise attack	conflict, rival, colony, dispute, sovereignty, marine, sailor, overwhelmed, assault craft, capture, barrack, casualty, amphibious, inhabitant, mainland, compromise, withdrawal, momentum, airfield, logistical, submarine, cruiser, destroyer, missile, beachhead, vessel, outnumber, simultaneous, surrender, garrison, patrol ship

 听说指导

Alpha　P Coy (Pre-parachute Selection)

【军事知识】

1. The Parachute Regiment 伞兵团

The Parachute Regiment, colloquially known as the Paras, is the Airborne Infantry of the British Army. One battalion is permanently under the command of the Director Special Forces in the Special Forces Support Group. The other battalions are the parachute infantry component of the British Army's rapid response formation 16 Air Assault Brigade. It is the only line infantry regiment that has not been amalgamated with another unit since the end of the Second World War. Members of the Parachute Regiment are known to the rest of the army and the British public by the nickname the Paras. The regiment took part in five major parachute assault operations in North Africa, Italy, Greece, France, the Netherlands and Germany, often landing ahead of all other troops.

2. RAF Brize Norton 普莱斯诺顿皇家空军驻地

Royal Air Force station Brize Norton or RAF Brize Norton in Oxfordshire, about 105 km west north-west of London, is the largest station of the Royal Air Force. It is close to the settlements of Brize Norton, Carterton and Witney. The station is home to Air Transport, Air-to-Air refuelling and Military Parachuting. RAF Brize Norton was opened in 1937 as a training base. Major infrastructure redevelopment began in 2010 ahead of the closure of RAF Lyneham in 2012, at which point Brize Norton became the sole air point of embarkation for British troops. By the end of June 2011 all flying units from RAF Lyneham had moved to RAF Brize Norton.

【词汇点拨】Words and Expressions

1. airborne /ˈeəbɔːn/ *adj.* [only before noun] (of soldiers 士兵) trained to jump out of aircraft onto enemy land in order to fight 空降的

During the second world war he served with 92nd Airborne.

2. camouflage /ˈkæməflɑːʒ/ *n.* [U] a way of hiding soldiers and military equipment, using paint, leaves or nets, so that they look like part of their surroundings（军事上的）伪装，隐蔽

They were dressed in camouflage and carried automatic rifles.

3. march /mɑːtʃ/ *v.* [usually + adv./prep.] to walk with stiff regular steps like a soldier 齐步走；行进

Soldiers were marching up and down outside the government buildings.

4. parade /pəˈreɪd/ *n.* to move or travel in a particular direction 行进；前往

They stood as straight as soldiers on parade.

5. parachutist /ˈpærəʃuːtɪst/ *n.* a person who jumps from a plane using a parachute 跳伞者

He was an experienced parachutist who had done over 150 jumps.

6. emblem /ˈembləm/ *n.* a design or picture that represents a country or an organization（代表国家或组织的）徽章，标记，图案

America's national emblem 美国的国徽

the club emblem 俱乐部的徽章

【词汇点拨】Proper Names

Brize Norton 普莱斯诺顿

【长难句解读】

1. To earn their place in the Regiment, recruits complete the 24-week Combat Infantryman's Course. (9-1)

> 句子分析：该句中 earn their place 意为获得一席之地，recruits 表示新兵，combat Infantryman's Course 意为战斗步兵课程。

> 翻译：为了能跻身该团，新兵要用24周完成战斗步兵课程。

2. The first 12 weeks of the course is Basic Training and in addition to the core skills, recruits learn first aid, signal communications, map reading, drill and nuclear, biological and chemical defence. (9-1)

> 句子分析：该段对话中 drill 意为队列，"nuclear, biological and chemical defence" 意为核生化防护。

> 翻译：前12周课程是基础训练，除了核心技能外，新兵还要学会急救、信号通信、识图、队列及核生化防护。

3. In weeks 19 to 20, soldiers complete the Pre-parachute Selection, or P Company, a series of physical tests designed to ensure that an individual has the self-discipline and motivation required for service with airborne forces. (9-1)

> 句子分析：Pre-parachute Selection 意为空降前选拔。

> 翻译：第19到20周，士兵们要完成"空降前选拔"或者叫P连，也就是一系列身体检查，以确保新兵个人具备符合参加空降部队要求的自律能力和动机。

【口语输出】

1. Describe the training of officers or soldiers in China.

【参考词汇】adapting to life in the military, schools and courses, military training, physical training

2. What expressions have you learned from the radio when it comes to the core skills taught on the Combat Infantryman's Course?

【参考词汇】weapons training with the SA80 and the Milan anti-tank weapon; learning to live outdoors, camouflage and movement in the field; cross-country runs, marches in battle order and assault courses

Bravo Personal equipment

【军事知识】

1. SA80 SA80 突击步枪

The SA-80 is a family of standard issue weapons used by the British Armed Forces and Jamaican Defence Force. All weapons in the SA80 (Small Arms for the 1980s) range uses the 5.56mm NATO round.

The development of the SA80 (Small Arms for 1980s) system, which included two weapons—SA80 IW (Infantry Weapon) assault rifle and SA80 LSW (Light Support Weapon) light machine gun, began in the late 1960s when British army decided to develop a new rifle, which will eventually replace the venerable 7.62 mm L1 SLR (British-made FN FAL rifle) in the 1980s.

2. G-36 G-36 突击步枪

The Heckler und Koch G-36 assault rifle had been born as HK-50 project in early 1990s. Some German soldiers complained about position of dual optical sights and those sights being easily fogged in bad weather (rain or snow). Otherwise it is a good rifle, accurate, reliable, simple in operations and maintenance, and available in a wide variety of versions—from the short-barreled Commando (some even said that it's a submachine gun) G-36C and up to a standard G-36 rifle.

【词汇点拨】Words and Expressions

1. grenade /grəˈneɪd/ *n.* a small bomb that can be thrown by hand or fired from a gun 榴弹；手榴弹；枪榴弹

A hand grenade was thrown at an army patrol.

2. ammunition /ˌæmjuˈnɪʃn/ *n.* [U] a supply of bullets, etc. to be fired from guns 弹药

Government forces are running short of ammunition and fuel.

3. ration /ˈræʃn/ *n.*

① [C] a fixed amount of food, fuel, etc. that you are officially allowed to have when there is not enough for everyone to have as much as they want, for example during a war （食品、燃料等短缺时的）配给量，定量

the weekly butter ration 每周的黄油配给量

② [pl.] a fixed amount of food given regularly to a soldier or to sb. who is in a place where there is not much food available （给战士或食品短缺地区的人提供的）定量口粮

We're on short rations until fresh supplies arrive.

4. webbing /ˈwebɪŋ/ *n.* [U] strong strips of cloth that are used to make belts, etc., and to support the seats of chairs, etc. （用以制作带子等的）带状结实织物

webbing with pouches for grenades

5. strip /strɪp/ *v.* to separate a machine, etc. into parts so that they can be cleaned or repaired 拆卸；拆开

They taught us how to strip down a car engine and put it back together again.

6. reassemble /ˌriːəˈsembl/ *v.* to fit the parts of sth. together again after it has been taken apart 重新装配（或组装）

We will now try to reassemble pieces of the wreckage.

7. chamber /ˈtʃeɪmbə(r)/ *n.* [C] a space in the body, in a plant or in a machine, which is separated from the rest （人体、植物或机器内的）腔，室

Her instructor plugged live bullets into the gun's chamber.

8. eject /ɪˈdʒekt/ *v.* to push sth. out suddenly and with a lot of force 喷出；喷射；排出

Used cartridges are ejected from the gun after firing.

9. catch /kætʃ/ *n.* [C] a device used for fastening sth. 扣拴物；扣件

She fiddled with the catch of her bag.

10. muzzle /ˈmʌzl/ *n.* the open end of a gun, where the bullets come out 枪口；炮口

Mickey felt the muzzle of a rifle press hard against his neck.

【词汇点拨】Proper Names

1. Powers /ˈpaʊəz/ 鲍尔斯

2. York /jɔːk/ 约克

3. Brice /braɪs/ 布赖斯

4. Ahmed /ˈɑːmed/ 艾哈迈德

【长难句解读】

1. Depending on the tactical situation, riflemen wear either fighting order or marching order. (9-2)

➢ 句子分析：句中的 tactical situation 意为战术情况，fighting order 和 marching order 分别为战斗行装和行军行装。

➢ 翻译：根据战术情况，步兵要负载战斗行装和行军行装。

2. The SA80 has a calibre of 5.56 mm and weighs approximately 5 kilos with a loaded magazine. (9-4)

➢ 句子分析：该句中 calibre 意为枪炮的口径，magazine 此处指弹匣。

➢ 翻译：SA80 口径为 5.56 毫米，重约 5 千克（装满弹匣）。

3. LSW has a calibre of 5.56 mm and has the weight of 6.58 kg with loaded magazine and sight. (9-4)

➢ 句子分析：该句中的 LSW 意为轻型支援武器，此处指 SA80 轻型机枪，sight 意为枪或炮上的瞄准器。

➢ 翻译：LSW 口径为 5.56 毫米，重量为 6.58 千克（含满载弹匣和瞄准具）。

4. It is ready to fire when it is cocked and there is a live round in the chamber. (9-5)

➢ 句子分析：该句中的 cock 表示上扳机准备射击，live round 表示实弹。

➢ 翻译：当拉上枪机，膛内有子弹时，枪就处于待击发状态。

5. When the magazine is off the weapon, pull the cocking handle back to eject the round from the chamber. (9-6)

➢ 句子分析：该句中 cocking handle 指枪机柄。

➢ 翻译：弹匣与枪身分离后，向后拉枪机柄，将膛内子弹退出。

【口语输出】

1. Compare SA80 with LSW and say which things are the same and which are different.

【参考词汇】standard individual weapon, a light machine gun, calibre, weigh, ammunition, range

2. What is the weapon or a piece of equipment you know well? Share your ideas of its operation with your partners.

【参考词汇】SA80, LSW, G-36, gun, rifle

3. How to load and unload the assault rifle?

【参考词汇】safety catch, change lever, insert, magazine , cocking handle

Charlie The platoon training programme

【军事知识】

1. NBC 核生化

NBC stands for nuclear, biological, chemical. It is a term used in the armed forces and in health and safety, mostly in the context of weapons of mass destruction (WMD) clean-up in overseas conflict or protection of emergency services during the response to a terrorist attack, though there are civilian and common-use applications (such as recovery and clean up efforts after industrial accidents).

In military operations, NBC suits are intended to be quickly donned over a soldier's uniform and can continuously protect the user for up to several days. Most are made of impermeable material such as rubber, but some incorporate a filter, allowing air, sweat and condensation to slowly pass through.

An NBC (Nuclear, Biological, Chemical) suit is a type of military personal protective equipment designed to provide protection against direct contact with and contamination by radioactive, biological or chemical substances, and provides protection from contamination with radioactive materials and some types of radiation, depending on the design. It is generally designed to be worn for extended periods to allow the wearer to fight (or generally function) while under threat of or under actual nuclear, biological, or chemical attack. The civilian equivalent is the Hazmat suit.

2. Colour Sergeant 上士

Colour sergeant or color sergeant (CSgt or C/Sgt) is a rank of non-commissioned officer in the Royal Marines and infantry regiments of the British Army, ranking above sergeant and below warrant officer class 2. It has a NATO ranking code of OR-7 and is equivalent to the rank of staff sergeant in other branches of the Army, flight sergeant or chief technician in the Royal Air Force, and chief petty officer in the Royal Navy. The insignia is the monarch's crown above three downward pointing chevrons.

The rank was introduced into the British Army during the Napoleonic Wars to reward long-serving sergeants. By World War I all colour sergeants held the appointments of company sergeant major (who were promoted to the new rank of warrant officer class II from February 1915) or company quartermaster sergeant and the rank itself was never used. It appears to have been reintroduced in the 1920s.

Historically, colour sergeants of British line regiments protected ensigns, the most junior officers who were responsible for carrying their battalions' Colours (flag or insignia) to rally troops in battles. For this reason, to reach the rank of colour sergeant was considered a prestigious attainment, granted normally to those sergeants who had displayed courage on the field of battle. This tradition continues today as colour sergeants form part of a colour party in military parades.

【词汇点拨】Words and Expressions

1. marksmanship /ˈmɑːksmənʃɪp/ *n.* [U] skill in shooting 射击术

They practised marksmanship every day.

2. range /reɪndʒ/ *n.* the distance over which a gun or other weapon can hit things 射程；射击距离

We are now within range of enemy fire.

3. instructor /ɪnˈstrʌktə(r)/ *n.* a person whose job is to teach sb. a practical skill or sport 教练；导师

She found an instructor and started taking lessons.

4. installation /ˌɪnstəˈleɪʃn/ *n.* a place where specialist equipment is kept and used 设施

a military installation 军事设施

5. fitness /ˈfɪtnəs/ *n.* the state of being physically healthy and strong 健壮；健康

She was never interested in fitness before but now she's been bitten by the bug.

6. tunnel /ˈtʌn(ə)l/ *n.* a passage built underground, for example to allow a road or railway/railroad to go through a hill, under a river, etc. 地下通道；地道；隧道

The tunnel project has already fallen behind schedule.

7. cargo /ˈkɑːɡəʊ/ *n.* the goods carried in a ship or plane（船或飞机装载的）货物

Customs men put dynamite in the water to destroy the cargo, but most of it was left intact.

8. ditch /dɪtʃ/ *n.* a long channel dug at the side of a field or road, to hold or take away water 沟；渠

The truck overturned and precipitated us into the ditch.

9. log /lɒɡ/ *n.* a thick piece of wood that is cut from or has fallen from a tree 原木

I shut the shed door and wedged it with a log of wood.

10. ramp /ræmp/ *n.* a slope that joins two parts of a road, path, building, etc. when one is higher than the other 斜坡；坡道

It took five strong men to heave it up a ramp and lower it into place.

【词汇点拨】Proper Names

1. Hutchinson /ˈhʌtʃɪnsən/ 哈钦森

2. Jarvis /ˈdʒɑːvis/ 贾维斯

3. Peters /ˈpiːtəz/ 彼得斯

4. Parks /pɑːks/ 帕克斯

【长难句解读】

1. A: I'd like to have a look at the platoon training plan for next week. What activities do you have planned?

B: At the moment, marksmanship, the NBC test and the obstacle course. (9-7)

➢ 句子分析：该句中 platoon training plan 指的是排训练计划。

➢ 翻译：

A：我想看一看你们排下周的训练计划。你打算搞哪些活动？

B：目前有射击、三防考核，还有过障碍。

2. A: No, I'd like to have a look at personnel first. And then we can look at facilities.

B: Right, sir. I'll start with the PTIs. Colour Sergeant Hutchinson is away all week, so Sgt Jarvis will have to take the obstacle course. (9-8)

➢ 句子分析：该句 facilities 指训练设施。PTIs 指体育训练教练员。Colour Sergeant 是上士。

➢ 翻译：

A：不，我想先看一看人员。然后咱们再看看训练设施。

B：好的，长官。我先从 PTI 开始说吧。哈钦森上士周末不在，杰维斯中士不得不参加过障碍。

3. A: What about installations? When is the obstacle course free?

B: It's free all week. There's no problem there.

A: And what about the range?

B: 2 Platoon and 3 Platoon have the range most of the week. It's only free on Monday morning and Tuesday morning. (9-8)

> **句子分析**：句中 range 表示靶场。

> **翻译**：

A：训练设施怎么样？障碍场地什么时间有空闲？

B：一周全空，没问题。

A：靶场呢？

B：2 排、3 排一周大部分时间都占着靶场。只有周一、周二上午有空闲。

【口语输出】

1. When will the platoon have marksmanship training, obstacle course training, and the NBC test according to track 9-8?

【参考词汇】installations, range, free

2. When will your platoon have climbing, fitness test, military law, and practice for parade? Share your weekly training programme to the class.

【参考词汇】first, then, after that, then, next, and finally, run, jump, crawl, climb, report

Delta Military experience

【军事知识】

1. NATO 北大西洋公约组织

NATO is an international organization which consists of the U.S., Canada, Britain, and other European countries, all of whom have agreed to support one another if they are attacked. NATO is an abbreviation for "North Atlantic Treaty Organization."

The North Atlantic Treaty Organization or NATO is an intergovernmental military alliance based on the North Atlantic Treaty which was signed on 4 April 1949. The NATO headquarters are in Brussels, Belgium, and the organization constitutes a system of collective defence whereby its member states agree to mutual defense in response to an attack by any external party. Part of its 10 year mission statement reads: "It confirms the North Atlantic Treaty Organization's core task of defending its territory and its commitment to collective defense of its members."

2. Royal Military Academy Sandhurst 桑德赫斯特皇家军事学院

The Royal Military Academy Sandhurst (RMAS), commonly known simply as Sandhurst, is the British Army officer initial training centre located adjacent to the village of Sandhurst, Berkshire, about 55 kilometres (34 mi) southwest of London. The Academy's stated aim is to be "the national centre of excellence for leadership." All British Army officers, including late entry officers who were previously Warrant Officers, as well as many from elsewhere in the world, are trained at Sandhurst. The Academy is the British Army equivalent of the Britannia Royal Naval College Dartmouth, Royal Air Force College Cranwell and the Commando Training Centre Royal Marines.

3. Yuri Gagarin 尤里·加加林

Yuri Alekseyevich Gagarin (9 March 1934–27 March 1968) was a Soviet pilot and cosmonaut. He was the first human to journey into outer space, when his Vostok spacecraft completed an orbit of the Earth on 12 April 1961.

Gagarin became an international celebrity, and was awarded many medals and titles, including Hero of the Soviet Union, the nation's highest honour. Vostok 1 marked his only spaceflight, but he served as backup crew to the Soyuz 1 mission (which ended in a fatal crash). Gagarin later became deputy training director of the Cosmonaut Training Centre outside Moscow, which was later named after him. Gagarin died in 1968 when the MiG-15 training jet he was piloting crashed.

【词汇点拨】Words and Expressions

1. curriculum /kəˈrɪkjələm/ *n.* the subjects that are included in a course of study or taught in a school, college, etc.（学校等的）全部课程

Teachers incorporated business skills into the regular school curriculum.

2. depot /ˈdepəʊ/ *n.* a place where large amounts of food, goods or equipment are stored（大宗物品的）贮藏处，仓库

an arms depot 军械库

The target have included air defense sites and ballistic missile depots.

3. commission /kəˈmɪʃn/

① *v.* to choose sb. as an officer in one of the armed forces 任命……为军官

He has just been commissioned (as a) pilot officer.

② *n.* an officer's position in the armed forces 军官职务

He accepted a commission as a naval officer.

4. **quartermaster** /ˈkwɔːtəmɑːstə(r)/ *n.* an officer in the army who is in charge of providing food, uniforms and accommodation 军需官；军需主任

He was promoted to company quartermaster from a cookhouse soldier rank by rank.

【词汇点拨】Proper Names

1. **Parry** /ˈpæri/ 帕里
2. **Yuri Gagarin** 尤里·加加林
3. **Cyprus** /ˈsaɪprəs/ 塞浦路斯
4. **Hampshire** /ˈhæmpʃɪə/ 汉普郡
5. **Northern Ireland** 北爱尔兰
6. **Belfast** /ˌbelˈfɑːst/ 贝尔法斯特
7. **Sandhurst** /ˈsændhəːst/ 桑德赫斯特（英国皇家军事学院所在地）
8. **Arctic** /ˈɑːktɪk/ 北极的

【长难句解读】

1. I did Northern Ireland training, and then I did a tour of duty in Belfast. (9-10)
   - ➢ 句子分析：该句中 tour of duty 意思为任期、服役期、值班。
   - ➢ 翻译：我参加北爱尔兰训练，然后到贝尔法斯特轮值。

2. Every year the British Army commissions about 300 offfcers from the ranks. This is called a late entry commission. (9-10)
   - ➢ 句子分析：第一句中 commission 是动词，意为任命。第二句中 commission 用作名词，late entry commission 意为晚进任命，是军士晋升军官制度。
   - ➢ 翻译：每年英国陆军从行伍中任命约 300 名军官。这个叫作"晚进任命"。

3. I served as motor transport offfcer in a line infantry regiment and later became quartermaster of 3 Para. (9-10)
   - ➢ 句子分析：句中 line infantry regiment 意为线列步兵团。3 Para，指的是第三空降。
   - ➢ 翻译：我在一个线列步兵团担任汽车运输官员，后来做了第三空降的军需官。

【口语输出】

1. Who is Yuri Gagarin? Give a brief introduction according to 9-11.
   【参考词汇】a Russian pilot, interested in aircraft, flying club, the Soviet Air Force Base, cosmonaut, in space, spaceship, MiG-15, accident

2. Describe your own career with the help of the sentence patterns in this part.

【参考词汇】my first posting was in...;

after that, I was promoted to...;

I did a tour of duty in...;

I deployed to...;

I served as...;

I was commissioned directly as...

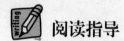 阅读指导

【文章导读】

This passage mainly focuses on the Malvinas War and its historical background, development and the Malvinas today. After explaining the surprise attack from Argentine troops, it elaborates how UK responded. Finally, what the Malvinas is like today is briefly introduced.

【军事知识】

1. Malvinas Islands 马尔维纳斯群岛

Malvinas Islands, also called Falkland Islands or Spanish Islas Malvinas, internally self-governing overseas territory of the United Kingdom in the South Atlantic Ocean. It lies about 300 miles (480 km) northeast of the southern tip of South America and a similar distance east of the Strait of Magellan. The capital and major town is Stanley, on East Falkland; there are also several scattered small settlements as well as a Royal Air Force base that is located at Mount Pleasant, some 35 miles (56 km) southwest of Stanley. In South America the islands are generally known as Islas Malvinas, because early French settlers had named them Malouines, or Malovines, in 1764, after their home port of Saint-Malo, France. Area 4,700 square miles (12,200 square km). Pop. (2012, excluding British military personnel stationed on the islands) 2,563.

2. United Nations Security Council Resolution 502 联合国安理会502号决议

On 31 March 1982, the Argentine ambassador to the UN, Eduardo Roca, tried garnering support against a British military build-up designed to thwart earlier UN resolutions calling for both countries to resolve their Falklands dispute through discussion. He did this because Argentina, based on inadequate intelligence gathering, was convinced a British task force was already on its way to the South Atlantic, and because of Britain's threat to use HMS Endurance to remove the scrap-metal workers from South Georgia. Any Argentine military action could then be justified as trying to counter Britain's use of force to evade complying with an earlier UN resolution. This Argentine approach to portray Britain as the aggressor came to nothing. On 1 April, London told the UK ambassador to the UN, Sir Anthony Parsons, that an invasion was imminent and he should call an urgent meeting of the Security Council to get a favourable resolution against Argentina. Parsons had to get nine affirmative votes from the 15 Council members (not a simple majority) and to avoid a blocking vote from any of the other four permanent members. The meeting took place at 11.00 am on 3 April, New York time (4.00 pm in London). Resolution 502 was adopted by 10 to 1, with 4 abstentions.

This was a significant win for the UK, giving it the upper hand diplomatically. The draft resolution Parsons submitted had avoided any reference to the sovereignty dispute (which might have worked against the UK): instead it focused on Argentina's breach of Chapter VII of the UN Charter which forbids the threat or use of force to settle disputes. The resolution called for the removal only of Argentine forces: this freed Britain to retake the islands militarily, if Argentina did not leave, by exercising its right to self-defence, that was allowed under the UN Charter.

【词汇点拨】Words and Expressions

1. conflict /ˈkɒnflɪkt/ *n.* a violent situation or period of fighting between two countries（军事）冲突；战斗

The conflict has now reached a new level of intensity.

2. rival /ˈraɪv(ə)l/ *n.* a person, company, or thing that competes with another in sport, business, etc. 竞争对手

The two teams have always been rivals.

The Japanese are our biggest economic rivals.

3. colony /ˈkɒləni/ *n.* [C] a country or an area that is governed by people from another, more powerful, country 殖民地

In France's former North African colonies, anti-French feeling is growing.

4. dispute /dɪˈspjuːt/

① *v.* to argue or disagree strongly with sb. about sth., especially about who owns sth. 争

论；辩论；争执

The issue remains hotly disputed.

② *n.* [C,U] an argument or a disagreement between two people, groups or countries; discussion about a subject where there is disagreement 争论；辩论；争端；纠纷

a dispute between the two countries about the border 两国间的边界争端

the latest dispute over fishing rights 最近关于捕鱼权的争端

5. sovereignty /ˈsɒvrənti/ *n.* complete power to govern a country 主权；最高统治权；最高权威

The country claimed sovereignty over the island.

6. casualty /ˈkæʒuəlti/ *n.* [C] a person who is killed or injured in war or in an accident（战争或事故的）伤员，亡者，遇难者

road casualties 交通事故伤亡人员

Both sides had suffered heavy casualties(= many people had been killed).

7. inhabitant /ɪnˈhæbɪtənt/ *n.* a person or an animal that lives in a particular place（某地的）居民，栖息动物

the oldest inhabitant of the village 这个村最老的居民

a town of 11000 inhabitants 有 11000 名居民的城镇

8. compromise /ˈkɒmprəmaɪz/ *n.* [C] an agreement made between two people or groups in which each side gives up some of the things they want so that both sides are happy at the end 妥协；折中；互让；和解

a compromise solution/agreement/candidate 折中的解决方案 / 协议 / 候选人

In any relationship, you have to make compromises.

9. withdrawal /wɪðˈdrɔːəl/ *n.* [U,C] the act of moving or taking sth. away or back 撤走；收回；取回

the withdrawal of support 不再支持

the withdrawal of the UN troops from the region 联合国部队从该地区的撤离

10. momentum /məˈmentəm/ *n.* the ability to keep increasing or developing 推进力；动力；势头

The fight for his release gathers momentum each day.

They began to lose momentum in the second half of the game.

11. beachhead /ˈbiːtʃhed/ *n.* a strong position on a beach from which an army that has just landed prepares to go forward and attack（军队的）滩头堡，滩头阵地

They were attacked unexpectedly from both sides as soon as they landed at a beachhead.

12. outnumber /ˌaʊtˈnʌmbə(r)/ *v.* to be greater in number than sb./sth.（在数量上）压倒，比……多

The demonstrators were heavily outnumbered by the police.

13. simultaneous /ˌsɪm(ə)lˈteɪnɪəs/ *adj.* happening or done at the same time as sth. else 同时发生（或进行）的；同步的

There were several simultaneous attacks by the rebels.

simultaneous translation/interpreting 同声传译

【词汇点拨】Proper Names

1. Malvinas War 马岛战争 / 福岛战争
2. South Atlantic Ocean 南大西洋
3. Great Britain 英国
4. Argentina /ˌɑːdʒənˈtiːnə/ 阿根廷
5. Viscount Falkland 福克兰子爵
6. Saint-Malo 圣马罗人
7. Leopoldo Galtieri 列奥波尔多·加尔铁里

【参考译文】

马岛战争

背景

马岛（西班牙语叫"马尔维纳斯群岛"）位于南大西洋，离阿根廷海岸不远。马岛战争指的是1982年3月至6月英国和阿根廷之间爆发的一次军事冲突。该群岛的政治历史比较复杂，可以追溯到16世纪西班牙和英国两国对该岛发现权的争议。1690年，英国以海军财务官福克兰子爵的姓氏命名该岛（西班牙名"马尔维纳斯"源自于定居在此地的圣马罗人的法国殖民地）。自发现之日起，围绕群岛的归属发生了多次争端。1982年英阿冲突之前，两国政府就群岛的主权问题进行了多次商议谈判，但是1982年1月，英阿谈判破裂。阿根廷当时的领导人列奥波尔多·加尔铁里将预谋发动突袭。

突袭

1982年4月2日，阿根廷特种部队92人分乘21艘冲锋艇发动突袭。阿军60人占领了皇家海军陆战队的营地，没有遇到抵抗，阿军另外30人直奔总督府。阿军有一些伤亡，他们接到命令，尽量不要流血。6：30，英总督府被这股阿军包围并遭枪弹攻击，此时阿军大队人马乘直升机和两栖舰登岛。9：30，英方总督莱克斯·汉特投降，英军无伤亡。当晚，总督及夫人还有大部英海军陆战队乘飞机离开马岛。

英方反应

阿根廷没有料到英国会为这几个岛开战,岛上仅有 1800 居民,离英国本土 8000 多英里。英国的确提出了折中方案,被阿方拒绝。4 月 3 日,联合国安理会通过决议,要求阿根廷从岛上撤军。4 月 5 日,英国舰队(包括两艘航空母舰)向马岛进发。

因距离遥远,到 4 月 22 日英舰才抵达马岛。同时,阿军加强了各位置的军力。4 月 25 日,英军夺回南乔治亚的小福克兰岛,取得初步胜利,鼓舞了士气。联合国继续和谈无果。5 月 1 日,英军对马岛首府斯坦利港的空军基地进行空袭。空袭说明英军在后勤补给上下了很大功夫。两架从英国本土来的中程轰炸机,带了九架加油机。只有一枚炸弹击中跑道,但是这一击让阿军把自己的战斗机召回本土,以防英军攻击。

英军取胜靠的是海军。英军担心阿军用导弹打军舰,于是 5 月 2 日,英军令一艘核动力潜艇击沉阿军的一艘巡洋舰,368 人丧生。两天后,阿根廷一枚导弹击中英军一艘驱逐舰,22 人丧生。5 月 20 日,联合国和谈失败,英国特混部队在东福克群岛的圣卡罗斯建立了滩头阵地。阿空军又摧毁了几艘英军舰船,使英军支援受阻。但是 5 月 28 日,英军第二空降团以少胜多,夺回了古斯格林居住区。此次战斗,阿军阵亡 150 人,英军阵亡 18 人。

6 月 8 日,英军准备占领斯坦利港,阿空军击毁两艘英军补给船,使 200 名英军丧生。但是,英军最终包围了斯坦利港。6 月 11 日晚,第三突击旅的几支部队,在海军炮火的支援下,对斯坦利港周边的三个阿军据点(阿里亚山、两姐妹、隆冬山)同时发动攻击。战斗十分激烈,但次日英军获胜。两天后,英军占无线崖和坦布尔荡山。6 月 14 日,阿军宣布投降。英军俘获阿军 1 万人。这场战事死亡 912 人,其中阿军死亡 655 人,英军死亡 254 人,还有 3 人是马岛居民。

今天的马岛

马岛战争后,英国很快加强了岛上的防御,对机场进行彻底修整,这样马岛可以迅速补充军力。现在,英国驻马岛部队包括陆海空三军,驻扎在离斯坦利港 35 英里处的欢乐山,那里有军港和空军基地。英军总兵力 500 人,支援飞机数架,喷气战斗机 4 架,驱逐舰 1 艘,巡逻艇 1 艘。当地的志愿支援也由马岛防御力量提供。

词汇表

Alpha P Coy (Pre-parachute Selection)
airborne force 空降部队
camouflage 伪装

cross-country run 野外跑步拉练
march in battle order 战斗序列行进
pass out parade 阅兵

Milan anti-tank weapon "米兰"反坦克武器
military parachutist wings (emblem) "伞兵飞翼"徽章

Bravo Personal equipment
beret 贝雷帽
radio headset 无线耳机
webbing 织带
grenade 手雷
ammunition 弹药
water bottle 水壶
individual weapon 单兵武器
protective clothing 防护服
combat rations 作战口粮
rucksack 背包
fieldcraft 野外作战技能；野外生存技能
the Royal Air Force (RAF) （英国）皇家空军
fighting order 战斗行装
marching order 行军行装
body armour 防弹服
calibre 口径
barrel 枪管
strip 拆卸
reassemble 重新装配
cocked 扣扳机准备射击
live round 实弹
chamber 枪膛
eject 弹出
cocking handle 机柄
change lever 挡杆
magazine catch 开关
sight 瞄准器

safety catch 保险；安全开关
muzzle 枪口
trigger 扳机

Charlie The platoon training programme
marksmanship 射击术
obstacle course 障碍课
range 靶场
instructor 教官
installation 训练设施
colour sergeant (CSgt) 上士
tunnel 隧道
ramp 斜坡
cargo net 攀爬网
wire 铁丝网
ditch 沟壕
fence 栅栏
low wall 矮墙

Delta Military experience
CV (curriculum vitae) 简历
posting 派驻
Cyprus 塞浦路斯
lance corporal 一等兵
tour of duty 服役期；值日勤务
company sergeant major 连军士长
regimental sergeant major 团军士长
depot 兵站；补给站
late entry commission 晚进任命（军士晋升军官制度）
line infantry regiment 线列步兵团
quartermaster 军需官

Echo Malvinas War

conflict 冲突
rival 对手
colony 殖民地
dispute 争议
sovereignty 主权
marine 海军陆战队员
sailor 水兵
overwhelmed 被击败
assault craft 突击艇
capture 占领；捕获
barrack 兵营
casualty 伤亡人员
amphibious 两栖的
inhabitant 居民

mainland 本土
compromise 折中；妥协
withdrawal 撤回
momentum 势头
airfield 机场
logistical 后勤上的
submarine 潜艇
cruiser 巡洋舰
destroyer 驱逐舰
missile 导弹
beachhead 滩头阵地
vessel 舰船
outnumber 数目超过
simultaneous 同时发生的
garrison 驻军

 拓展学习

The AK-47, M-16, M240, the PK and QBZ-95 Assault Rifle:5 Most Deadly Guns of War

Modern warfare has seen breathtaking advances in the last hundred years, as mortal competition between nations spawns successively deadlier weapons. Aircraft, missiles, tanks, submarines and other inventions—many of which did not exist in practical terms in 1914 - have quickly earned key positions in the militaries of the world.

Yet there is still one invention that, although conceived more than five hundred years ago, still has a vital place on today's battlefield: the infantry weapon and supporting arms. No matter how high tech the armed forces of the world have become, warfare since the end of the Second World War has consistently involved some form of infantry combat.

In his seminal work on the Korean War, This Kind of War, historian T.R. Fehrenbach wrote, "you may fly over a land forever; you may bomb it, atomize it, pulverize it and wipe it clean of life—but if you desire to defend it, protect it and keep it for civilization, you must do this on the ground, the way the Roman legions did, by putting your young men in the mud." With that in mind, here are five of the most deadly guns of modern war.

AK-47:

The undisputed king of the modern battlefield is the Avtomat Kalashnikova model 47, or AK-47. Extremely reliable, the AK-47 is plentiful on Third World battlefields. From American rap music to Zimbabwe, the AK-47 has achieved icon status, and is one of the most recognizable symbols—of any kind—in the world. The AK series of rifles is currently carried by fighters of the Islamic State, Taliban fighters in Afghanistan, various factions in Libya and both sides in the Ukraine conflict.

The AK-47, as the story goes, was the brainchild of the late Mikhail T. Kalashnikov. A Red Army draftee, Kalashnikov showed a talent for small-arms design while convalescing from battlefield injuries, and had assembled a prototype assault rifle by 1947. (There is some circumstantial evidence, however, that the AK-47 was at least partially designed by the German designer Hugo Schmeisser, who had created the similar Stg44 in 1942.)

The AK-47 was the world's first standard-issue assault rifle. The rifle used a new 7.62-millimeter cartridge that generated less recoil and was lighter than rounds used in traditional infantry rifles. In return, the 7.62×39 round offered more controllability when fired in full automatic and allowed the infantryman to carry more rounds into battle.

The AK-47 has endured because it is a weapon for the lowest common denominator. It requires little training to learn how to shoot, and as a result large armies or militias can be raised by simply handing out AK-47s. It is dead simple to use and requires little maintenance. Disassembly is quick and the weapon can run virtually without lubrication. All of these are important considerations when your soldiers or militiamen are often illiterate, untrained draftees.

An estimated one hundred million AK-47s of all varieties have been manufactured by countries including the Soviet Union, China, North Korea, Egypt, Yugoslavia and most of the former Warsaw Pact. As The Independent pointed out, that could be one AK for every seventy people on Earth. Even Finland and Israel, neither Soviet allies nor client states, built variants. The most recent version issued to the Russian Army is the AK-74M, chambered in the lighter 5.45-millimeter.

The M16 family of weapons:

The modern M16 rifle got its start in 1956, when inventor Eugene Stoner tested its predecessor, the AR-15, at the Infantry School at Fort Benning. The rifle would not enter U.S. service for another four years, and then with the U.S. Air Force. The U.S. Army would jump on the M16 bandwagon in 1965, with the U.S. Marines following in 1966.

The original AR-15 was a reliable, innovative rifle, but a last-minute change of gunpowder and misconceptions about the rifle's need to be cleaned contributed to a poor reliability rate in Vietnam. Exacerbating the problem was the M16's direct impingement self-loading system, in which gasses and carbon residue created when gunpowder is burned are cycled back into the weapon's internal mechanism.

The most recent version, the M16A4, weighs 8.79 pounds loaded with a 30-round magazine. The rifle is effective to 550 meters, with a sustained rate of fire of 12-15 rounds per minute. The 5.56-millimeter SS109/M855 bullet, which emphasized armor-piercing capability over lethality on NATO battlefields, is being phased out in favor of the M855A1 round.

The original M16 led to the improved M16A1 by 1967, and the M16A2 by 1986. The M16A3 was a short-run rifle built for Navy SEALs, while the M16A4 has become standard issue in the U.S. Marine Corps. The M4A1 carbine, currently the standard-issue infantry weapon for the U.S. Army, is identical except for a shorter barrel, collapsible stock and the ability to be fired fully automatic.

The M16 has evolved into a reliable rifle. Modular and highly adaptable, variants have fulfilled roles from carbine to infantry rifle, squad automatic weapon and designated marksman rifle. The civilian version, again dubbed AR-15, has enjoyed explosive growth in the last ten years with the sunset of the Federal assault-weapons ban. The author owns two.

M240 Machine Gun:

The M240 Machine Gun is the current medium machine gun for the U.S. Army and U.S. Marine Corps. The M240 is in service with another sixty-eight countries and has served long enough that at least one adopter—Rhodesia—is no longer in existence.

The M240 is the American version of the FN-MAG designed in the 1950s by the Belgian arms maker Fabrique Nationale (FN). Utilizing features from both Axis and Allied infantry weapons, the MAG, as it was known, became wildly popular and standard issue with many NATO countries. In the years since introduction, the MAG has served from South Africa to the Falklands, to Afghanistan and Iraq.

The M240 can be used to engage point targets, such as individual enemy troops and light vehicles, or provide suppressive fire. The M240's heavier 7.62-millimeter bullet gives it a maximum effective range of up to 1,800 meters.

The M240 weighs 27 pounds, and with spare barrel, tripod, and other accessories and spare parts can weigh in at up to 47 pounds. The M240 can fire 100 rounds per minute sustained fire, meaning the weapon will not overheat. It can fire up to 650 rounds per minute, but overheating is imminent.

Obviously, too many countries use the M240 to run down its use in every case, but the state of issue in the U.S. military is typical of use worldwide. In the U.S. Army, M240 machine guns are found on armored vehicles and issued at a rate of two per infantry platoon. In the U.S. Marine Corps, six guns are issued to an infantry company, allowing the company commander flexibility in their deployment.

PK Machine Gun

The PK (Pulemyot Kalashnikova, or "Kalashnikov's machine gun") light machine gun was the Soviet Union's solution to high firepower at the squad level. Like the AK-47, the PK has seen extensive use around the world—where one finds an AK, a PK is never far behind.

The PK was also invented by Mikhail Kalashnikov. Although it resembles the M240, it is in the same class as the U.S. Marine Corps' M-27 Infantry Automatic Rifle, or the NATO Minimi/M-249 Squad Automatic Weapon.

The PK fills an important role as a Squad Automatic Weapon (SAW). Although most infantrymen throughout the world carry weapons capable of fully automatic or burst fire, trained soldiers rarely fire full auto. Full-auto fire from lightweight assault rifles is generally inaccurate and rapidly consumes ammunition.

Instead, a single gun like the PK is designated as the squad automatic weapon. The PK is equipped with a heavier barrel and frame to absorb heat and recoil from sustained fire. Accuracy, particularly mounted on a tripod, is reported to be excellent out to 800 meters.

The PK uses the same sights as the AK-47 to ease training. The PK uses a heavier hitting round than the AK-47, the 7.64×54r round originally used in the Moisin Nagant infantry rifle and Degtyaryov machine gun in World War II.

The PK has some anti-aircraft capability while the gun is mounted to a bipod or tripod, although realistically that is limited to low-flying helicopters and perhaps drones.

QBZ-95 Assault Rifle

The QBZ-95, or "Type 95 Automatic Rifle" is the standard-issue assault rifle of China. Designed to replace the Chinese copy of the AK-47, the QBZ-95 is unlike any other Chinese rifle. The rifle is issued to all arms of the People's Liberation Army, as well as the People's Armed Police.

The QBZ-95 is a bullpup design, with the magazine inserted behind the trigger. Like other bullpup rifles, this shortens the overall length of the rifle. The result is a rifle with a barrel 3.5" longer than the M4 carbine, but shorter in overall length. The rifle features a built-in carrying handle, although the use of such a handle creates issues when attaching scopes and other optics.

An entire line of infantry weapons has grown up around the QBZ-95. A carbine version with a shorter overall length is available for vehicle crews and special forces, while a heavier barrel variant is available to boost squad firepower. Unfortunately, the heavier version cannot accept belt-fed ammunition and only takes 30-round magazines, limiting its ability to provide high-volume firepower.

The rifle uses a unique 5.8-millimeter round developed by China and not used outside of its borders. The justification for the round is a bit of a mystery. It seemingly does not provide any significant improvement over existing NATO and Russian cartridges, both of which have seen extensive research and development and the development of a wide variety of subrounds. One possible explanation for the Chinese round is that it makes the QBZ-95 unable to accept externally sourced ammunition.

Although the rifle has so far not seen widespread use outside of China, China's status and the size of China's Army earn it a place on this list. (1678 words)

(*Selected from https://nationalinterest.org/blog/the-buzz/the-ak-47-m-16-m240-the-pk-qbz-95-assault-rifle-5-most-23512?page=0%2C1*)

【Notes】

1. T.R. Fehrenbach, (January 12, 1925–December 1, 2013) was an American historian, columnist, and the former head of the Texas Historical Commission (1987–1991). He graduated from Princeton University in 1947, and had published more than twenty books, including the bestseller Lone Star: A History of Texas and Texans and This Kind of War, about the Korean War. This Kind of War is seen by many senators and generals as "perhaps the best book ever written on the Korean War" (John McCain, The Wall Street Journal). Secretary of Defense James Matthis said "There's a reason I recommended T.R. Fehrenbach's book...that we all pull it out and read it one more time."

2. Eugene Stoner, Eugene Morrison Stoner (November 22, 1922–April 24, 1997), is the man most associated with the design of the AR-15. It was adopted by the US military as the M16. He is regarded by most historians, along with John Browning and Mikhail Kalashnikov, as one of the most successful firearms designers of the 20th century.

3. Fort Benning is a United States Army post outside Columbus, Georgia. Fort Benning supports more than 120,000 active-duty military, family members, reserve component soldiers, retirees, and civilian employees on a daily basis. It is a power projection platform, and possesses the capability to deploy combat-ready forces by air, rail, and highway. Fort Benning is the home of the United States Army Maneuver Center of Excellence, the United States Army Armor School, United States Army Infantry School, the Western Hemisphere Institute for Security Cooperation (formerly known as the School of the Americas), elements of the 75th Ranger Regiment, 3rd Brigade–3rd Infantry Division, and many other additional tenant units. Since 1918, Fort Benning, has served as the Home of the Infantry. Since 2005, Fort Benning has been transformed into the Maneuver Center of Excellence, as a result of the 2005 Base Realignment and Closure (BRAC) Commission's decision to consolidate a number of schools and installations to create various "centers of excellence". Included in this transformation was the move of the Armor School from Fort Knox to Fort Benning.

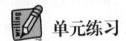 单元练习

I. Label the picture

1. _____   2. _____   3. _____   4. _____
5. _____   6. _____   7. _____   8. _____
9. _____

II. Multiple choice

1. Infantry sections are also issued with the Light Support Weapon (LSW), a version of the _____ designed as a light machine gun.

 A. AK-47   B. G-36   C. SA80   D. M-16

2. _____ refers to a design or picture that represents a country or an organization.

 A. Emblem   B. Camouflage C. NBC   D. NCO

3. Depending on the _____ situation, riffemen wear either fighting order or marching order.

  A. tactical   B. offensive   C. strategy   D. defensive

4. The political history of the Malvinas islands is complex and goes back to rival claims of discovery by _____ and _____ government in the 16th century.

  A. Argentina; British   B. Argentina; Spanish

  C. Spanish; France   D. Spanish; British

5. Which of the following doesn't belong to the SA80?

  A. trigger   B. muzzle   C. tracks   D. magazine

6. Every year the British Army commissions about 300 offfcers from the ranks. This is called ___.

  A. an army training regiment   B. a late entry commission

  C. a tour of duty   D. recruit training

7. The Parachute Regiment (Para Reg) is the British Army's ___ force.

  A. armoured   B. aviation   C. artillery   D. airborne

8. SA80 is ___ when there is no magazine and no round in the chamber.

  A. unloaded   B. loaded   C. line   D. live

9. Leading up to the 1982 conflict, the Argentine and UK governments had been engaged in negotiations over the islands' ___.

  A. sovereignty   B. echelon right   C. momentum   D. victory

10. On 28 May,1982, the Second Parachute Regiment managed to re-take the settlement of Goose Green, despite being greatly ___.

  A. overwhelmed   B. compromise   C. simultaneous   D. outnumbered

III. Translation: English to Chinese

1. My first posting abroad was in Cyprus in 1975. I deployed to Cyprus with my unit. We were there as part of a UN peacekeeping force.

2. The lifestyle was fantastic although we were there at a difficult time.

3. On exercises and operations, soldiers carry their individual weapon, ammunition, water, food and protective clothing.

4. Fighting order weighs about ten kilos and includes all the equipment the soldier needs to survive for two to three days: individual weapon, extra ammunition, and grenades, webbing, digging tool, water bottle, combat rations and washing and shaving kit.

5. Make sure the weapon is pointing in a safe direction, put the safety catch to F and pull

the trigger. Finally, put the safety catch in the S position.

IV. Translation: Chinese to English

19 到 20 周，士兵们要完成"空降前选拔"或者叫 P 连，也就是一系列身体检查，以确保新兵个人具备符合参加空降部队要求的自律能力和动机。测试包括在 1 小时 50 分钟内负重 35 磅并携带武器行军 10 英里。通过 P 连之后，新兵便完成了步兵训练的最后环节，可以参加新训结业阅兵了。之后，新兵要在普莱斯诺顿的皇家空军驻地进行为期四周的基础跳伞课程。该训练结束时，士兵被授予"伞兵飞翼"徽章，等待去伞兵营。该团下辖三个伞兵营。

V. Oral practice

1. What is the weapon or the piece of equipment you know well? Can you introduce this weapon or equipment in terms of its history and formation?

2. Who is Yuri Gagarin? Can you give a brief introduction in terms of his story?

附录：Unit 9 练习答案

I. Label the picture

1. cocking handle   2. change lever   3. magazine catch   4. magazine
5. sights   6. safety catch   7. muzzle   8. rounds
9. trigger

II. Multiple choice

1-5 CAADC   6-10 BDAAD

III. Translation: English to Chinese

1. 1975 年首次驻扎海外塞浦路斯，我和所在单位被派驻到塞浦路斯，参加联合国维和部队。

2. 虽然我们到那儿的时候正好遇到困难时期，但那儿的生活方式还很迷人的。

3. 在演习和作战中，士兵们常携带单兵武器、弹药、水、食物和防护服。

4. 战斗行装重约十千克，包括士兵生存两至三天所需的装备：单兵武器、弹药和手雷、织带、挖掘工具、水壶、作战口粮和洗漱及剃须套装。

5. 一定要把枪口指向安全方向，将保险扳到 F 位置，扳动扳机。最后，将保险扳回 S 位置。

IV. Translation: Chinese to English

In weeks 19 to 20, soldiers complete the Pre-parachute Selection, or P Company, a series of physical tests designed to ensure that an individual has the self-discipline and motivation required for service with airborne forces. The tests include a ten-mile march carrying a 35 lb pack and weapon in one hour and 50 minutes. After passing P Company, recruits complete the final phase of infantry training and prepare for the Pass Out parade. Soldiers then move on to the four-week Basic Parachute Course at the Royal Air Force station at Brize Norton. At the end of this course, soldiers are presented with their Military Parachutist Wings and they are ready to take their place in one of the Regiment's three operational Parachute Battalions.

V. Oral practice

1. The SA80 is the standard individual weapon issued to British troops. The SA80 has a calibre of 5.56 mm and weighs approximately 5 kilos with a loaded magazine. The magazine holds 30 rounds of ammunition. The SA80 is very accurate and has a range of about 400 metres. Infantry sections are also issued with the Light Support Weapon (LSW), a version of the SA80 designed as a light machine gun. LSW has a calibre of 5.56 mm and has the weight of 6.58 kg with loaded magazine and sight. It is 900 mm long and its barrel length is 646 mm. Its effective range is 1,000 metres and its rate of ffre is from 610 to 775 rounds per minute.

2. Yuri Gagarin was a Russian pilot. He was born near Moscow in Russia in 1934. When he was a boy, he was interested in aircraft and when he was eighteen or nineteen, he joined a local ffying club. In 1955, when he was twenty-one, he became a pilot in the Soviet Air Force. His ffrst assignment was at the Soviet Air Force Base in the Arctic. Gagarin was very intelligent, and soon he was a test pilot for new aircraft. And then he became a cosmonaut at Star City. In 1961, he was the ffrst human in space. The name of his spaceship was Vostock 1. Vostock 1 was in space for about 100 minutes. Yuri was only twenty-seven years old. After this he was a hero. But Gagarin wasn't happy when he wasn't in a plane and soon he was a test pilot again. He died in an accident in a new MiG-15 in 1968. He was thirty-four years old.

Unit 10 The 3rd Armoured Cavalry Regiment

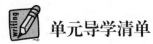
单元导学清单

模块	主题	学习目标	任务	军事知识	参考词汇
Alpha	新兵报到	1）了解新兵报到流程和第三装甲骑兵团基本知识；2）掌握指路的常用词汇和表达；3）能够识别地图位置。	Task 1 Task 2 Task 3 Task 4 Task 5	■ The 3rd ACR ■ Fort Carson	squared away, quarters, bend, proceed, lane, garrison, intersection, installation
	新兵与首长初次见面	1）了解新兵与首长初次见面的注意事项；2）掌握请求进入和离开的表达。	Task 6 Task 7	■ report for duty	disciplinary, assignment, deployment, fill one's shoes, accommodation, dismiss
Bravo	第三装甲骑兵团的历史	1）了解第三装甲骑兵团的历史；2）掌握介绍该骑兵团历史的相关词汇和表达。	Task 1 Task 2 Task 3 Task 4 Task 5	■ "Brave Rifles" ■ the Regiment of Mounted Riflemen ■ transformation from a horse cavalry to a mechanised one	barbed wire, designate, engagement, screen, station, squadron, trench, insignia, escort, settler, rebel, comrade, frontier, convert, mchanised, rage, rear
	第三装甲骑兵团的现状	1）了解第三装甲骑兵团的编制构成、装备力量等；2）掌握与第三装甲骑兵团编制和力量构成相关的词汇与表达。	Task 6 Task 7 Task 8 Task 9 Task 10	■ the formation of the 3rd ACR today ■ TO&E ■ combined arms	reconnaissance, offensive, defensive, cutting edge, asset, scout
Charlie	坦克构造	1）了解坦克构造；2）熟练掌握与坦克结构相关的词汇与术语；3）介绍坦克、坦克乘员等。	Task 1	■ Tank ■ Tank suspension ■ Tank crew	barrel, cupola, engine deck, periscope, gunsight, road wheel, side skirt, toe plate, tracks, loader, driver, gunner

续表

模块	主题	学习目标	任务	军事知识	参考词汇
Charlie	坦克队形	1）了解坦克编队队形；2）熟练掌握与坦克队形相关的词汇和句型；3）讨论不同坦克队形的应用情境。	Task 2 Task 3 Task 4	■ Restricted and open terrain ■ Tactics and strategy	column, echelon, line, staggered column, vee, wedge, terrain, flank, screen
Delta	新式坦克	1）了解不同的军事装备及其特点；2）熟练掌握与装备相关的词汇和表达；3）介绍我军的新型主战坦克。	Task 1 Task 2 Task 3	■ Main Battle Tank ■ Transport aircraft ■ Amphibious ships ■ the Leopard ■ Light armoured vehicle	consumption, amphibious, territory, land, steep, LAV, transport aircraft, Stryker
Echo (Review)	合成部队	1）了解合成部队和美国陆军第三装甲骑兵团的相关知识；2）熟练掌握相关的词汇和表达；3）运用所学知识介绍合成部队与单兵种部队的异同。	Task 1 Task 2	■ the U.S. armed forces ■ Military intelligence ■ Reconnaissance ■ Mechanised infantry ■ Mortar	temporary, segregate, flexibility, co-ordinate, intelligence, reconnaissance, brigade, squadron, troop, mechanise, integrate, parent, aviation, saber, skinner, howitzer

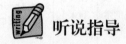 听说指导

Alpha New duty station

【军事知识】

1. The 3rd ACR 美国第三装甲骑兵团

The 3rd Cavalry Regiment, formerly 3rd Armoured Cavalry Regiment, is a regiment of the United States Army currently stationed at Fort Hood, TX. The Regiment has a history in the United States Army that dates back to 19 May 1845, when it was constituted in the Regular Army as the Regiment of Mounted Riflemen at Jefferson Barracks, Missouri. This unit was reorganized at the start of the American Civil War as the 3rd U.S. Cavalry Regiment on 3 August 1861. In January 1943, the Regiment was re-designated as the 3d Cavalry Group. Today they are equipped with Stryker vehicles. The 3d Armoured Cavalry Regiment was the last heavy armoured cavalry regiment in the U.S. Army until it officially switched over to a Stryker regiment on 16 November 2011. It will retain its lineage as the 3d Cavalry Regiment.

2. Fort Carson 卡森堡

Fort Carson is a United States Army installation located near Colorado Springs, primarily in El Paso County, Colorado. It is 40 miles (64 km) north of Pueblo, Colorado in Pueblo County. Fort Carson is the home of the 4th Infantry Division, the 10th Special Forces Group, the 71st Ordnance Group (EOD), the 4th Engineer Battalion, the 759th Military Police Battalion, the 10th Combat Support Hospital, the 43rd Sustainment Brigade, and the 13th Air Support Operations Squadron of the United States Air Force. The post also hosts units of the Army Reserve, Navy Reserve and the Colorado Army National Guard.

【词汇点拨】Words and Expressions

1. squared away (informal) all the necessary arrangements for something or someone have been completed; be ready to go or in good shape 一切就绪；状态优秀

I've got my tickets and hotel squared away.

2. quarters /ˈkwɔːtəz/ *n.* [pl.] living accommodation on a military base 营房；驻地

He was confined to his quarters.

3. bend /bend/ *n.* a curve or turn, especially in a road or river（尤指道路或河流的）拐弯；弯道

to round the bend 绕过弯道

a sharp bend in the road 道路的急拐弯

4. proceed /prəˈsiːd/ *v.* to move or travel in a particular direction 行进；前往

The marchers proceeded slowly along the street.

5. lane /leɪn/ *n.* a section of a wide road, that is marked by painted white lines, to keep lines of traffic separate 车道

a four-lane highway 四车道公路

to change lanes 变换车道

6. garrison /ˈɡærɪsn/ *n.* troops who occupy a fortress or town in order to defend it 驻军；卫戍区；卫戍部队；守备部队

a garrison of 5000 troops 有 5000 士兵驻守的防地

The garrison held out for three weeks.

7. intersection /ˌɪntəˈsekʃn/ *n.* a place where roads or other lines meet or cross 十字路口；交叉路口

We crossed at a busy intersection. 我们穿过了一个繁忙的道路交叉口。

8. installation /ˌɪnstəˈleɪʃn/ *n.* a building, complex or other permanent structure, which contains some form of technical equipment (such as communications equipment, radar, weapons system, etc.) 设施

Our target was the radar installations along the north coast.

9. disciplinary /ˌdɪsəˈplɪnəri/ *adj.* designed to enforce discipline 有关纪律的；加强纪律的

The company will be taking disciplinary action against him.

10. assignment /əˈsaɪnmənt/ *n.* a task or job 任务；工作

My first assignment was to update the brigade security orders.

11. deployment /dɪˈplɔɪmənt/ *n.* the movement of troops to a war zone or area of operations 部署

The deployment to Germany was completed in 72 hours.

12. fill one's shoes [idiom] to take someone's place, often by doing the job they have just left 取代（某人）；接替（某人）的工作

Who do you think will fill Sarah's shoes when she goes?

13. accommodation /əˌkɒməˈdeɪʃn/ *n.* a place to live; an act of providing shelter 住处；提供住宿

The officer inspected the soldiers' accommodation.

You are responsible for the accommodation of refugees.

14. dismiss /dɪsˈmɪs/

① *v.* to send someone away 让（某人）离开

② *v.* to remove someone from their job 解雇；免职；开除

③ *v.* to release servicemen at the end of a parade 解散

He dismissed the clerk.

The brigade commander has been dismissed.

Company, dismiss!

【词汇点拨】Proper Names

1. Fort Carson 卡森堡（美军基地，科罗拉多州）

2. Colorado Springs 科罗拉多斯普林斯（又名科罗拉多泉，位于美国科罗拉多州）

3. Denver /ˈdenvə(r)/ 丹佛（位于美国科罗拉多州）

4. Sergeant Chambers 中士钱伯斯

5. Captain Paige 佩奇上尉

【长句解读】

1. After rounding the bend, cross under highway I-25 and proceed west along highway 83. (10-1)

➢ 句子分析：该句中 round 作动词，bend 意为拐弯、弯道，round the bend 表示绕过弯道，cross under 是指在……下面穿过，proceed 意为继续前进。

➢ 翻译：绕过弯道以后，从 I-25 号高速下穿过，继续沿着 83 号高速朝西行驶。

2. A: Good morning, welcome to Fort Carson. State your business please.

B: Good morning, I'm Sergeant Jones. I'm here to process in. (10-2)

➢ 句子分析：该段对话中有两个表达值得注意。第一句中 state your business 意为"表明来意"，第二句中 process in 意为办理报到。

➢ 翻译：

A：早上好，欢迎来到卡森堡。请陈述您的来意。

B：早上好，我是中士琼斯。我是来这里办理报到手续的。

3. It's a pleasure to have you on board. (10-5)

➢ 句子分析：on board 原意为"在船上（或飞机上、火车上）"，在该句中 have you on board 意为"和我们一起携手合作"。

➢ 翻译：很高兴有你加入。

4. A: Sergeant Frank Jones requests permission to enter, sir.

C: Permission granted.

…

A: Request permission to be dismissed, sir.

C: Dismissed. Sergeant, one more thing. (10-5)

➢ 句子分析：该对话中的表达 request permission to enter 和 request permission to be dismissed 分别用于向上级请求"进入"和"离开"。

➢ 翻译：

A：长官，中士弗兰克·琼斯请求允许进入。

C：准许进入。

……

A：长官，中士弗兰克·琼斯请求允许离开。

C：去吧。中士琼斯，还有一件事。

5. C: I see here you played some basketball in school and a little in college before you joined the army.

A: Yes, sir, I was a forward and guard. (10-5)

> **句子分析**：该对话中的 forward 和 guard 分别表示篮球比赛中的"前锋"和"后卫"。

> **翻译**：

C：我看到你参军前在学校时打过篮球。

A：是的，长官，我当时打前锋和后卫。

【口语输出】

1. Which route would you suggest for your foreign friends to come to your base from the nearest train station or airport?

【参考词汇】exit, roundabout, overpass, flyover, highway, locate, proceed, approach, lane, head towards, turn south onto, round the bend, cross under, go past, carry on, take a left

2. What should you pay attention to when reporting for new duty?

【参考词汇】paperwork, housing, process in, in-process, accommodation

3. Suppose you were Captain Paige, what would you say to Sergeant Jones in your first meeting?

【参考词汇】accommodation, previous assignment, deployments, family, interests

Bravo   Regimental history

【军事知识】

1. TO&E 编制与制备表

A table of organization and equipment (TOE or TO&E) is a document published by the U.S. Department of Defense which prescribes the organization, staffing, and equipage of units. It also provides information on the mission and capabilities of a unit as well as the unit's current status. A general TOE is applicable to a type of unit (for instance, infantry) rather than a specific unit (the 3rd Infantry Division). In this way, all units of the same branch (such as infantry) follow the same structural guidelines.

2. Combined arms 联合作战

Combined arms is an approach to warfare which seeks to integrate different combat arms of a military to achieve mutually complementary effects (for example, using infantry and armour in an urban environment, where one supports the other, or both support each other).

Combined arms doctrine contrasts with segregated arms where each military unit is composed of only one type of soldier or weapon system. Segregated arms is the traditional method of unit/force organisation, employed to provide maximum unit cohesion and concentration of force in a given weapon or unit type.

【词汇点拨】Words and Expressions

1. barbed wire wire with sharp spikes attached to it, used as an obstacle 带刺铁丝网

a barbed wire fence 带刺铁丝网围栏

2. designate /ˈdezɪgneɪt/ *v.* to say officially that sth has a particular character or name 命名

This area has been designated (as) a National Park.

3. engagement /ɪnˈgeɪdʒmənt/ *n.* an exchange of fire between opposing forces 战斗；交战；交火

The engagement lasted just over an hour.

4. screen /skriːn/

① *v.* (~ sth./sb. from sth./sb.) to hide or protect sth./sb. by placing sth. in front of or around them 警戒；屏蔽；掩护

Dark glasses screened his eyes from the sun.

② *n.*[C] (~ of sth.) something that prevents sb. from seeing or being aware of sth., or that protects sb./sth. 掩蔽物；掩护物；屏障；庇护

All the research was conducted behind a screen of secrecy.

5. station /ˈsteɪʃn/ *v.* to send a serviceperson to serve in a particular location 派驻；使驻扎

I was stationed in Germany.

6. squadron /ˈskwɒdrən/ *n.* an air force unit consisting of two or more flights; a battalion-sized armoured cavalry grouping, consisting of three cavalry troops, one tank company and one battery（空军）中队；（美）骑兵中队

The squadron was on a reconnaissance mission.

7. trench /trentʃ/ *n.* a narrow hole or channel dug into the ground in order to provide protection from enemy fire 战壕；堑壕

trench warfare 堑壕战

8. insignia /ɪnˈsɪgniə/ *n.* a decorative symbol (used to denote the identity of a unit, specialist qualification, rank, etc.) 徽章

His uniform bore the insignia of a captain.

9. escort

① /ˈeskɔːt/ *n.* a person, vehicle or aircraft or ship which accompanies an individual or group in order to protect them 护送者；护卫队；护卫舰（或车队、飞机）

② /ɪˈskɔːt/ *v.* to act as an escort 护卫；护送

Armed escorts are provided for visiting heads of state.

The convoy was escorted by two destroyers.

10. settler /ˈsetlə(r)/ *n.* a person who goes to live in a new country or region 移民；殖民者

white settlers in Africa 非洲的白人移民

11. rebel

① /ˈrebl/ *n.* a person who uses armed force to oppose the established government 造反者，反抗者

② /rɪˈbel/ *v.* to oppose the established government with armed force 造反；反抗

The rebels have captured the barracks.

Some mountain tribes have rebelled against the provincial government.

12. comrade /ˈkɒmreɪd/ *n.* a fellow, soldier, worker, etc. 战友；同志

They were old army comrades.

13. frontier /ˈfrʌntɪə(r)/ *n.* a region on the border between two states 国界；边界；边境

a customs post on the frontier with Italy 与意大利的交界边境上的海关关卡

14. convert /kənˈvɜːt/ *v.* change or make sth. change from one form, purpose, system, etc. to another （使）转变；转换；转化

The hotel is going to be converted into a nursing home.

15. mechanised /ˈmekənaɪzd/ *adj.* equipped with machinery, especially transport 机械化的

In large facilities, slaughtering is carried out in fully mechanised lines.

大型设施的屠宰过程是完全机械化的。

16. rage /reɪdʒ/ *v.* (of a storm, a battle, an argument, etc.) to continue in a violent way 猛烈地继续；激烈进行

The blizzard was still raging outside.

17. rear /rɪə(r)/ *adj.* moving or located at the back of a formation or position 后部；后方

They set off, two men out in front as scouts, two behind in case of any attack from the rear.

他们出发了，两人在前作为侦查员，两人殿后以防后方攻击。

18. reconnaissance /rɪˈkɒnɪsns/ *n.* the activity of getting information about an area for military purposes, using soldiers, planes, etc. 侦查

Time spent on reconnaissance is seldom wasted.

19. offensive /əˈfensɪv/ *adj.* connected with the team that has control of the ball; connected with the act of scoring points 攻方的；进攻型的；攻击型的

offensive play 进攻打法

20. defensive /dɪˈfensɪv/ *adj.* connected with trying to prevent the other team or player from scoring points or goals 防守的

defensive play 防守型打法

21. cutting edge the newest, most advanced stage in the development of sth.（处于某事物发展的）尖端；最前沿；领先阶段

working at the cutting edge of computer technology 从事计算机技术的最前沿工作

22. asset /ˈæset/ *n.* a person or thing that is valuable or useful to sb./sth. 有价值的人（或事物）；有用的人（或事物）

In his job, patience is an invaluable asset.

23. scout /skaʊt/

① *n.* a person, an aircraft, etc. sent ahead to get information about the enemy's position, strength, etc. 侦查员；侦查机

The scouts haven't returned yet.

② *v.* to search an area or various areas in order to find or discover sth. 侦查；搜寻

Platoon is scouting the enemy position.

【词汇点拨】Proper Names

1. **Oregon** /ˈɒrəɡən/ 俄勒冈州（美国）
2. **Fort Lewis** 路易斯堡（美军基地，华盛顿州）
3. **Washington** /ˈwɒʃɪŋtən/ 华盛顿州（美国）
4. **Fort Bliss** 布利斯堡（美军基地，得克萨斯州）
5. **Texas** /ˈteksəs/ 得克萨斯州（美国）

【长句解读】

1. Originally designated the Regiment of Mounted Riflemen, the Regiment first saw action during the Mexican War in 1847 and it was during this conflict that they earned the title, "The Brave Rifles". (10-7)

➢ **句子分析**：过去分词短语 originally designated... 用于修饰主语 the Regiment。saw action 意为"经历战斗"。The Brave Rifles 为第三装甲骑兵团的绰号。

➢ **翻译**：第三装甲骑兵团，最初被命名为"马背上的骑兵团"，第一次参与的战斗是 1847 年墨西哥战争，也正是在这场冲突期间，他们获得了"勇士"头衔。

2. This was a difficult time, as many officers and men joined the rebels and soldiers found themselves fighting their former comrades. (10-7)

➢ **句子分析**：该句中 rebel 作名词，此处意为叛军，comrade 指战友。句中 as 引导原因状语从句。

➢ **翻译**：这是一个艰难的时期，因为许多官兵加入了叛军，士兵们发现他们在和以前的战友战斗。

3. At the outbreak of World War I, the 3rd Cavalry was deployed to Europe but the use of trenches, barbed wire, gas and machine guns meant that horse cavalry fought in few engagements during that conflict. (10-7)

➢ **句子分析**：该句中的 outbreak 意为战争爆发，horse cavalry 意为马背上的骑兵，在这里指代当时的第三装甲骑兵团。

➢ **翻译**：第一次世界大战爆发时，第三骑兵团被部署到欧洲，但是战壕、铁丝网、毒气和机枪的使用意味着在那场冲突中，"马背上的骑兵团"很少参与战斗。

4. In the '60s and '70s—while war raged in Vietnam—the Regiment was stationed in West Germany, providing rear area security for the 7th US Army and patrolling the East / West German border. (10-7)

➢ **句子分析**：该句中的 rage 用于表示战争的激烈程度，East / West German 分别表示当时的"东德"和"西德"，patrol the border 意思为"巡逻边境"。

➢ **翻译**：在 20 世纪 60 年代和 70 年代——当时越南战争肆虐——该团驻扎在西德，为美国第 7 集团军提供后方安全保障，并在东德／西德边境巡逻。

5. The Squadron provides the Regiment's air assets and its authorized table of organisation and equipment (TO&E) includes scout, transport and attack helicopters or birds. (10-8)

➢ **句子分析**：该句中的 the Squadron 指上句中的 the 4th Longknife Air Cavalry Squadron，是美国第三装甲骑兵团的第四长刀中队，assets 原意为财产、资产，在此处引申为"军事力量"。句中的 birds 与 helicopters 同义。

➢ **翻译**：该中队为该团提供空中力量，其编制与装备表中包括侦查、运输和攻击直升机。

【口语输出】

1. What can you learn from the transformation of the 3rd ACR from a horse cavalry to a mechanised one?

【参考词汇】convert, be equipped with, designated, armoured vehicles, deployment

2. What is the cutting edge of the 3rd ACR today?

【参考词汇】three armoured cavalry squadrons, Tiger, Sabre, Thunder, MBT (main battle tank), CFV (cavalry fighting vehicle), howitzer

3. How many squadrons are there in the 3rd ACR and what are they equipped with?

【参考词汇】Tiger, Sabre, Thunder, Longknife, attack helicopter, CFV, MBT, SPH (self-propelled howitzer), scout helicopter, transport helicopter

Charlie How tanks fight

【军事知识】

1. Tank 坦克

A tank is an armored fighting vehicle intended as a primary offensive weapon in front-line ground combat. Tank designs are a balance of heavy firepower, strong armor, and good battlefield mobility provided by tracks and a powerful engine; usually their main armament is mounted in a turret. They are a mainstay of modern 20th and 21st century ground forces and a key part of combined arms combat.

The first tank prototype, "Little Willie", was unveiled in September 1915. Following its underwhelming performance—it was slow, became overheated and couldn't cross trenches–a second prototype, known as "Big Willie", was produced.

By 1916, this armored vehicle was deemed ready for battle and made its debut at the First Battle of the Somme near Courcelette, France, on September 15 of that year. Known as the Mark I, this first batch of tanks was hot, noisy and unwieldy and suffered mechanical malfunctions on the battlefield; nevertheless, people realized the tank's potential. Further design improvements were made and at the Battle of Cambrai in November 1917, 400 Mark IV's proved much more successful than the Mark I, capturing 8,000 enemy troops and 100 guns.

Tanks rapidly became an important military weapon. During World War II, they played a prominent role across numerous battlefields.

2. Tank suspension 坦克悬挂装置

A tank's suspension, also known as "tracks" or "treads", determines the load limit of the tank as well as the traverse speed of the tank. In battle, each tank has a set of two tracks that get damaged independently; a few tanks have double tracks (four total tracks). To be able to mount a new module to your vehicle, its weight must be supported by the suspension. If one of the tracks is destroyed, the vehicle is immobilized until the track is repaired. If a track is damaged, the vehicle will move at a reduced speed. Upgrading the suspension also upgrades a number of hidden characteristics, including track HP, terrain resistance, and aim dispersion while on the move. While upgrading suspension cannot increase your maximum speed, it can have an impact on acceleration and maintaining speed while on rough terrain.

3. Tank crew 坦克乘员

No matter how sophisticated and well armored a tank may be, it will only perform as well as the crew controlling it. This is perhaps better illustrated if one thinks of the crew as an integral, indispensible component of the overall machine.

In the past, especially in the early days, the tank had a large crew, as many as eighteen men in the case of the German A7V, and often six, seven or eight was considered necessary with most heavy tanks of World War I.

As the art of tank battle and tank design progressed into World War II, the average-sized tank would have a crew of four men: a driver, a gunner, a gun-loader/radio operator and a tank commander, all working as a team in their special positions within the tank. The driver was at the front, usually with a little viewing hatch and escape hatch right in front of him, and would sometimes have control of a forward-pointing machine gun, which in some cases would be handled by a fifth crewman, or co-driver.

In very modern tanks some of the crew jobs are being automated, taken over by robots and computers. One of the first and most obvious choices one might think of for replacement is the weapons loader.

4. Restricted terrain and open terrain 受限制地形和开阔地形

Restricted terrains often refer to unfavorable terrain, such as mountains, littorals, jungles, subterranean areas, and urban areas that hinder ground movement. Effort is needed to enhance mobility.

Open terrains are mostly flat and free of obstructions such as trees and buildings. Examples include farmland, grassland and specially cleared areas such as an airport.

5. Tactics and strategy 战术和战略

Strategy refers to the movement of armies in order to achieve the overall objectives of a campaign or war (for example, the capture of a port, which can be used to land supplies and reinforcements for future operations), while tactics refers to the movement of battalions, brigades, divisions and equivalent-sized groupings, in order to achieve local objectives (for example, the destruction of an enemy battalion, which is defending one of the approaches to the port).

About 2,500 years ago, Chinese military strategist Sun Tzu wrote *The Art of War*, in which he said, "Strategy without tactics is the slowest route to victory. Tactics without strategy is the noise before defeat."

Strategy defines your long-term goals and how you're planning to achieve them. In other words, your strategy gives you the path you need toward achieving your organization's mission.

Tactics are much more concrete and are often oriented toward smaller steps and a shorter time frame along the way. They involve best practices, specific plans, resources, etc. They're also called "initiatives".

【词汇点拨】Words and Expressions

1. cupola /ˈkjuːpələ/ *n.* a revolving turret housing a gun or machine-guns, which is fitted to a warship, aircraft or fighting vehicle 指挥塔

2. barrel /ˈbærəl/ *n.* the tube part of a gun, down which the bullet or shell slides when it is fired 炮筒（管）；枪管

He spent hours cleaning the barrel of his rifle.

3. periscope /ˈperɪskəʊp/ *n.* an optical instrument, which enables an observer on a lower level (e.g. in a submerged submarine or at the bottom of a trench) to see things on a higher level (such as on the surface of the sea or ground)（火炮的）周视瞄准镜；（坦克的）周视指挥观瞄镜；（潜艇的）潜望镜

Submarine motion at the periscope depth will suffer the interference that comes from waves.

4. gunsight /ˈɡʌnsaɪt/ *n.* a part of a gun that you look through in order to aim it accurately（枪炮的）瞄准具（器）；观瞄镜

The gunsight would have been switched on during the search period and adjusted for brightness.

5. road wheel（坦克、装甲车辆等的）负重轮

The road wheel is connected with the soleplate through an axle and a bearing bracket.
负重轮通过轴和轴承支架与底板连接。

6. track /træk/ *n.* a moving band of metal links fitted around the wheels of a tank or other armoured vehicle, enabling it to move over soft or uneven ground 履带

The tank came off the road when it lost a track.

7. loader /ˈləʊdə/ *n.* a person who loads a gun or other firearm 填装手；装弹机

The PLZ04/ 04 use an automatic loader, thus a crew of only four is required.

8. gunner /ˈɡʌnə(r)/ *n.* a member of the armed forces who is trained to use large guns 火炮瞄准手；炮手；枪手

The crew was composed of the commander, gunner, loader, driver and radio-operator.

9. armament /ˈɑːməmənt/ *n.* [C usually pl.]

① a general term for a weapon 武器；装备

② the process of equipping with weapons 武装；战备

armaments factory 军工厂

The main innovation in the new vehicle was the turret and its armament.

10. suspension /səˈspenʃn/ *n.* the system by which a vehicle is supported on its wheels and which makes it more comfortable to ride in when the road surface is not even 承载装置；悬挂装置

The car's improved suspension gives you a smoother ride.

11. turret /ˈtʌrɪt/ *n.* a small metal tower on a ship, plane or tank that can usually turn around and from which guns are fired（战舰、飞机或坦克的）回转炮塔；旋转枪架

She began to talk to the man in the turret of the car.

12. column /ˈkɒləm/ *n.* a tactical formation consisting of several files of soldiers moving forward together with one behind the other 纵队

Two columns of infantry advanced across the desert.

13. echelon /ˈeʃəlɒn/ *n.* a tactical formation in which troops, vehicles or aircraft are deployed in a series of parallel lines, each of which is longer than the one in front 梯形编队；梯队

14. staggered /ˈstæɡəd/ *adj.* arranged in such a way that not everything happens at the same time 交错的；错开的

staggered column 交错纵队编队

15. vee /viː/ *n.* a tactical formation in the shape of a letter V V字形编队；V形队形

16. wedge /wedʒ/ *n.* a tactical formation in the shape of a triangle (e.g. one sub-unit leading as point, with the other two sub-units following abreast of each other) 楔形编队

17. flank /flæŋk/ *n.* the left-hand or right-hand side of a military force which is deployed in a defensive position or tactical information 侧翼

The enemy's right flank was exposed.

18. tactics /ˈtæktɪks/ *n.*[pl.] the art of moving soldiers and military equipment around during a battle or war in order to use them in the most effective way 战术；兵法

This was just the latest in a series of delaying tactics.

【词汇点拨】Proper Names

1. Sergeant Ambrose 中士安布罗斯
2. Private Parks 列兵帕克斯
3. Lance Corporal Jones 上等兵琼斯

【长句解读】

1. The column formation permits excellent fires to the platoon's flanks but only the lead track can fire to the front. (10-9)

➤ 句子分析：该句中有两个并列分句，由表示转折的并列连词 but 连接，介绍纵队队形的优缺点。其中，the lead track 指的是坦克队形中的第一辆坦克。

➤ 翻译：纵队队形可以对排的侧翼进行完美射击，但只有领头坦克才能朝前方开火。

2. This formation is when the platoon is moving across open terrain, when contact with the enemy is likely, and when the platoon needs to protect or screen an exposed flank or the flank of another moving force. (10-9)

➤ 句子分析：该句为主系表结构，其中表语包含三个 when 引导的表语从句，介绍采用该坦克阵型的三种常见情况。其中，exposed flank 指"暴露的侧翼"。

➤ 翻译：当坦克排穿越开放地形，或者可能与敌人相遇，或者需要保护或掩护暴露的侧翼或另一个行进部队的侧翼时，采用该队形。

3. This formation gives maximum firepower to the front, but the platoon is open to ambush from the flanks. (10-9)

➤ 句子分析：句中 maximum 表示"最高的；最多的"，修饰 firepower，maximum firepower 意思是"最大火力"。该句中的 be open to sth. 是固定搭配，表示"易受……的"，相似表达结构有 be vulnerable to sth.。

> **翻译**：这个队形可以提供最大的正面火力，但容易遭到侧翼伏击。

【口语输出】

1. What is the basic structure of a tank?

【参考词汇】barrel, cupola, engine deck, periscope/gunsight, road wheel, side skirt, toe plate, hull, tracks, main armament, suspension, turret, be composed of, consist of, belong to

2. Can you introduce the members of the tank crew?

【参考词汇】commander, gunner, loader, driver

3. What are the basic formations for tanks and armoured vehicles?

【参考词汇】column, echelon, line, staggered column, vee, wedge

4. What are the advantages and disadvantages of the column formation?

【参考词汇】**advantages:** speed, move along a route, through a restricted terrain, advance and keep position by following the leader, easy to maintain the formation, excellent fire to the platoon's flanks

disadvantages: not a good battle formation, only the lead track, fire to the front

5. Which is the best tank formation? And why?

【参考词汇】the wedge, meet every possible situation, contact with enemy, move in open terrain, excellent firepower to the front and the flanks, fire straight ahead, three tanks, cover either flank

Delta　New tanks

【军事知识】

1. Main Battle Tank 主战坦克

A main battle tank, also known as a battle tank or universal tank, is a subset of tank that fills the armor-protected direct fire and maneuver role of many modern armies. Cold War-era development of more powerful engines, better suspension systems and lighter weight composite armor allowed the design of a tank that had the firepower of a super-heavy tank, the armor protection of a heavy tank, and the mobility of a light tank, in a package with the weight of a medium tank. Through the 1960s, the MBT replaced almost all other types of tanks, leaving only some specialist roles to be filled by lighter designs or other types of armored fighting vehicles.

2. Transport aircraft 运输飞机

A military transport aircraft, military cargo aircraft or airlifter is a military-owned transport aircraft used to support military operations by airlifting troops and military equipment. Transport aircraft are crucial to maintaining supply lines to forward bases that are difficult to reach by ground or waterborne access, and can be used for both strategic and tactical missions. They are also often used for civilian emergency relief missions by transporting humanitarian aid.

3. Amphibious ships 两栖舰船

An amphibious warfare ship is an amphibious vehicle warship employed to land and support ground forces, such as marines, on enemy territory during an amphibious assault. Specialized shipping can be divided into two types, most crudely described as ships and craft. In general, the ships carry the troops from the port of embarkation to the drop point for the assault and the craft carry the troops from the ship to the shore. Amphibious assaults taking place over short distances can also involve the shore-to-shore technique, where landing craft go directly from the port of embarkation to the assault point. Some tank landing ships may also be able to land troops and equipment directly onto shore after travelling long distances, such as the Ivan Rogov-class landing ship.

4. The Leopard 德制"豹"2型主战坦克

The Leopard tank, designed and produced in Western Germany, has over the years proven to be one of the very best Main Battle Tanks (MBTs) ever, functioning at almost the same level as the M1 Abrams. The development process of the first Leopards took off in late 1956, in a bid to replace the American-built M-47 and M-48 Patton tanks which were rapidly becoming outdated, based on the technological trends of the time. The Leopard first entered service in 1965.

5. Light armoured vehicle 轻型装甲车

The Light Armored Vehicle or LAV is a kind of armored military transport capable of equipping a small squad into tight situations, combining speed, maneuverability and firepower to fulfill a variety of missions.

【词汇点拨】Words and Expressions

1. consumption /kənˈsʌmpʃn/ *n.* the act of using energy, food or materials; the amount used（能量、食物或材料的）消耗，消耗量

Gas and oil consumption always increases in cold weather.

2. amphibious /æmˈfɪbɪəs/ *adj.* suitable for use both on water and on land 两栖的

amphibious assault 两栖突击（突袭、攻击）

amphibious craft 两栖舰船

amphibious landing 两栖登陆

amphibious operation 两栖（登陆）行动

amphibious vehicle 两栖车辆

3. territory /ˈterətri/ *n.* [C,U] land that is under the control of a particular country or ruler 领土；版图；领地

They have refused to allow enemy troops to be stationed in their territory.

4. land /lænd/ *v.*

① to bring a flying aircraft back onto the ground 着陆；降落

The squadron has just landed.

② to leave a ship and go back onto dry land territory 乘飞机或船着陆；登陆

The troops landed at dawn.

③ to deploy troops from aircraft or ships （使）着陆；降落；靠岸；登陆

The troops were landed by helicopter.

5. steep /stiːp/ *adj.* rising or falling quickly, not gradually 陡的；陡峭的

The path grew steeper as we climbed higher.

6. transport aircraft 运输机

【词汇点拨】Proper Names

1. **Defence Department**（澳大利亚）国防部

2. **MBT**（Main Battle Tank）主战坦克

3. **Australian Air force** 澳大利亚空军

4. **Australian Army** 澳大利亚陆军

5. **Leopard AS1** 德制"豹 AS1"型主战坦克

6. **Stryker** /ˈstraɪkə/ 斯特瑞克装甲车

7. **ASLAV**（Australian Light Armoured Vehicle）/ˈæslæv/ 澳大利亚轻型装甲车

8. **Australia** /ɒˈstreɪlɪə/ 澳大利亚

9. **Canada** /ˈkænədə/ 加拿大

10. **Canadian** /kəˈneɪdɪən/ 加拿大人

【长句解读】

1. So if our current planes can't lift the Leopard, it will be impossible for them to transport a newer, heavier tank with all the latest equipment and shooting systems. (10-10)

> 句子分析：该句为主从复合句，由 if 引导的条件状语从句和主句构成。其中 with all the latest equipment and shooting systems 为后置定语，修饰前面的名词 tank，意思为"配备了全部最新设备和射击系统"。

> 翻译：因此，如果我们现在的飞机不能吊运豹式坦克的话，它们就更无法运输配备了全部最新设备和射击系统的更新、更重的坦克。

2. So that makes going overseas with all that heavy equipment a bit of a problem. (10-10)

> 句子分析：该句主语为指示代词 that，指代上文提到的内容，谓语动词 makes，宾语为动名词 going overseas with all that heavy equipment，a bit of a problem 为宾语补语。整个结构"make+ 宾语 + 宾语补语"意为"使、让某人 / 某物成为……"。其中 a bit of 后面可以接不可数名词，也可以接可数名词，可数名词通常用单数，而且其前面应该有不定冠词。

> 翻译：所以，携带这些重型设备到海外执行任务会有些问题。

3. But unlike the ASLAV, the Stryker is armed with 105 mm guns. (10-10)

> 句子分析：句中 be armed with 为固定搭配，意思为"使用……武器；装备有……"。105 mm，指的是枪炮的口径为 105 毫米。

> 翻译：但与澳大利亚轻型装甲车不同的是，斯特瑞克装甲车配备了 105 毫米口径的坦克炮。

【口语输出】

1. What are the two Australian officers talking about?

【参考词汇】two Australian officers, discuss, the Defence Department, replace the Leopard, new MBTs

2. According to one of the speakers, why are tanks still so important in a combined arms team?

【参考词汇】one of the most important elements, more exposed, attack, helicopters, can't provide enough support, offer greater protection, make forces more effectively

3. What problems will the Australian Army have if they buy new tanks?

【参考词汇】first of all, have higher fuel consumption, parts for new tanks, more expensive, another problem, none of the transport aircraft, lift, a Leopard AS1, impossible, transport a newer, heavier tank with the latest equipment and shooting system, moreover, don't have enough amphibious ships to move them

4. Where do they think the Australian Army is more likely to deploy, overseas or domestically?

【参考词汇】be conducted overseas, the UN on peace-keeping mission

5. Which country has the similar problem? How did this country solve the problem?

【参考词汇】Canada, buy the Strykcr, moves on wheels, much lighter than a tank, armed with 105mm guns, faster and much lighter

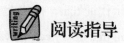 阅读指导

【文章导读】

This passage mainly focuses on combined arms and its differences from traditional segregated units. After explaining the cavalry terminology, it elaborates the differences and similarities between an armoured cavalry regiment and a mechanised infantry brigade. Finally, the formations of US Army 3rd Armoured Cavalry Regiment and Armoured Cavalry Squadron are briefly introduced.

【军事知识】

1. Six service branches of the U.S. armed forces 美军军种

Get a brief overview of the six service branches of the U.S. armed forces:

U.S. Air Force (USAF)

U.S. Army (USA)

U.S. Coast Guard (USCG)

U.S. Marine Corps (USMC)

U.S. Navy (USN)

U.S. Space Force (USSF)

The Air Force is part of the Department of Defense (DOD). It's responsible for aerial military operations, defending U.S. air bases, and building landing strips. Its service members are airmen. The reserve components are Air National Guard and Air Force Reserve.

The Army is part of the DOD and is the largest of the military branches. It handles significant ground combat missions, especially operations that are ongoing. Army Special Forces are called Green Berets for their headgear. The Army's members are its soldiers. The reserve components are the Army Reserve and Army National Guard.

The Coast Guard is part of the Department of Homeland Security (DHS). It provides national security and search and rescue for America's waterways, seas, and coast. It's responsible for stopping drug smugglers and others breaking maritime law. It enforces marine environmental protection laws. Service members are Coast Guardsmen and nicknamed Coasties. The reserve component is the Coast Guard Reserve.

The Marine Corps is part of the DOD. It provides land combat, sea-based, and air-ground operations support for the other branches during a mission. This branch also guards U.S. embassies around the world and the classified documents in those buildings. Marine Corps Special Operations Command (MARSOC) members are known as Raiders. All service members are called Marines. The reserve component is the Marine Corps Reserve.

The Navy is part of the DOD. It protects waterways (sea and ocean) outside of the Coast Guard's jurisdiction. Navy warships provide the runways for aircraft to land and take off when at sea. Navy SEALs (sea, air, and land) are the special operations force for this branch. All service members are known as sailors. The reserve component is Navy Reserve.

The Space Force is a new branch, created in December 2019 from the former Air Force Space Command. The Space Force falls within the Department of the Air Force. It organizes, trains, and equips space forces to protect U.S. and allied interests in space and provides space capabilities to the joint force.

2. Military intelligence 军事情报

Military intelligence is a military discipline that uses information collection and analysis approaches to provide guidance and direction to assist commanders in their decisions. This aim is achieved by providing an assessment of data from a range of sources, directed towards the commanders' mission requirements or responding to questions as part of operational or campaign planning. To provide an analysis, the commander's information requirements are first identified, which are then incorporated into intelligence collection, analysis, and dissemination.

3. Reconnaissance 侦查

In military operations, reconnaissance or scouting is the exploration of an area by military forces to obtain information about enemy forces, terrain, and other activities. Examples of reconnaissance include patrolling by troops, ships or submarines, crewed or uncrewed reconnaissance aircraft, satellites, or by setting up observation posts. Espionage is usually considered to be different from reconnaissance, as it is performed by non-uniformed personnel operating behind enemy lines.

4. Mechanised infantry 机械化步兵

Mechanised infantry is infantry equipped with armoured personnel carriers (APCs) or infantry fighting vehicles (IFVs) to move soldiers during heavy fire fights. They often work together with tanks in a symbolic setting, providing shelter and fire support for other troops.

Mechanised infantry is distinguished from motorized infantry in that its vehicles provide a degree of protection from hostile fire, as opposed to "soft-skinned" wheeled vehicles (trucks or jeeps) for motorized infantry. Most APCs and IFVs are fully tracked, or are all-wheel drive vehicles (6×6 or 8×8), for mobility across rough ground.

5. Mortar 迫击炮

A mortar is usually a simple, lightweight, man-portable, muzzle-loaded weapon, consisting of a smooth-bore (although some models use a rifled barrel) metal tube fixed to a base plate (to spread out the recoil) with a lightweight bipod mount and a sight. They launch explosive shells (technically called bombs) in high-arcing ballistic trajectories. Mortars are typically used as indirect fire weapons for close fire support with a variety of ammunition. The defining features of a mortar are its responsiveness, high-angle arching trajectory (above 45°), low velocity and relatively short range. In the theatre of combat, mortars are used to kill military personnel, harass adversaries and prevent the use of ground with interdiction fire. They can also be used to lay smoke screens.

Mortars can engage targets at less than 70 meters to 9,000 meters from the firer's position. Medium mortars (61-99 mm) can fire at ranges of 100 meters to 5,500 meters, while heavy mortars (100-120 mm) have a range of some 500 meters to 9,000 meters. The range of any given mortar systems depends on the cartridge which possesses the maximum range.

【词汇点拨】Words and Expressions

1. temporary /ˈtemprəri/ *adj.* lasting or intended to last or be used only for a short time; not permanent 短暂的；暂时的；临时的

They had to move into temporary accommodation.

2. segregate /ˈseɡrɪɡeɪt/ *v.* ~ **sth. (from sth.)** to keep one thing separate from another（使）分开；分离；隔离

A large detachment of police was used to segregate the two rival camps of protesters.

3. flexibility /ˌfleksəˈbɪləti/ *n.* the property of being flexible; easily bent or shaped 灵活性；弹性

The flexibility of distance learning would be particularly suited to busy managers.

4. co-ordinate /kəʊˈɔːdɪneɪt/ *v.* to manage the actions of two or more people or groups so that they work towards a common goal 使协调；使相配合

This contrasts sharply with the coordinated national programs of most other countries.

5. intelligence /ɪnˈtelɪdʒəns/ *n.*

① any information which may be useful (especially information about the enemy) 情报

We have received some fresh intelligence on the enemy artillery.

② people and equipment involved in the gathering, analysis and dissemination of intelligence 情报人员；情报机构

We are feeding false information to the enemy's intelligence.

6. brigade /brɪˈɡeɪd/ *n.* a subdivision of an army, typically consisting of a small number of infantry battalions and/or other units and typically forming part of a division 旅

A third brigade is at sea, ready for an amphibious assault.

7. troop /truːp/ *n.* 部队；（英陆军装甲兵、工兵、通信兵的）排；（美陆军的）骑兵连

8. mechanise /ˈmekənaɪz/ *v.* [usually passive] to change a process, so that the work is done by machines rather than people 机械化；使机械化

The production process is now highly mechanised.

9. integrate /ˈɪntɪɡreɪt/ *v.* ~ (A) (into/with B) /~ A and B to link up several things to form a whole（使）合并；成为一体

10. parent /ˈpeərənt/ *n.* (often used as an adjective) an organization that produces and owns or controls smaller organizations of the same type 创始机构；母公司；总部

Each unit including the parent company has its own, local management.

11. aviation /ˌeɪviˈeɪʃn/ *n.* the designing, building and flying of aircraft 航空

The airline's biggest headache is the increase in the price of aviation fuel.

12. saber /ˈseɪbə(r)/ *n.* (BrE sabre) a heavy sword with a curved blade（弯刃）军刀；马刀；佩剑

13. skinner /ˈskɪnə(r)/ *n.* a person who prepares or deals in animal skins 皮革商；剥皮工；（美）赶牲口的人

14. howitzer /ˈhaʊɪtsə(r)/ *n.* a short-barrelied artillery piece designed to fire shells at high trajectories 榴弹炮

【词汇点拨】Proper Names

US Army's 3rd Armoured Cavalry Regiment 美国陆军第3装甲骑兵团

【参考译文】

合成部队

"合成部队"这个术语是指军事建制，即把不同兵种单位比如步兵、坦克、炮兵和空中支援力量合成诸如团或营这样的一个独立单位。这样的建制可能是永久的，比如美国陆军第3装甲骑兵团，也可能是为一个特定任务而临时组建的，比如把一个坦克连暂时编入一个步兵营中。

合成部队不同于单一兵种部队。单一兵种部队主要由一个兵种类型组成，因而火力更为集中。合成部队既有较高的灵活性，也能独立于其他作战单位单独行动。这个特点使之能够快速、协同地应对多种战场需求。例如，合成部队自身就能进行情报搜集和侦查，制订自己的后勤保障计划，执行自己的战斗和防御计划，同时还可以提供空中和炮火支援。这些都需要协调指挥才能完成。我们将会看到一个具体的例子，但先了解几个新的术语会很有帮助。

美国骑兵部队术语

军队建制术语根据不同的部门和国家会有很大差别。例如，美国陆军骑兵部队通常使用以下术语：

团（regiment）：通常由四到五个营组成，规模等同于一个常规旅；

营（battalion）：通常由四到五个骑兵连组成，规模等同于一个常规营；

骑兵连（troop）：规模等同于一个常规连。

装甲骑兵与机械化步兵

作为合成部队和单一兵种部队区别的一个例子，我们可以看一下装甲骑兵团和机械化步兵旅之间的异同点。两支部队都使用装甲运兵车运输兵力，且都列装主战坦克。但是装甲骑兵团的一体化深入到连一级，常规的装甲骑兵连包括2个坦克排、2个步兵排、1个迫击炮排和1个排部。装甲骑兵团也有自己的空中支援营。然而，常规的机械化步兵旅有2~3个机械化步兵营、1个装甲营和多个支援中队，各营都有很强的独立性。机械化步兵旅通常没有自己的空中支援力量，都是由师里的另一个团来为其提供。

装甲骑兵营的一体化使它能够独立行动，比传统的必须依靠旅来支援的单一兵种营反应更加迅速。这意味着，装甲骑兵更易作为先遣进攻部队，而机械化步兵会部署在战斗最激烈的地方。

第三装甲骑兵团（勇敢步枪手）

来看一个特别的例子，美国陆军第3装甲骑兵团下设3个装甲骑兵营、1个航空营和 1 个支援保障营，具体如下：
- 一营（老虎营）（装甲骑兵营）
- 二营（佩剑营）（装甲骑兵营）
- 三营（雷电营）（装甲骑兵营）
- 四营（长刀营）（航空营）
- 支援保障营（骡夫营）

装甲骑兵营

一个典型的装甲骑兵营包括：
- 连部
- 三个骑兵连
- 一个坦克连
- 一个炮（榴弹炮）兵连（约连级规模）
- 附属单位（一个后备连队，比如炮兵连、工兵连、情报连或者医疗连）

 词汇表

Alpha New duty station
squared away 状态优秀；一切就绪［美军口语］
bend （尤指道路或河流的）拐弯；弯道
highway （尤指城镇间的）公路，干道，交通要道
proceed 行进；前往
approach （在距离或时间上）靠近，接近
lane 车道
garrison 驻军；卫戍区；卫戍部队；守备部队
identification 身份证明
process in 报到
car registration 车辆登记
driving license 驾照
speed limit （道路上的）限速
temporary quarters 临时宿舍

reservation 预订；预约
intersection 十字路口；交叉路口；交点
installation 设施
disciplinary 有关纪律的；执行纪律的；惩戒性的
assignment （分派的）工作，任务
accommodation 住宿
forward 【篮球】前锋
guard 【篮球】后卫
dismissed 准予……离去

Bravo Regimental history
TO&E 编制与制备表
barbed wire 带刺铁丝网（尤用作围栏）
engagement 战斗；交火；交战
squadron （空军）中队；（美陆军工兵、装

甲兵）营；（英陆军）连
designate 命名；提名；任命；（将部队）定番号为……
full strength 满员；满编；全员
screen 警戒；屏蔽（物）；掩护
trench 战壕
station 驻扎
rear area 后方（地域）
insignia 标识；徽章；军装服饰
escort 护卫（队）；护航（队）
rename 重命名
rebel 叛乱；反叛；叛军；叛乱分子
comrade 伙伴，战友；同志
frontier 边境，国界
deploy 部署
outbreak （战争、疾病、暴力等的）爆发，突然发生
convert （使）转变，转换，转化
rage （暴风雨、战斗、争论等）猛烈地继续；激烈进行
base 基地
attack helicopter 攻击直升机
cavalry fighting vehicle (CFV) 骑兵（装甲兵）战斗车辆
main battle tank (MBT) 主战坦克
scout helicopter 侦查直升机
self-propelled howitzer (SPH) 自行榴弹炮
transport helicopter 运输直升机
security mission 警戒任务；安全保卫任务
reconnaissance mission 侦查任务
defensive mission 防御任务
offensive mission 进攻任务
tactical 战术的
combined arms 合成部队

mobile 机动的
asset 有价值的人（或事物）；有用的人（或事物）
authorized 授权的

Charlie How tanks fight
barrel 炮筒（管）；枪管
cupola 指挥塔
engine deck 发动机甲板
periscope （火炮的）周视瞄准镜；（坦克的）周视指挥观瞄镜；（潜艇的）潜望镜
gunsight （枪炮的）瞄准具（器）；观瞄镜
road wheel （坦克、装甲车辆等的）负重轮
side skirt 侧裙
toe plate 前甲板
track 履带
loader 填装手；装弹机
driver 驾驶员
gunner 火炮瞄准手；炮手；枪手
hull 壳体（坦克、舰艇等的主要结构）
armament 武器；装备
suspension 承载装置；悬挂装置
turret （战舰、飞机或坦克的）回转炮塔；旋转枪架
column 纵队
echelon 梯形编队；梯队
line 横队
staggered column 交错纵队编队
vee V字形编队；V形队形
wedge 楔形编队
tactics 战术
scout 侦查
terrain 地形
flank 侧翼

modified 改良的；改进的
rear 后部；后面
variation 变体
assault 袭击；攻击
maximum 最高的；最多的；最大极限的
firepower 火力
ambush 伏击；埋伏
wooded 树木繁茂的；长满树木的；树木覆盖的

Delta New tanks

cargo（船或飞机装载的）货物
consumption（能量、食物或材料的）消耗，消耗量
amphibious 两栖的
armour 装甲
territory 领土；版图；领地
firepower 火力
land（使）着陆；降落；靠岸；登陆
steep 陡的；陡峭的

Echo Combined arms

arms 兵种
temporary 短暂的；暂时的；临时的

segregate（使）分开；分离；隔离
flexibility 灵活性；弹性
reconnaissance 侦查
co-ordinate 使协调；使相配合
intelligence 情报；情报人员；情报机构
logistical 后勤的
terminology 术语
brigade 旅
squadron（空军）中队；（美陆军工兵、装甲兵）营；（英陆军）连
troop 部队；（英陆军装甲兵、工兵、通信兵的）排；（美陆军的）骑兵连
mechanise 机械化；使机械化
integrate（使）合并；成为一体
mortar 迫击炮
division 师
parent 创始机构；总公司；总部
advance 先行的；先遣的
concentrate 使……集中（或集合、聚集）
aviation 航空
saber（弯刃）军刀；马刀；佩剑
skinner 皮革商；剥皮工；（美）赶牲口的人
howitzer 榴弹炮
engineer 工兵

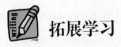 拓展学习

Responsibilities of a Tank Platoon

The tank crew is a tightly integrated team. Though all members have primary duties, success depends on their effectiveness as a crew. They must work together to maintain and service their tank and equipment, and they must function as one in combat. Crews must cross-train so each member can function at any of the other crew positions.

Platoon Leader

The platoon leader is responsible to the commander for the discipline and training of his platoon, the maintenance of its equipment, and its success in combat. He must be proficient in the tactical employment of his section and of the platoon in concert with a company team or troop. He must have a solid understanding of troop-leading procedures and develop his ability to apply them quickly and efficiently on the battlefield.

The platoon leader must know the capabilities and limitations of the platoon's personnel and equipment; at the same time, he must be well versed in enemy organizations, doctrine, and equipment. He must serve as an effective tank commander (TC). Most important of all, the platoon leader must be flexible, using sound judgment to make correct decisions quickly and at the right times based on his commander's intent and the tactical situation.

Platoon leaders must know and understand the task force mission and the task force commander's intent. They must be prepared to assume the duties of the company commander in accordance with the succession of command.

Platoon Sergeant

The PSG is second in command of the platoon and is accountable to the platoon leader for the training, discipline, and welfare of the soldiers in the platoon. He coordinates the platoon's maintenance and logistics requirements and handles the personal needs of individual soldiers. The PSG is the most experienced TC in the platoon. His tactical and technical knowledge allow him to serve as mentor to crewmen, other NCOs, and the platoon leader. His actions on the battlefield must complement those of the platoon leader. He must fight his section in concert with the platoon leader's section.

Tank Commander

The TC is responsible to the platoon leader and signed equipment, the reporting of logistical needs, and the tactical employment of his tank. He briefs his crew, directs the movement of the tank, submits all reports, and supervises initial first-aid treatment and evacuation of wounded crewmen. He is an expert in using the tank's weapon systems, requesting indirect fires, and executing land navigation.

The TC must know and understand the company mission and company commander's intent. He must be prepared to assume the duties and responsibilities of the platoon leader or PSG in accordance with the succession of command. These requirements demand that the TC maintain situational awareness by using all available optics for observation, by eavesdropping on radio transmissions, and by monitoring the intervehicular information system (IVIS) or appliqué digital screen (if available).

Gunner

The gunner searches for targets and aims and fires both the main gun and the coaxial machine gun. He is responsible to the TC for the maintenance of the tank's armament and fire control equipment. The gunner serves as the assistant TC and assumes the responsibilities of the TC as required. He also assists other crewmembers as needed. Several of his duties involve the tank's communications and internal control systems: logging onto and monitoring communications nets; maintaining digital links if the tank is equipped with the IVIS or appliqué digital system; inputting graphic control measures on digital overlays; and monitoring digital displays during the planning and preparation phases of an operation.

Driver

The driver moves, positions, and stops the tank. While driving, he constantly searches for covered routes and for covered positions to which he can move if the tank is engaged. He maintains his tank's position in formation and watches for visual signals. If the tank is equipped with a steer-to indicator, the driver monitors the device and selects the best tactical route. During engagements, he assists the gunner and TC by scanning for targets and sensing fired rounds. The driver is responsible to the TC for the automotive maintenance and refueling of the tank. He assists other crewmen as needed.

Loader

The loader loads the main gun and the coaxial machine gun ready box; he aims and fires the loader's machine gun (if the vehicle is equipped with one). He stows and cares for ammunition and is responsible to the TC for the maintenance of communications equipment. Before engagement actions are initiated, the loader searches for targets and acts as air or antitank guided missile (ATGM) guard. He also assists the TC as needed in directing the driver so the tank maintains its position in formation. He assists other crewmembers as necessary. Because the loader is ideally positioned both to observe around the tank and to monitor the tank's digital displays, platoon leaders and TCs should give strong consideration to assigning their second most experienced crewman as the loader. (819 words)

(Selected from http://www.bits.de/NRANEU/others/amd-us-archive/FM-17-15%2896%29.pdf)

【Notes】

1. **cross-train** 交叉训练
2. **eavesdrop** 偷听；窃听

3. appliqué（织物的）缝饰，嵌花，贴花

4. coaxial 同轴的；共轴的

5. automotive 汽车的；机动车辆的

6. stow 妥善放置；把……收好

7. TC (tank commander) 坦克指挥官

8. PSG (Platoon Sergeant) 坦克排长

9. **IVIS:** The Intervehicular Information System (IVIS) is a computer driven communications network embedded in the vehicle electronics of the M1A2 main battle tank now entering international service. The system uses multifunction controls and displays to provide a graphics user interface for human-machine interaction, embedded sensors for the information integration, a distributed multiprocessor computer system for information processing, and a digital communications channel.

10. **ATGM:** An anti-tank missile (ATM), anti-tank guided missile (ATGM), anti-tank guided weapon (ATGW) or anti-armour guided weapon, is a guided missile primarily designed to hit and destroy heavily armoured military vehicles. ATGMs range in size from shoulder-launched weapons, which can be transported by a single soldier, to larger tripod-mounted weapons, which require a squad or team to transport and fire, to vehicle and aircraft mounted missile systems.

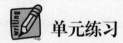

单元练习

I. Label the picture

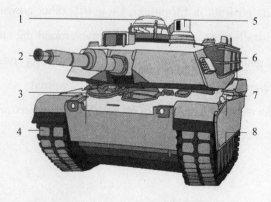

1._____ 2._____ 3._____ 4._____
5._____ 6._____ 7._____ 8._____

II. Multiple choice

1. In the US Cavalry, a regiment is typically composed of 4 to 5 squadrons, the equivalent of a normal ___.

   A. corps   B. division   C. brigade   D. battalion

2. ___ refers to a unit with two or more arms and elements of a military service.

   A. Combined arms unit   B. Squadron
   C. TO&E   D. Segregated unit

3. An operation designed to cause an enemy attack to fail is a/an ___ mission.

   A. defensive   B. offensive   C. security   D. reconnaissance

4. ___ is another word for battle; ___ is another word for helicopter.

   A. Asset; bird   B. Engagement; trench
   C. Scout; CFV   D. Engagement; bird

5. Which of the following doesn't belong to the tank crew?

   A. Loader.   B. Driver.   C. Track.   D. Gunner.

6. The ___ is the main armament of a tank.

   A. cupola   B. barrel   C. engine deck   D. side skirt

7. The ___ formation has two variations: left and right.

   A. column   B. line   C. wedge   D. echelon

8. The ___ formation is used when speed is important and when the platoon is moving along a route, such as a road or track, or through restrictive terrain.

   A. column   B. vee   C. line   D. echelon

9. The ___ formation gives maximum firepower to the front, but the platoon is open to ambush from the flanks.

   A. line   B. echelon right
   C. staggered column   D. wedge

10. Compared with a traditional segregated battalion, the armoured cavalry may be used easily as a/an ___.

    A. amphibious force   B. logistical support force
    C. intelligence force   D. advance attack force

III. Translation: English to Chinese

1. Although the Regiment was still a horse cavalry unit at the outbreak of World War II, it was soon converted to mechanised cavalry and re-equipped with armoured vehicles before deployment to Europe.

2. It is a highly mobile force that can conduct reconnaissance, security, offensive and defensive operations.

3. The cutting edge of the Regiment is the three armoured cavalry squadrons: the 1st Tiger Squadron, the 2nd Sabre Squadron and the 3rd Thunder Squadron.

4. The wedge is the best formation to meet every possible situation, including when contact with the enemy is expected.

5. I suppose heavy armour and big guns were perhaps the most important things when we were thinking more about defending our own country on our own territory.

IV. Translation: Chinese to English

美国陆军第3装甲骑兵团以科罗拉多州的卡森堡为基地，是一支高度机动的机械化部队。该团成立于1846年，是美国陆军最老的部队之一，曾参加过第一、二次世界大战等7次大的战争和36场战役。第3装甲骑兵团擅长沙漠作战，平时十分注重沙漠实战训练，被称为"陆军沙漠作战专家"。1991年的海湾战争中，该团曾深入伊拉克境内300千米，击溃伊拉克共和国卫队3个师。第3装甲骑兵团下辖3个坦克营、1个炮兵营及支援分队等，拥有M1A1坦克、M3A2战车和AH-64A攻击直升机等武器装备。

V. Oral practice

1. What are the features of the combined arms? How is it different from the traditional segregated units?

2. What do you know about US Army's 3rd Armoured Cavalry Regiment? Can you introduce this unit in terms of its history and formation?

附录：Unit 10 练习答案

I. Label the picture

1. cupola　　2. barrel　　3. toe plate　　4. tracks
5. periscope　6. engine deck　7. side skirt　8. road wheels

II. Multiple choice

1-5 CAADC　　6-10 BDAAD

III. Translation: English to Chinese

1. 虽然该团在第二次世界大战爆发时仍然是一个骑兵部队，但很快就转变为机械化骑兵，并在部署到欧洲之前重新装备了装甲车。

2. 这是一支高度机动的部队，可以执行侦查、安全、进攻和防御作战任务。

3. 该团最前沿的力量是三个装甲骑兵营：一营（老虎营）、二营（佩剑营）和三营（雷电营）。

4. 楔形队形是应对所有可能情况的最有利队形，包括遭遇敌军的情况。

5. 我想当我们想在自己的领土上保卫我们自己的国家时，重型装甲和大炮可能是最重要的武器装备。

IV. Translation: Chinese to English

The US Army's 3rd Armoured Cavalry Regiment, based in Fort Carson, Colorado, is a highly mobile mechanised unit. Founded in 1846, the regiment is one of the oldest units of the US Army. It has participated in seven major wars and 36 battles, including World War I and World War II. The 3rd Armoured Cavalry Regiment is good at desert combat, usually pays great attention to actual combat training in desert, and is called the "desert combat expert" of the Army. In the Gulf War of 1991, the regiment penetrated into Iraq for 300 kilometers and defeated three divisions of the Iraqi Republican Guard. The 3rd Armoured Cavalry Regiment has three tank squadrons, one artillery squadron and one support squadron, and has weapons and equipment such as M1A1 tanks, M3A2 fighting vehicles and AH-64A attack helicopters.

V. Oral practice

1. The term combined arms refers to military formations which join different arms categories—such as infantry, tank, artillery, and air support—into a single unit. This organization may be either permanent or temporary. A combined arms unit is flexible and can act independently of other combat units. It can respond in a rapid, co-ordinated manner to a wide range of battlefield needs. It can carry out many tasks by itself, such as reconnaissance, intelligence, logistical support, defence, air and artillery support. A typical example is the 3rd Armoured Cavalry Regiment, which is made up of three armoured cavalry squadrons, one aviation squadron and one support squadron.

2. Formed nearly 160 years ago, the 3rd Armoured Cavalry Regiment (the 3rd ACR) is the second oldest unit in the United States Army. Although the Regiment was still a horse cavalry unit at the outbreak of World War II, it was soon converted to mechanised cavalry and re-equipped with armoured vehicles before deployment to Europe. The Regiment was redesignated as the 3rd Armoured Cavalry Regiment in 1948. The 3rd ACR today is one of the largest and most powerful tactical units in the US Army. It is a highly mobile force that can conduct reconnaissance, security, offensive and defensive operations. The cutting edge of

the Regiment is the three armoured cavalry squadrons: the 1st Tiger Squadron, the 2nd Sabre Squadron and the 3rd Thunder Squadron. Each squadron is equipped with M1A2 Abrams MBTs, M3A2ODS Bradley CFVs and M109A6 howitzers. The 4th Longknife Air Cavalry Squadron is organised and equipped to conduct highly mobile reconnaissance and screening operation.

Unit 11　International HQ

单元导学清单

模块	主题	学习目标	任务	军事知识	核心词汇
Alpha	联合司令部	1）了解欧洲军团相关知识及联合司令部组织架构及其分部门任务；2）掌握相关词汇和表达。	Task 1 Task 2 Task 3 Task 4 Task 5 Task 6	■ Eurocorps ■ NATO ■ High Readiness Forces	discharge, liaise, spokesperson, executive, protocol, CHOD, CIS, COS, DOS, SNR, CIMIC
Bravo	参观拜访	1）了解军事访问相关内容及北约任务清单的主要内容；2）掌握军事访问的相关词汇和表达。	Task 1 Task 2 Task 3 Task 4	■ Protocol (Office) ■ Vin d'honneur	agenda, office call, welcome package, catering, suspense, tasker, Commander's Mess, POC
Charlie	重新安排	1）了解英语中打电话及电话留言相关知识；2）熟练掌握打电话相关的词汇与表达。	Task 1 Task 2 Task 3	■ Liaison officer ■ Defence Attaché Germany	confirm, accompany, installation, maintenance, slide
Delta	将军来访	1）了解军事接待礼仪中介绍规则；2）熟练掌握军事接待介绍相关词汇及表达；3）运用所学知识进行军事接待英语对话。	Task 1 Task 2 Task 3 Task 4	■ Military Assistant	accommodation, subordinate, venue
Echo (Review)	北约总部内部	1）了解北约总部内部组织架构及11个分部门的任务；2）熟练掌握相关的词汇和表达；3）运用所学知识介绍北约总部内部。	Task 1 Task 2	■ NATO Headquarter Structure ■ The "G" division	equivalent, manpower, logistical, counterintelligence, supervision, appropriate, restoration, accounting, billet, prefix, symbolic

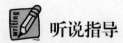 听说指导

Alpha Joint headquarters

【军事知识】

1. Eurocorps 欧洲军团

Created by France and Germany in 1992, Eurocorps is composed of 6 Framework Nations and 5 Associated Nations. Since 2009, all those nations are solidly linked together by the Treaty of Strasbourg which confers Eurocorps its own legal capacity and makes it a unique truly multinational army corps.

Located in the Eurometropole Strasbourg, Eurocorps is close to the decision-making centers of the European Union and NATO, allowing it to maintain close relations on a political and military level.

2. NATO 北约

Its full form is North Atlantic Treaty Organization, a political and military alliance of countries from Europe and North America. It provides a unique link between these two continents, enabling them to consult and cooperate in the field of defence and security, and conduct multinational crisis-management operations together.

The foundations of NATO were officially laid down on 4 April 1949 with the signing of the North Atlantic Treaty, more popularly known as the Washington Treaty. The most important players in NATO are the member countries themselves. There are currently 32 members.

3. High Readiness Forces 高度戒备（战备）部队

In order to provide flexibility for conducting the full range of missions, as well as describing the availability of Allied Forces to NATO commanders, HQs and forces can be further sub-divided into two types of forces reflecting readiness levels: High Readiness Forces (HRF) and Forces of Lower Readiness (FLR). Together, HRF and FLR form the Graduated Readiness Forces (GRF). Graduated Readiness Forces Headquarters (GRF HQs) provide these forces with the appropriate command and control.

HRF readiness should range from 0 to 90 days and include capabilities for an immediate response (from 0 to 30 days and in the framework of the NATO Response Force). FLR should be reported with readiness ranges from 91 to 180 days and normally used to sustain deployed HQs and forces.

【词汇点拨】Words and Expressions

1. staff /stɑːf/ *n.* a group of officers and other ranks, who assist the commander of a large tactical grouping (such as a brigade, division, corps, etc.), and who form the headquarters 参谋

A group of military staff officers were honored last week.

2. discharge /dɪsˈtʃɑːdʒ/ *n.* the release of a person from duty 退役；退伍

He was accused of obstructing the sergeant in the discharge of his duty.

3. liaise /liˈeɪz/ *v.* ~ **(with sb.)** (especially BrE) to work closely with sb. and exchange information with them（与某人）联络；联系

His job is to liaise with other similar organizations and to plan a joint campaign.

4. infrastructure /ˈɪnfrəˌstrʌktʃə/ *n.* basic amenities and facilities upon which a modern society relies in order to function properly (such as electricity, roads and railways, telecommunications, water, etc.) 基础设施

Israel would continue its efforts to destroy the terrorist infrastructure, arrest those behind terrorist activities and harshly punish those who perpetrate these murderous activities.

5. principal /ˈprɪnsəpl/ *adj.* [only before noun] most important; main 最重要的；主要的

Their principal job was to further the policy agenda of the White House.

6. spokesperson /ˈspəʊkspɜːsn/ *n.* a person who speaks on behalf of a group or an organization 发言人

Spokesperson Zhao Lijian made the remarks at a daily press briefing when responding to a query on a recent purchasing contract of a field information and communication system between the United States and Taiwan, which was a follow-up to the US arms sale plan to Taiwan on December 2020.

7. report to (not used in the progressive tenses，即不用于进行时) if you report to a particular manager in an organization that you work for, they are officially responsible for your work and tell you what to do 被……领导；隶属；从属

The Secretary General is responsible for all UN peacekeeping operations and he reports to the UN Security Council.

8. executive /ɪɡˈzekjətɪv/ *n.* a person who has an important job as a manager of a company or an organization（公司或机构的）主管领导；管理人员

The executive officer needs a way to synchronize warfighting functions, understand all plans and keep the staff task oriented.

9. deputy /ˈdepjuti/ *n.* a person who is the next most important person below a business manager, a head of a school, a political leader, etc. and who does the person's job when he or she is away 副手；副职；代理

Deputy Chief of Staff 副参谋长

The plane completely vanished from radar one minute before it entered Vietnam's air traffic space, according to Lt. Gen. Vo Van Tuan, deputy chief of staff of the Vietnamese army.

10. protocol /ˈprəʊtəkɒl/ *n.* a system of fixed rules and formal behaviour used at official meetings, usually between governments 礼仪；外交礼节

He has become something of a stickler for the finer observances of royal protocol.

11. budget /ˈbʌdʒɪt/ *n.* the money that is available to a person or an organization and a plan of how it will be spent over a period of time 预算

The House of Representatives approved a new budget.

12. finance /ˈfaɪnæns/ *n.* the activity of managing money, especially by a government or commercial organization 财政；金融；财务

The accounts were certified (as) correct by the finance department.

13. take a look round 四处看看

It's an interesting place. Do you want to take a look round?

14. in store 准备着；将要发生

Those soldiers don't know what life holds in store for them.

The president would need all his strength and bravery to cope with what lay in store.

15. briefer /ˈbriːfə(r)/ *n.* an official who has the job of giving information about something, for example a war 信息通报官

Military briefers said no planes were shot down today.

【词汇点拨】Proper Names

1. CHOD (Chief of Defense) 国防部长

2. CIS (Communications and Information Systems) 通信与信息系统处

3. CIMIC (Civil-military Co-operation) 军民合作处

4. COS (Chief of Staff) 参谋长

5. DOS (Director of Staff) 参谋部主任

6. **SNR (Senior National Representative)** 多国部队各国最高军事代表

7. **Chief of Public Information** 新闻处处长

8. **Public Information Office** 新闻处

9. **Strasbourg** /ˈstræsbɜːg/ 斯特拉斯堡（位于法国东北部下莱茵省，欧盟委员会所在地）

10. **Belgium** /ˈbeldʒəm/ 比利时（西欧国家）

11. **Belgian** /ˈbeldʒən/ 比利时人；比利时（人）的

12. **Command Group** 指挥部

13. **NGO (non-governmental organization)** 非政府组织

14. **Colonel Reiter** 赖特上校

15. **Colonel Shapiro** 夏皮罗上校

16. **Lieutenant Colonel Esteban** 埃斯特班中校

17. **Major Algin** 阿尔根少校

18. **Captain Evans** 埃文斯上尉

【长难句解读】

1. It is commanded by the Chief of Staff, who is Belgian, with a Deputy Chief of Staff Operations, from Spain, and a French Deputy Chief of Staff Support and a Director of Staff – myself – (we are both French) all reporting to him. (11-1)

➤ 句子分析：该句中 be commanded by 意为"由……指挥"，who is Belgian 为定语从句，其先行词为 the Chief of Staff；with...all reporting to him 则是由 with 引导的复合结构，作伴随状语。

➤ 翻译：（欧洲军团）由比利时籍的参谋长指挥，另有一名来自西班牙的作战副参谋长和一名来自法国的支援副参谋长和一名参谋部主任——我本人——（我们都是法国人）都向他汇报工作。

2. G3 is in charge of Operations. It will have the task of writing the operation order to achieve the Commander's objective if Eurocorps is deployed. (11-2)

➤ 句子分析：该句中的 Operations 值得注意，首字母大写，表示作战处；operation order 意为"作战命令"。

➤ 翻译：G3 负责作战处。如果欧洲军团被部署，其任务就是编写作战命令，实现指挥官的目标。

3. This is the eyes and ears of the Commander, tasked with helping him lead his operation. (11-2)

> 句子分析：eyes and ears 原意为"眼和耳"，在该句中意为"耳目"，特指该部门对于指挥官的重要性。

> 翻译：该部门就是指挥官的千里眼、顺风耳，任务是协助指挥官领导行动。

4. This branch has to adapt the logistics support to each possible mission. (11-2)

> 句子分析：adapt...to 意为"调整，使适应"。

> 翻译：该部门必须根据每个可能的任务调整后勤支援。

5. I'm afraid today's tour does not include a visit to G5, which is the branch responsible for long-term planning and includes the sections for policy, OPS analysis, and international relations, or to G6, which is tasked with planning and organizing Communications and Information Systems.(11-2)

> 句子分析：该句中有两个 which 引导的非限制性定语从句，先行词分别是 G5 和 G6，用以说明两者的具体职责。

> 翻译：

恐怕今天的行程不包括 G5 和 G6。G5 是负责长期规划的部门，下设政策、行动分析和国际关系等子部门；G6 的任务是负责规划和组织通信和信息系统。

【口语输出】

1. What do you know about the organization of Eurocorps?

【参考词汇】restructure, traditional NATO army corps, is commanded by, report to

2. What's the mission of G1?

【参考词汇】responsible for, personnel, administrative aspects, protocol, ceremonies

3. What's the last stop on the tour? What's the mission of it?

【参考词汇】G9, CIMIC, establishing and maintaining the necessary contact with government and civilian agencies, NGOs, plan operations, the civilian population

4. How many briefers are there in this morning's briefing and which branch are they from respectively?

【参考词汇】four, G1, G4, G6, G9

5. What expressions have you learned from the listening materials when it comes to describing the tasks of your unit?

【参考词汇】be tasked with.../to do..., have the mission of.../to do..., be responsible for...

Bravo　The visit

【军事知识】

1. Protocol (Office) 礼宾处

It is an office that is responsible for the forms of ceremony and etiquette observed by diplomats and heads of state.

2. Vin d'honneur 简餐（酒水、点心、小食等）；小型酒会

This is a French word that translates to "Wine Reception", referring to a reception for a visitor with drinks and snacks. People usually stand up.

【词汇点拨】Words and Expressions

1. agenda /əˈdʒendə/ *n.* a list of items to be discussed at a meeting（会议的）议程表；议事日程

Can we move on to the next item on the agenda?

2. office call a visit to the Commander's office（在办公室）拜会；会见

The Protocol Chief is responsible for co-ordinating the office call with the Commander's Office.

3. welcome package /ˈpækɪdʒ/ helpful documents for visitors and new personnel 报到资料

The commander asked him to prepare a welcome package.

4. catering /ˈkeɪtərɪŋ/ *n.* the work of buying food and drink and cooking meals for people 饮食服务；餐饮供应

I wondered if you could do some catering for us next week.

The Chief asked the catering to prepare something for lunch in the Commander's Mess.

5. IAW an abbreviation which stands for "in accordance with" 依照；与……一致

This is also in accordance with international laws and international principles.

6. NLT an abbreviation which means "no later than" 不迟于

The elections promised for April were supposed to be held no later than July, 2023.

7. suspense /səˈspens/ *n.* (usually in the military) the date (and sometimes the hour) by which an activity must be done 最后期限

suspense date 最后期限

What is the suspense on completing the operation order?

8. tasker *n.* a term used in NATO to mean document detailing the individual(s)/department(s) tasked to carry out the duty/duties listed for a specific event and the suspense 任务清单

The sergeant was asked to write up the tasker immediately.

9. snack /snæk/ *n.* (informal) a small meal or amount of food, usually eaten in a hurry 点心；小吃；快餐

It is customary to offer a drink or a snack to guests.

10. focal point a thing or person that is the centre of interest or activity 集中点；焦点（指人或事物）

He quickly became the focal point for those who disagreed with government policy.

11. etiquette /ˈetɪket/ *n.* the formal rules of correct or polite behaviour in society or among members of a particular profession （社会或行业中的）礼节，礼仪；规矩

medical/legal/professional etiquette 医学界的/法律界的/行业规矩

In the code of military etiquette silence and fixity are forms of deference.

12. observe /əbˈzɜːv/ *v.* to obey rules, laws, etc. 遵守（规则、法律等）

Will the rebels observe the ceasefire?

13. diplomat /ˈdɪpləmæt/ *n.* a person whose job is to represent his or her country in a foreign country, for example, in an embassy 外交官

He started his official career as a diplomat.

14. head of state (pl. heads of state) the official leader of a country who is sometimes also the leader of the government 国家元首

The first African female head of state may just get elected for a second term.

15. detail /ˈdiːteɪl/ *v.* to give a list of facts or all the available information about sth. 详细列举；详细说明；详述

The report detailed the outbreak of the war.

16. departure /dɪˈpɑːtʃə(r)/ *n.* (~ from...) the act of leaving a place; an example of this 离开；起程

His sudden departure threw the office into chaos.

17. unclassified /ʌnˈklæsɪfaɪd/ *adj.* (of documents, information, etc.) not officially secret; available to everyone 非机密的；公开的

Many of the details surrounding last week's event are classified—but even the unclassified information confirms that something extraordinary occurred.

18. pass /pɑːs/ *n.* an official document that shows that you have the right to enter or leave a place 通行证

Without the pass, they cannot enter the barracks.

19. Commander's Mess a building in which commanders take their dinner 司令部餐厅；指挥官（首长）餐厅

They will treat the general in the Commander's Mess.

【词汇点拨】Proper Names

1. Polish /ˈpɒlɪʃ/ 波兰人；波兰语；波兰（人）的

2. Pablo /ˈpæbləʊ/ 巴勃罗（人名）

3. Lieutenant General Wojak 沃贾克中将

4. Madrid /məˈdrɪd/ 马德里（西班牙首都）

5. Exercise Mountain Fury "山怒"演习

6. Dieter /ˈdaɪətə(r)/ 迪特尔（人名）

【长难句解读】

1. Pablo's been the point of contact, so I'll ask him to bring us up to date. (11-4)

➢ 句子分析：该句中有两个短语，point of contact 意为"联络员"，bring sb. up to date 意为"将最新消息告诉某人"。

➢ 翻译：巴勃罗是联络人，所以我会请他向我们介绍最新情况。

2. His visit is quite short, so there's no point in including a lot of information about Madrid. (11-4)

➢ 句子分析：该句中 there is no point in doing... 为固定搭配，意为"做……是没有意义的"。

➢ 翻译：他的访问时间很短，所以没有必要包括很多关于马德里的资料。

3. I've made a note of all this and I'll have someone write up the tasker this afternoon. (11-4)

➢ 句子分析：该句中有两个需要注意的短语。make a note of 意为"把……记下来"，write up 意为"把……整理成文；详细写出（所做的事或所说的话）"。

➢ 翻译：我已经把这些都记下来了，今天下午我会叫人把任务清单整理出来。

【口语输出】

1. What can you learn at the very beginning of the conversation?

【参考词汇】visit, Polish CHOD, 11 November, Pablo, point of contact

2. What information is usually included in the tasker?

【参考词汇】tasked organization, situation, tasks, suspense

Charlie　Rearrangements

【军事知识】

1. Liaison officer　联络军官；联络官

A liaison officer or LO is a person that liaises between two organizations to communicate and coordinate their activities. Generally, they are used to achieve the best utilization of resources or employment of services of one organization by another. In the military, liaison officers may coordinate activities to protect units from collateral damage. They also work to achieve mutual understanding or unity of effort among disparate groups. For incident or disaster management, liaison officers serve as the primary contact for agencies responding to the situation. Liaison officers often provide technical or subject matter expertise of their parent organization. Usually an organization embeds liaison officers in other organizations to provide face-to-face coordination.

2. Defence Attaché Germany　德国国防武官

A defence attaché is a member of the military serving in an overseas embassy, representing their country's defence abroad.

The term "defence attaché" covers personnel from all branches of the military, and those in the role have diplomatic immunity and status.

The Defense Attaché Office in Germany traces its origin to the establishment of US Military Attachés in the 1880s. Excepting the period during and immediately after the two World Wars (1917-1921 and 1941-1954, respectively) when there was no US diplomatic presence in Germany, there has been a military attaché assigned to Germany since 1880.

The Defense Attaché is the primary military advisor to the Ambassador and Country Team on military issues and developments within Germany. Additionally he represents the Secretary of Defense, the Chairman of the Joint Chiefs of Staff, and greater Department of Defense (DoD) elements; plans and coordinates US military activities with the German Armed Forces throughout Germany (including coordination with DoD, Joint Staff, and USEUCOM); observes and reports on German military developments; oversees US military training programs (including Foreign Area Officer in-country training and the Personnel Exchange Programs); and supports DoD and other VIP visits.

【词汇点拨】Words and Expressions

1. confirm /kənˈfɜːm/ *v.* to state or show that sth. is definitely true or correct, especially by providing evidence（尤指提供证据来）证实；证明；确认

It has not been confirmed that the ceasefire agreement is to be signed.

2. accompany /əˈkʌmpəni/ *v.* to travel or go somewhere with sb. 陪同；陪伴

The White House will announce in coming months the delegates who will accompany Obama to London.

3. inconvenience /ˌɪnkənˈviːniəns/ *n.* trouble or problems, especially concerning what you need or would like yourself 不便；麻烦；困难

We apologize for the delay and regret any inconvenience it may have caused.

4. cancel /ˈkænsl/ *v.* to decide that sth. that has been arranged will not now take place 取消；撤销

The commander decided to cancel the briefing this afternoon.

5. flu /fluː/ *n.* an infectious disease like a very bad cold, that causes fever, pains and weakness 流行性感冒；流感

The soldiers were diagnosed as having flu.

6. slide /slaɪd/ *n.* a small piece of film held in a frame that can be shown on a screen when you shine a light through it 幻灯片

a slide show/projector 幻灯放映；幻灯机

This military briefing with slides is very successful.

7. electrician /ɪˌlekˈtrɪʃn/ *n.* a person whose job is to connect, repair, etc. electrical equipment 电工；电器技师

He worked as an electrician in this military academy.

8. maintenance /ˈmeɪntənəns/ *n.*

① the act of keeping sth. in good condition by checking or repairing it regularly 维护；保养

② a department with staff who are in charge of maintain sth. 维修部

The school pays for heating and the maintenance of the buildings.

9. installation /ˌɪnstəˈleɪʃn/ *n.*

① a piece of equipment or machinery that has been fixed in position so that it can be used 安装的设备（或机器）

② a place where specialist equipment is kept and used 设施

a heating installation 供暖装置

a military installation 军事设施

10. **lamp** /læmp/ *n.* a device that uses electricity, oil or gas to produce light 灯

a table/desk/bicycle, etc. lamp 台灯、自行车灯等

【词汇点拨】Proper Names

1. **Thomas Schneider** 托马斯·施耐德

2. **Captain Keller** 凯勒上尉

3. **General Heidemann** 海德曼上将

4. **Major Kurz** 库尔茨少校

5. **Berlin** /bəːˈlin/ 柏林（德国首都）

6. **Captain González** 冈萨雷斯上尉

7. **Corporal Watts** 下士沃茨

8. **Sergeant Christakis** 中士克里斯塔基斯

9. **Major Frutos** 弗鲁托斯少校

10. **Lieutenant Adler** 阿德勒中尉

11. **Lieutenant Colonel Devreux** 德弗罗中校

12. **Staff Sergeant Sanz** 上士桑兹

【长难句解读】

1. Er... Good morning. This is Thomas Schneider, the SNR for Germany. I'm phoning in connection with the visit to your base tomorrow. There has been a change of plan and instead of Captain Keller, I have to inform you that General Heidemann will now be accompanied by Major Kurz—that's Kurz–kilo–uniform–romeo–zulu–Major Kurz. I repeat: General Heidemann, the German Chief of Defence will be accompanied on his visit by Major Kurz, and not Captain Keller. Thank you. Goodbye. (11-5)

➢ **句子分析**：此部分内容为电话语音留言，请注意基本格式。先是打招呼（Good morning），然后自我介绍（This is Thomas Schneider, the SNR for Germany.），紧接着说明打电话的目的（I'm phoning in connection with the visit to your base tomorrow.），接下来就是具体内容（There has been a change of plan and instead of Captain Keller, I have to inform you that General Heidemann will now be accompanied by Major Kurz—that's Kurz–kilo–uniform–romeo–zulu–Major Kurz. I repeat: General Heidemann, the German Chief of Defence will be accompanied on his visit by Major Kurz, and not Captain Keller.），最后结

尾（Thank you. Goodbye.）。

> **翻译**：早上好。我是德国的军官代表团团长托马斯·施耐德。我打电话是有关明天参观你们基地的事宜。计划发生了变化，我得通知你们现在将由库尔茨少校而不是凯勒上尉陪同海德曼将军——是 Kurz–kilo–uniform–romeo–zulu– 库尔茨少校。我再重复一遍，德国国防部长海德曼将军将由库尔茨少校陪同访问，而不是凯勒上尉。谢谢，再见。

2. A: Support Group. How may I help you?

B: Hello, this is Captain González. Corporal Watts, please.

A: Corporal Watts speaking.

B: This is Captain González. (11-6)

> **句子分析**：该部分为英文电话对话，请注意格式。一般来说，接电话的人先说话，然后是打电话方告知对方来电何人。如果想找某人通话，接下来可以简单表明要找的通话对象：Can I speak to sb., please? 或如对话中直接说 Corporal Watts, please。一般来说，如果对方就是打电话方要找的人，就会听到简洁的回答：Speaking 或 sb. speaking（我就是，请讲）。

> **翻译**：A：这是支援组。我能为您做些什么？
> B：你好，我是冈萨雷斯上尉。我想请沃茨下士接电话。
> A：我是沃茨下士。
> B：我是冈萨雷斯上尉。

3. A: This is Staff Sergeant Sanz in the catering section. We have a small problem.

B: Sorry, the line is bad. (11-6)

> **句子分析**：该组对话中 the line is bad 为电话对话中常用表述，表明"线路信号不好"。类似表述有 the line is busy/engaged（电话占线）。

> **翻译**：A：我是餐饮部门的桑兹上士。我们遇到了一个小问题。
> B：对不起，线路信号不好。

4. A: I've spoken to maintenance, sir. They say they can't fix it till tomorrow.

B: How come they need 24 hours to change a light bulb?

A: It's not that simple, sir. It seems there's a big problem with the installation. (11-6)

> **句子分析**：该组对话中 How come... 为口语常用语，意为"怎么会，怎么会这样"，就是 why 的意思。但要注意 how come 的句型是：How come+Subject（主词）+Verb（动词）+Object（受词）？另 It's not that simple 中，that 为副词，修饰形容词，not that 意为 not very, or not as much as has been said（不很；不那么）。

> **翻译**：A：先生，我已经和维修部门说过了。他们说明天才能修好。

B：为什么他们需要24小时才能换一个灯泡？

A：没那么简单，先生。看起来安装设备有很大问题。

【口语输出】

1. What are the original arrangements of the visit? And what are the changes to the original arrangements?

【参考词汇】January 15th, ETA, 1035 hours, Captain Keller, lunch, Major Kurz, new ETA, 1445 hours, cancel

2. If you were Major Frutos, would you be angry and what would you do?

【参考词汇】a little, calm down, get ready immediately

3. What do you think of the solution? In terms of the problem, what other solutions can you think of?

【参考词汇】wonderful, special, vin d'honneur, playground

Delta General's visit

【军事知识】

1. Military Assistant 军事副手

In the British Armed Forces and many of those derived from them, it refers to an officer appointed to the personal office of a general officer. They are similar to aides-de-camp (French for field assistant) but generally have a more overtly administrative role. In the United States Armed Forces and the Canadian Forces, the equivalent position is that of Executive Assistant.

【词汇点拨】Words and Expressions

1. accommodation /əˌkɒməˈdeɪʃn/ *n.* somewhere to live or stay, often also providing food or other services 住宿；膳宿

The officer inspected the soldiers' accommodation.

2. subordinate /səˈbɔːdɪnət/ *n.* a person who has a position with less authority and power than sb. else in an organization 下级；部属

Sixty of his subordinate officers followed his example.

3. bow /baʊ/ *v.* ~ **(down) (to/before sb./sth.)** to move your head or the top half of your body forwards and downwards as a sign of respect or to say hello or goodbye 鞠躬；点头

He bowed low to the assembled soldiers.

4. embrace /ɪmˈbreɪs/ *v.* (formal) to put your arms around sb. as a sign of love or friendship 抱；拥抱

On hearing the news that the war was over, people were sort of crying for joy and embracing each other.

5. venue /ˈvenjuː/ *n.* a place where people meet for an organized event, for example a concert, sporting event or conference 地点；聚会地点（如音乐厅、体育比赛场馆、会场）

The venue for tomorrow's military briefing remains undecided.

6. luxury /ˈlʌkʃəri/ *n.* [U] the enjoyment of special and expensive things, particularly food and drink, clothes and surroundings 奢侈的享受；奢华

By all accounts that captain leads a life of considerable luxury.

【词汇点拨】Proper Names

1. Major Lowca 洛卡少校

2. Carlos /ˈkɑːlɒs/ 卡洛斯（人名）

3. Valencia /vəˈlenʃɪə/ 瓦伦西亚（位于西班牙瓦伦西亚省）

4. Lieutenant García 加西亚中尉

5. Spanish Civil Guard 西班牙国民警卫队

6. Lieutenant Colonel Okur 奥库中校

7. Turkish /ˈtɜːkɪʃ/ 土耳其人；土耳其语；土耳其（人）的

8. Barcelona /ˌbɑːsɪˈləʊnə/ 巴塞罗那（位于西班牙巴塞罗那省）

9. Poland /ˈpəʊlənd/ 波兰

10. Major Ortega 奥尔特加少校

11. The Hotel Palace 皇宫酒店

12. Warsaw /ˈwɔːsɔː/ 华沙（波兰首都）

【长难句解读】

1. D: Pleased to meet you, sir.

D: Sir. Gentlemen, if you would like to come this way. We have a car waiting outside. (11-7)

➢ 句子分析：该对话中，would like to do 意为"想要做某事"，表示意愿、喜爱，常用于有礼貌提出请求或者建议，语气较为委婉。

➢ 翻译：D: 见到您很高兴，先生。

D：先生，先生们，请这边走。我们的车在外面等着。

2. The airport is quite close to the city and traffic is quite light at this time of the day. (11-7)

> 句子分析：该句中的 light 一词为形容词，意为 not great in amount, degree, etc.（少量的；程度低的），也就是说交通畅通，不堵车。相应地，heavy traffic 是指交通拥挤。

> 翻译：机场离市区很近，每天这个时候交通都很畅通。

【口语输出】

1. How many speakers are there in the conversation? What is Major González tasked to do? And what about Lieutenant García?

【参考词汇】four, escort, hotel, Spanish Civil Guard, look after, transport

2. Should we follow some rules for introductions? What rules for introductions can you learn from the conversations?

【参考词汇】Yes, a younger person, an older person, in the military, a subordinate, a senior

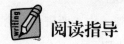 阅读指导

【文章导读】

This passage mainly focuses on the structure of a NATO headquarters and the job descriptions of the 11 typical divisions in a NATO headquarters. Moreover, the "G" division prefix is illustrated and we know that NATO divisions are often denoted with the prefix "G".

【词汇点拨】Words and Expressions

1. **equivalent** /ɪˈkwɪvələnt/ *n.* a person or thing that is equal to or corresponds with another in value, amount, function, etc.（在数量、功能、意义等方面）对等的人（或事物）

That branch can be headed by a major or the equivalent.

2. **manpower** /ˈmænpaʊə(r)/ *n.* the number of workers needed or available to do a particular job 人力；劳动力

He thinks that Israel itself has long been superior in weaponry and manpower.

3. **logistical** /ləˈdʒɪstɪkəl/ *adj.* of or relating to logistics 后勤上的

For logistical and political reasons, they have only recently been able to gain access to the area.

4. counterintelligence /ˈkaʊntərɪnˌtelɪdʒəns/ *n.* actions that a country takes in order to find out whether another country is spying on it and to prevent it from doing so 反情报

Within months he was being trained in counterintelligence.

Counterintelligence seeks to nullify the enemy's espionage efforts.

5. supervision /ˌsuːpəˈvɪʒ(ə)n/ *n.* the supervising of people, activities, or places 监督

The FCC should be put under strict supervision.

6. appropriate /əˈprəʊpriət/ *adj.* suitable, acceptable or correct for the particular circumstances 合适的；恰当的

In the circumstances, his plan for the operation looked highly appropriate.

7. formulate /ˈfɔːmjuleɪt/ *v.* to create or prepare sth. carefully, giving particular attention to the details 制订；规划；构想；准备

to formulate a policy/theory/plan/proposal 制订政策；创立理论；构想计划；准备建议

Little by little, he formulated his plan for defeating the enemy.

8. guideline /ˈgaɪdlaɪn/ *n.* [pl. guidelines] rules or instructions that are given by an official organization telling you how to do sth., especially sth. difficult 指导方针；指导原则；行动纲领；准则

Now it's your chance to take part in this first free massive online open course (MOOC) which delves into what the UN Guidelines look like in practice.

9. deployable /dɪˈplɔɪəbl/ *adj.* (of soldiers or equipment) able to be moved to a place where they can be used when they are needed 可部署的

Strategically deployable joint forces must be able to conduct operational and tactical maneuver.

10. restoration /ˌrestəˈreɪʃn/ *n.* the work of repairing and cleaning an old building, a painting, etc. so that its condition is as good as it originally was 整修；修复

His visit is expected to lead to the restoration of diplomatic relations.

11. accounting /əˈkaʊntɪŋ/ *n.* [U] the process or work of keeping financial accounts 会计

Luckily, they offered him a new position in the accounting department.

12. facility /fəˈsɪləti/ *n.* [pl. facilities] buildings, services, equipment, etc. that are provided for a particular purpose 设施；设备

sports/conference facilities 运动 / 会议设施

This network is widely used in military facilities, communications and sensor matrices.

13. day-to-day *adj.* [only before noun]

① planning for only one day at a time 按日计划的；逐日的；每天的

② involving the usual events or tasks of each day 日常工作的；例行的

I have organized the cleaning on a day-to-day basis, until our usual cleaner returns.

As adjutant he was responsible for his day-to-day programme.

14. billet /ˈbɪlɪt/

① *v.* [usually passive] to send soldiers to live somewhere temporarily, especially in private houses during a war 使（部队）临时设营（常在民宅里）

② *n.* a place, often in a private house, where soldiers live temporarily 部队临时营舍（常设在民宅里）

Many citizens were happy to billet the soldiers in their homes.

15. prefix /ˈpriːfɪks/ *n.*

① (grammer 语法) a letter or group of letters added to the beginning of a word to change its meaning, such as un-in unhappy and pre- in preheat 前缀（缀于单词前以改变其意义的字母或字母组合）

② a word, letter or number that is put before another 前置代号（置于前面的单词或字母、数字）

In the word "unimportant", "un-" is a prefix.

Car insurance policies have the prefix MC (for motor car).

16. denote /dɪˈnəʊt/ *v.* to mean sth. to represent 表示

The red triangle denotes danger.

17. symbolic /sɪmˈbɒlɪk/ *adj.* ~ **(of sth.)** containing symbols, or being used as a symbol 使用象征的；作为象征的；象征性的

The president's visit is loaded with symbolic significance.

Cyberspace has become symbolic of the computing devices, networks, fibre-optic cables, wireless links and other infrastructure that bring the internet to billions of people around the world.

【参考译文】

北约总部内部

　　无论是在冲突时期还是和平时期，北约总部都是一个繁忙的运作中心。在冲突期间，北约总部将根据北约政策和决策，规划并协调具体行动。它还将同民政当局和非政府组织联络，帮助提供人道主义援助并确保民用基础设施（例如道路和水）的安全。在和平时期，总部将计划训练演习，为其部队制定长期政策，并与民政当局和非政府组织协调，以确保在未来发生冲突时能够进行合作。

总部结构

北约总部由指挥部领导，指挥部包括总部指挥官（通常为四星上将）、副指挥官（通常为三星上将）、参谋长（二星上将）以及负责人员管理、规划、礼宾、公共信息、联络和法律顾问的办公室。

在指挥部之下，总部下设多个部门（分部）。这些部门通常由上校或同等级别的文职人员领导。以下是北约总部 11 个具有代表性的部门（分部）的工作描述：

（1）负责文职和军事人员的招聘、管理、薪资和退伍工作。协调人力需求，为参谋提供建议，并评估人员绩效报告。

（2）规划和协调重大后勤保障，包括运输、物资和医疗保障。在和平时期，协调后勤演习。

（3）管理通信和信息系统，包括电话、计算机和视频系统。

（4）负责情报和反情报工作，监督内部安全。收集和分析数据，并向相关部门报告相关信息。

（5）为所有以冲突为导向的行动提供指挥和控制人员，从和平支援行动到战斗。

（6）根据北约指导方针和任务协调行动规划，并为总部制定政策。

（7）计划和协调训练和演习，以确保总部指挥下的部队做好充分准备和部署。

（8）与民政当局和非政府组织联络。帮助规划人道主义援助，恢复民用基础设施，也处理平民文化问题。

（9）协调军事和市政工程保障。规划可能包括：建立运输路线，设置障碍以拖延敌人，清除雷区，恢复水电等重要服务设施。

（10）负责总部的会计和财务，包括管理训练、行政、差旅和设施维护的成本以及支付相关费用。

（11）负责总部的日常事务，包括餐饮、供应、设施维护、运输和车辆维护以及住宿。

"G"部门前缀

北约各部门通常用前缀"G"表示（如 G2 情报部门）。这个"G"用于象征性地指代由一名将军指挥的"地面"军事组织。每个 G 部门由一名上校或同等级别的军官领导。尽管象征性地使用了"G"，但为了确保联合规划能力，总部参谋人员还包括陆军、海军、空军和海军陆战队所有四个主要军事部门的成员。

(10) 预算与财务处	(8) 军民合作处	(3) 通信处
(9) 工程处	(11) 总部支援组	(4) 情报处
(2) 后勤处	(5) 作战处	(1) 人事处
(6) 计划政策处	(7) 训练与发展处	

词汇表

Alpha Joint headquarters
CHOD 国防部长
CIS 通信与信息系统处
CIMIC 军民合作处
COS 参谋长
DOS 参谋部主任
manpower 人力
SNR 多国部队各国最高军事代表
staff 参谋
discharge 退役；退伍
liaise （与某人）联络；联系
infrastructure 基础设施
principal 最重要的；主要的
spokesperson 发言人
executive （公司或机构的）主管领导；管理人员
deputy 副手；副职；代理
Eurocorps 欧洲军团
budget 预算
finance 财政；金融；财务
protocol 礼仪；外交礼节
briefer 信息通报官

Bravo The visit
catering 饮食服务；餐饮供应
Commander's Mess 司令部餐厅；指挥官（首长）餐厅
IAW 依照；与……一致
ILT 不迟于

office call （在办公室）拜会；会见
POC 满员；满编；全员
suspense 最后期限
tasker 任务清单
vin d'honneur 简餐（酒水、点心、小食等）；小型酒会
welcome package 报到资料
snack 点心；小吃；快餐
focal point 集中点；焦点（指人或事物）
etiquette （社会或行业中的）礼节，礼仪；规矩
observe 遵守（规则、法律等）
diplomat 外交官
head of state 国家元首
detail 详细列举；详细说明；详述
departure 离开；起程
unclassified 非机密的；公开的
security 保安
pass 通行证
agenda （会议的）议程表；议事日程

Charlie Rearrangements
confirm （尤指提供证据来）证实；证明；确认
accompany 陪同；陪伴
inconvenience 不便；麻烦；困难
cancel 取消；撤销
flu 流行性感冒；流感
slide 幻灯片

electrician 电工；电器技师
maintenance 维护；保养；维修部
installation 安装的设备（或机器）；维修部
lamp 灯

Delta General's visit

accommodation 住宿；膳宿
subordinate 下级；部属

bow 鞠躬；点头
embrace 抱；拥抱
venue 地点；聚会地点（如音乐厅、体育比赛场馆、会场）
luxury 奢侈的享受；奢华
escort 护送
present 正式介绍

 拓展学习

NATO Headquarters

NATO Headquarters is the political and administrative centre of the Alliance. It is located at Boulevard Leopold III in Brussels, Belgium. It offers a venue for representatives and experts from all member countries to consult on a continuous basis, a key part of the Alliance's consensual decision-making process, and to work with partner countries.

Highlights

◆ NATO Headquarters is the political and administrative centre of the Alliance.

◆ It is the permanent home of the North Atlantic Council—NATO's senior political decision-making body.

◆ It is also home to national delegations of member countries and to liaison offices or diplomatic missions of partner countries.

◆ The work of these delegations and missions is supported by NATO's International Staff and International Military Staff, also based at the Headquarters.

◆ The Headquarters hosts roughly 6,000 meetings every year.

◆ Initially based in London, the Headquarters was moved to Paris in 1952 before being transferred to Brussels, Belgium in 1967.

Role, responsibilities and people

NATO Headquarters is where representatives from all the member states come together to make decisions on a consensus basis. It also offers a venue for dialogue and cooperation between partner countries and NATO member countries, enabling them to work together in their efforts to bring about peace and stability.

Roughly 4,000 people work at NATO Headquarters on a full-time basis. Of these, some 2,000 are members of national delegations and supporting staff members of national military representatives to NATO. About 300 people work at the missions of NATO's partners countries. Some 1,000 are civilian members of the International Staff or NATO agencies located within the Headquarters and about 500 are members of the International Military Staff, which also includes civilians.

Working mechanism

With permanent delegations of NATO members and partners based at the Headquarters, there is ample opportunity for informal and formal consultation on a continuous basis, a key part of the Alliance's decision-making process.

Meetings at NATO Headquarters take place throughout the year, creating a setting for dialogue among member states. More than 5,000 meetings take place every year among NATO bodies, involving staff based at the Headquarters as well as scores of experts who travel to the site.

Evolution

In 1949, Allied countries established NATO's first Headquarters in London, the United Kingdom, at 13 Belgrave Square.

As NATO's structure developed and more space was needed, its Headquarters moved to central Paris in April 1952. At first it was temporarily housed at the Palais de Chaillot, but then moved to a purpose-built edifice at Porte Dauphine in 1960.

In 1966, however, France decided to withdraw from NATO's integrated military command structure, which called for another move—this time to Brussels. The new site in Belgium was constructed in a record time of six months and was inaugurated on 16 October 1967.

By 1999, NATO Heads of State and Government came to realise that, with NATO's enlargement and transformation, the facilities no longer met the requirements of the Alliance. They agreed to construct a new Headquarters situated across the road from the existing Headquarters, Boulevard Léopold III, Brussels. The construction of the building was finalised in 2017 and the move took place in 2018.

The design of the new building reflects the unity and adaptability of the Alliance. Its unity is manifest through the concept of interlocking fingers, while its adaptability is ensured by state-of-the-art facilities, allowing the building to adapt to the Alliance's evolving needs. It is also equipped for the 21st century with cutting—edge information and communications technologies. Furthermore, the building helps reduce NATO's environmental footprint by reducing energy consumption (geothermal heating in the winter and cooling in the summer), making full use of natural light via huge glass surfaces, and reducing water consumption (rain water collection via the sloped wings of the construction).

(Retrieved from https://www.nato.int/cps/en/natohq/topics_49284.htm)

【Notes】

1. **the Alliance** 联盟
2. **Boulevard Leopold III** 利奥波德三世大道
3. **Brussels** 布鲁塞尔（比利时首都）
4. **consult** 咨询；（与某人）商议，商量
5. **on a ... basis** 在……基础上
6. **consensual** 一致同意的
7. **highlight** 最重要 / 最精彩的部分
8. **North Atlantic Council** 北大西洋理事会
9. **delegation** 代表团
10. **NATO's International Staff** 北约国际参谋部
11. **International Military Staff** 国际军事参谋部
12. **be based at** 设在；建立在
13. **transfer**（使）转移，搬迁
14. **consensus** 一致的意见；共识
15. **ample** 足够的；充足的
16. **scores of** 许多；大量
17. **the United Kingdom** 英国
18. **Belgrave Square** 贝尔格雷夫广场
19. **be housed at** 入驻
20. **Palais de Chaillot**（法国）夏乐宫
21. **edifice** 大厦
22. **Porte Dauphine**（法国）皇太子妃门站
23. **inaugurate** 为……举行落成典礼
24. **enlargement** 扩大
25. **finalise** 使结束；使完结
26. **interlocking** 紧紧相扣的
27. **state-of-the-art**（技术上）最先进的

28. **cutting-edge** 尖端的

29. **geothermal heating** 地热采暖

30. **sloped** 倾斜的

 单元练习

I. Match the words and make terms

1. DOS	Director	National	Representative
2. IAW	in	later	staff
3. CHOD	Chief	of	with
4. SNR	Senior	Information	contact
5. COS	Public	accordance	staff
6. PIO	Chief	of	Office
7. NLT	no	of	defense
8. POC	point	of	than

II. Multiple choice

1. _____ is the principal advisor to the Commander.

 A. SNR   B. DOS   C. CHOD   D. COS

2. _____ is the department that is in charge of telephone, computer and video systems.

 A. POC   B. CIS   C. CIMIC   D. Logistics

3. Manpower and Personnel is responsible for recruitment, administration, payment and _____ of civilian and military personnel.

 A. dismiss   B. discipline   C. discharge   D. disagreement

4. _____ is another word for kitchen.

 A. Ceiling   B. Captain   C. Catering   D. Commander

5. The date by which a task must be completed is called _____.

 A. sustain   B. suspense   C. suspect   D. suspension

6. One of the tasks of the Protocol office is to prepare _____.

 A. welcome package   B. meals   C. a briefing   D. snacks

7. The highest authority in a country's armed forces is _____.

 A. HOM   B. Chief of public information

C. CHOD   D. Protocol chief

8. Usually, commanders will take their dinner in _____.

A. Protocol Office   B. office call

C. vin de'honneur   D. Commander's Mess

9. _____ is a term used in NATO to mean a document detailing the individual(s) / department(s) tasked to carry out the duty / duties listed for a specific event and the suspense.

A. Task   B. Tasker   C. Venue   D. Pass

10. The Chief of Public Information leads the Public Information Office and serves as the _____ of headquarters.

A. spokesperson   B. highest authority

C. personnel   D. civilian

III. Translation: English to Chinese

1. I also coordinate support to the Command Group and as executive for Chief of Staff, I am also responsible for all headquarters staff work.

2. This is the branch responsible for personnel and deals with all administrative aspects including personnel, protocol and ceremonies, and military visits—including this one!

3. G4 is responsible for organising, coordinating and assisting with logistics supplies within the EC, from planning to movement and transport, including materials, services and medical support.

4. G6 has the mission to plan and organise our communications and information systems. They are also responsible for coordinating security regulations for CIS systems.

5. And could you call catering and ask them if they can prepare something for lunch in the Commander's Mess?

IV. Translation: Chinese to English

北约是由北美和欧洲的多个国家根据1949年4月4日签署的条约组成的联盟。其根本目标是利用政治和军事手段维护其成员的自由和安全。多年来，北约在国际灾害管理和维持和平任务中发挥了越来越大的作用。

北约拥有一支由成员国军队组成的军事力量。虽然他们协同作战，但军队始终处于本国政府的控制之下。尽管北约成员国在政策方面存在分歧，但北约仍是有历史记载以来持续时间最长的防御同盟，多名国际分析人士认为，北约也是最成功的同盟。

V. Oral practice

1. Can you talk about the structure in a NATO headquarters?

2. What are the 11 typical divisions in a NATO headquarters? Can you talk about their jobs respectively?

附录：Unit 11 练习答案

I. Match the words and make terms

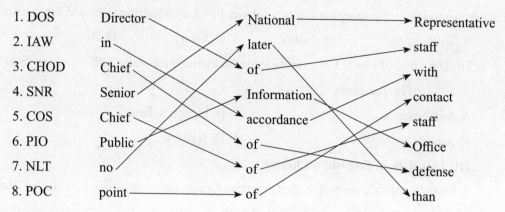

II. Multiple choice

1-5 DBCCB   6-10 ACDBA

III. Translation: English to Chinese

1. 我还协调对指挥部的支援，担任参谋长的执行官，我还负责所有司令部的参谋工作。

2. 这是负责人事的部门，负责所有行政事务，包括人事、礼宾和仪式，以及军事访问——包括这次！

3. 第四小组负责组织、协调和协助欧洲军团内的后勤供应，从规划到运输，包括物资、服务和医疗支援。

4. G6 的任务是规划和组织我们的通信和信息系统。他们还负责协调通信与信息系统的安全规定。

5. 你能不能给餐饮部门打个电话，问问他们能不能在指挥官餐厅准备点午餐？

IV. Translation: Chinese to English

NATO is an alliance of many countries from North America and Europe formed by a treaty signed on Apr. 4, 1949. Its fundamental goal is safeguarding its members' freedom and security using political and military means. Over the years, NATO has taken on an increasing role in international disaster management and peacekeeping missions.

NATO maintains a military force made up of member countries' troops. Although they work in concert, troops always remain under the control of their home nation's government.

Although NATO countries have their disagreements over its policies, NATO has lasted longer than any other defense alliance in recorded history, and several international analysts say it's the most successful alliance too.

V. Oral practice

1. A NATO headquarters is led by the Command Group, which will include the HQ Commander (typically a 4-star general), the Deputy Commander (typically a 3-star general), the Chief of Staff (a 2-star general), and offices for staff management, planning, protocol, public information, liaison and legal advisors. Under the Command Group, the headquarters is divided into several divisions, usually headed by a colonel or the civilian equivalent.

2. The 11 typical divisions in a NATO headquarters are:

G1: Personnel. It is responsible for recruitment, administration, payment and discharge of civilian and military personnel. Coordinating manpower requirements, advising staff, and evaluating personnel performance reports.

G2: Logistics. It is in charge of planning and coordinating major logistical support, including transportation, materials and medical support. During peacetime, coordinating logistical exercises.

G3: Communications. It is responsible for managing communications and information systems, including telephone, computer and video systems.

G4: Intelligence. It is tasked with intelligence and counterintelligence, and supervision of internal security. Collecting and analysing data, and reporting relevant information to appropriate divisions.

G5: Operations. Its mission is to provide command and controlling staff for all conflict-oriented operations, from peace support to combat.

G6: Plans and Policy. It is in charge of coordinating planning for operations in accordance with NATO guidelines and missions, and formulating policy for the headquarters.

G7: Training and Exercises. Its mission is to plan and coordinate training and exercises to make sure units under the headquarters' command are well prepared and deployable.

G8: Civil-military Co-operation. It liaises with civilian authorities and NGOs. Helping plan for humanitarian support and restoration of civilian infrastructure. Also dealing with civilian cultural issues.

G9: Engineer. It is responsible for coordinating military and civil engineer support. Planning may include: creation of transportation routes, creation of obstacles to delay enemy, clearing minefields, restoration of vital services like power and water.

G10: Budget, Contracting and Finance. It is tasked to provide accounting and finance support for the headquarters, including managing costs and payments for training, administration, travel and facility maintenance.

G11: Headquarters Support Group. It is tasked with providing support for the day-to-day needs of the headquarters, including catering, supply, facilities maintenance, transportation and vehicle maintenance, and billeting.

Unit 12　Carrier

单元导学清单

模块	主题	学习目标	任务	军事知识	核心词汇
Alpha	舰船基础	1) 了解航母的基本构造；2) 掌握航母构造相关的词汇和表达；3) 能够识别航母各部分并用英语表达。	Task 1 Task 2 Task 3 Task 7	■ Flight deck ■ Island ■ Hangar bay	hull, starboard, waterline, port, stern, island, hangar bay, flight deck, bridge, elevator
	舰载机起降	1) 了解舰载机起降过程；2) 描述舰载机起降过程及关键装置的作用。	Task 4 Task 5 Task 6	■ Electromagnetic aircraft launch system (EMALS) ■ Optical landing system	vessel, warship, launch, catapult, recover, arresting wire, tailhook
Bravo	航母发展历史及航母战斗群	1) 了解航母的发展历程、航母战斗群的编成及任务类型；2) 掌握与航母发展历程、航母战斗群的编成及任务类型相关的词汇与表达。	Task 1 Task 2 Task 3	■ Eugene Ely ■ HMS Argus ■ Carrier battle group (CBG)	sortie, flotilla, anti-submarine, anti-aircraft, anti-surface
	英国皇家海军"无敌"号航母的规格及任务	1) 了解皇家"无敌"号航母的规格及任务；2) 掌握与航母规格及任务相关的词汇与表达。	Task 4 Task 5 Task 6 Task 7	■ Invincible class aircraft carrier ■ Ski-jump	invincible, specification, displacement, cruise, amphibious, assault
Charlie	航母上的生活	1) 了解航母上的生活；2) 熟练掌握与航母生活相关的词汇与术语；3) 运用明喻和暗喻的修辞手法介绍航母上的生活。	Task 1 Task 2 Task 3 Task 4	■ Underway ■ Flight deck shirt colors	petty officer, berthing, rack, shore station, sea duty

续表

模块	主题	学习目标	任务	军事知识	核心词汇
Delta	海军指令	1）了解海军指令；2）熟练运用海军指令进行对话。	Task 1 Task 2	■ Naval Language and Slang of the royal navy ■ Nautical metaphors in the English language	gangway, helmsman, ten-hut, belay that last order
Echo (Review)	海军术语及主战舰艇	1）了解海军主战舰艇的相关知识；2）熟练掌握相关的词汇和表达；3）运用所学知识介绍海军主战舰艇的分类、特点和功能。	Task 1 Task 2	■ Fujian aircraft carrier ■ Nimitz class aircraft carriers ■ Type 23 frigate ■ Spruance class destroyer ■ Vanguard class submarine	terminology, initial, anchor, versatile, embargo, backbone, vanguard, prime minister

 听说指导

Alpha　Boats for beginners

【军事知识】

1. Flight deck　飞行甲板

The flight deck of the aircraft carrier is mainly used for the parking and landing of carrier-based aircraft. In order to be able to park as many aircraft as possible and have less interference during take-off and landing, aircraft carriers need to use a flight deck that is much larger than the width of the hull. Generally speaking, the width of the flight deck of an aircraft carrier is almost 1.5-2 times the width of the hull.

2. Island　舰岛

An aircraft carrier's "island" is the command center for flight-deck operations, as well as the ship as a whole. The island is about 150 feet (46 m) tall, but it's only 20 feet (6 m) wide at the base, so it won't take up too much space on the flight deck. The top of the island, well above the height of any aircraft on the flight deck, is spread out to provide more room.

3. Hangar bay　机库

The flight-deck crew can keep a small number of aircraft up top, but there's not nearly enough room for the 80 to 100 aircraft stationed on a typical carrier. When they're not in use, most of the aircraft are secured in the hangar bay, the "carrier's garage". The hangar bay is located two decks below the flight deck, just below the galley deck. The bay itself is 110 feet (~34 m) wide, 25 feet (~8 m) high and 685 feet (~209 m) long — more than two-thirds the length of the entire ship. It can hold more than 60 aircraft, as well as spare jet engines, fuel tanks and other heavy equipment, in four zones divided by sliding doors (a safety precaution to stop a fire from spreading). The hangar is three decks high, and it's flanked by various single-deck compartments on both sides. There are also four giant elevators surrounding the hangar, which move the aircraft from the hangar to the flight deck. The high-speed, aluminum hydraulic elevators are big enough and powerful enough to lift two 74,000 pound (~34,000 kg) fighter jets.

4. Electromagnetic aircraft launch system (EMALS) 电磁弹射系统

The electromagnetic aircraft launch system (EMALS) is a type of aircraft launching system developed by General Atomics for the United States Navy. The system launches carrier-based aircraft by means of a catapult employing a linear induction motor rather than the conventional steam piston. Its advantages include lower system weight, cost, and maintenance; the ability to launch both heavier and lighter aircraft than conventional systems; and lower requirements for fresh water, reducing the need for energy-intensive desalination.

5. Optical landing system 光学着陆系统

The Fresnel lens optical landing system provides guidance for correctly landing on an aircraft carrier. The lens is located on the side of the runway so that it can be seen by the pilots throughout the entire landing process.

The optical landing system consists of a horizontal bar of green lights and a vertical bar of red lights on both sides of the "meatball". The "meatball" is the centerpiece that consists of five amber colored lenses. Certain lenses will light up one at a time depending on the angle the plane is in relation to the "meatball". This causes the center light to appear to be moving up and down in relation to the horizontal green bars on the sides. In order to safely land, the pilot tries to keep the center amber lens horizontal with the green bar throughout process. If the pilot gets too low, the amber light will turn red indicating that the aircraft is dangerously low and risks hitting the back end of the aircraft carrier. The red lights around the green horizontal bars will be flashing if the carrier is not able to receive the aircraft, and so the jet must keep circling or find another place to land.

【词汇点拨】Words and Expressions

1. hull /hʌl/ *n.* the main, bottom part of a ship, that goes in the water 船身；船体

They climbed onto the upturned hull and waited to be rescued.

2. starboard /ˈstɑːrbərd/ *n.* the side of a ship or an aircraft that is on the right when you are facing forward（船舶或飞机的）右舷；右侧

The hull had suffered extensive damage to the starboard side.

3. waterline /ˈwɔːtərlaɪn/ *n.* the level that the water reaches along the side of a ship（船的）吃水线；水线

The ship was holed along the waterline by enemy fire.

4. vessel /ˈvesl/ *n.* a large ship or boat 大船；轮船

Naval vessel electric power system is small but complicated.

5. warship /ˈwɔːʃɪp/ *n.* a ship used in war 军舰；舰艇

The Navy is to launch a new warship today.

6. launch /lɔːntʃ/ *v.* to send sth. such as a spacecraft, weapon, etc. into space, into the sky or through water 发射；把（航天器、武器等）发射上天；水中发射

to launch a missile/rocket/torpedo 发射导弹/火箭/鱼雷

The countdown has begun for the launch of the space shuttle.

7. recover /rɪˈkʌvər/ *v.* 返回（作战飞机返回机场或航母）

Flight deck crewmembers prepare to recover aircraft aboard amphibious assault ship USS Makin Island.

8. secure /sɪˈkjʊr/ *v.* to protect sth. so that it is safe and difficult to attack or damage 保护；保卫；使安全

The windows were secured with locks and bars. 窗户已经插上栓，上了锁，都关好了。

9. manoeuvre /məˈnuːvər/

① *n.* a movement performed with care and skill 细致巧妙的移动；机动动作

② *n.* military exercises involving a large number of soldiers, ships, etc. 军事演习；作战演习

You will be asked to perform some standard manoeuvres during your driving test.

The army is on manoeuvres in the desert.

10. catapult /ˈkætəpʌlt/ *n.* a machine used for sending planes up into the air from a ship 弹射器

On that type of aircraft carrier a catapult was used to help launch aircraft.

11. reactor /riˈæktər/ *n.* a large structure used for the controlled production of nuclear energy 核反应堆

The nuclear reactor was not damaged in the lightning storm that struck late last night.

12. accelerate / əkˈseləreɪt / *v.* to happen or to make sth. happen faster or earlier than expected 使加速，加快

Exposure to the sun can accelerate the ageing process.

13. naught /nɔːt/

① *n.* the figure 0 零

② *pron.* nothing 无；化为乌有

come to naught 成为泡影

My efforts came to naught.

14. cable / ˈkeɪbl / *n.* A cable is a kind of very strong, thick rope, made of wires twisted together. 缆绳

The miners rode a conveyance attached to a cable made of braided steel wire.

15. arresting wire a wire cable that is stretched across an airfield runway or the flight deck of an aircraft carrier and that can be engaged by an aircraft's arrester hook during landing in order to halt the forward motion of the aircraft within a limited space 阻拦绳

The ship is equipped with arresting wires which are capable of safely landing aircraft traveling at speeds of more than 200 miles per hour in about 300 feet.

【长难句解读】

1. An aircraft carrier is a type of warship fitted with a runway or flight deck which is used to launch and recover planes. (12-3)

➤ 句子分析：该句中 fitted with 意为"装有……的"，launch and recover planes 指舰载机起降。

➤ 翻译：航空母舰（简称航母）是一种装有跑道或者飞行甲板的战舰，供飞机起飞和降落。

2. Below the waterline, the hull is rounded and narrow but above the waterline the hull gets wider and forms the flight deck. (12-3)

➤ 句子分析：该句中 rounded 意为"圆形的"，比如 a surface with rounded edges 翻译成"带圆边的面"。

➤ 翻译：吃水线以下，船体圆而窄；吃水线以上，船体逐渐变宽，形成飞行甲板。

3. To prepare for launching an aircraft, the carrier sails into the wind. (12-3)

> 句子分析：该句中 sails into the wind 意为"航母逆风航行"。

> 翻译：为弹射起飞做准备，航母要逆风航行。

4. This manoeuvre reduces the plane's minimum take off speed by getting the wind moving over the flight deck and over the plane's wings. (12-3)

> 句子分析：该句中 manoeuvre 意为"细致巧妙的移动、机动动作"，指前句中提到的航母逆风航行。另外，manoeuvre 也有军事演习之意。

> 翻译：这一机动可使风掠过飞行甲板和机翼，从而减少飞机的最低起飞速度。

5. Thick metal cables, or arresting wires are stretched across the deck and the pilot's aim when he lands is to catch the cable with the tailhook – a long hook attached to the plane's tail. (12-3)

> 句子分析：该句中的 arresting wires 是指辅助舰载机着舰的阻拦索，tailhook 是指尾钩。

> 翻译：甲板上布设几根很粗的金属缆绳或者拦阻索，飞机着舰时，飞行员要让飞机尾部的长钩即尾钩钩住缆索。

【口语输出】

1. Suppose you are a Navy offier, can you introduce the different parts of a carrier to a group of foreign visiters to Liaoning?

【参考词汇】bow, stern, port, starboard, hull

2. What is the major function of a carrier?

【参考词汇】platform for operations, launch and recover the carrier-based aircraft

3. Can you introduce the launch and recovery process of carrier-based aircraft?

【参考词汇】flight deck, catapult, arresting wire, tailhook

4. What do you think makes a qualified carrier fighter pilot?

【参考词汇】perfect physical fitness, strong mentality and adaptability, flexibility, emotional stability, hard training

Bravo City at sea

【军事知识】

1. Eugene Ely 尤金·伊利

Eugene Ely's daring bravery and love of flying lead him to pioneer the world of flight from a ship. Born in Williamsburg, Iowa in 1886, Eugene became one of the premiere pilots during the early days of flying. Self-trained as a race car driver and mechanic who worked in the Davenport area, Eugene loved speed and anything mechanical.

In 1910 he was approached by the United States Navy about the challenge of flying an airplane off a ship. On November 14, 1910, Eugene Ely became the first pilot to make a successful unassisted airplane takeoff from the deck of a ship. Despite damaging the propeller in the takeoff he kept the airplane airborne for two and a half miles. He made history again on January 18, 1911, when he successfully landed an airplane on a ship, the U.S.S. Pennsylvania. He is also credited with designing the arresting gear that helped stop the airplane upon landing. This principle is still used today.

2. HMS Argus "百眼巨人"号航母

HMS Argus was a British aircraft carrier that served in the Royal Navy from 1918 to 1944. She was converted from an ocean liner that was under construction when the First World War began, and became the first example of what is now the standard pattern of aircraft carrier, with a full-length flight deck that allowed wheeled aircraft to take off and land. After commissioning, the ship was heavily involved for several years in the development of the optimum design for other aircraft carriers. Argus also evaluated various types of arresting gear, general procedures needed to operate a number of aircraft in concert, and fleet tactics.

3. Carrier battle group (CBG)

An aircraft carrier is extremely valuable. And without protection, an aircraft carrier is extremely vulnerable. That's why aircraft carriers never leave home alone. They are always escorted by an extensive flotilla of other ships. The aircraft carrier plus the flotilla is known as the carrier battle group. A modern carrier battle group is nearly invincible. The US Navy forms carrier battle groups on an as-needed basis and assigns ships to the group based on the mission. Therefore, no two carrier battle groups are the same. However, a typical carrier battle group consists of the following ships: the aircraft carrier itself, two guided-missile cruisers, two destroyers, one frigate, two submarines and a supply ship.

4. Invincible class aircraft carriers

The Invincible class is a class of light aircraft carrier operated by the Royal Navy. Three ships were constructed, HMS Invincible, HMS Illustrious and HMS Ark Royal. The vessels were built as aviation-capable anti-submarine warfare (ASW) platforms to counter the Cold War North Atlantic Soviet submarine threat, and initially embarked Sea Harrier aircraft and Sea King HAS.1 anti-submarine helicopters. With the cancellation of CVA-01, the three ships became the replacements for the Audacious and Centaur classes, and the Royal Navy's sole class of aircraft carrier.

5. Ski-jump 滑跃起飞（甲板）

A ski-jump is a curved ramp that allows aircraft to take off from a runway that is shorter than the aircraft's required takeoff roll. By launching the aircraft at a slight upward angle, the ski-jump gives the aircraft additional time to build up velocity and reach stable flight. ski-jumps are commonly used to launch airplanes from aircraft carriers that lack aircraft catapults.

【词汇点拨】Words and Expressions

1. sortie /ˈsɔːrti/ *n.* a flight that is made by an aircraft during military operations（在军事行动中飞机的）出动架次

In the execution of every major task, the division commander will always fly the first sortie and fire the first shot of live ammunition.

2. flotilla /floʊˈtɪlə/ *n.* a group of boats or small ships sailing together 船队；小型舰队

Greenpeace managed to assemble a small flotilla of inflatable boats to waylay the ship at sea.

3. operation /ˌɑːpəˈreɪʃ(ə)n/ *n.* military activity 军事行动

He was the officer in charge of operations.

4. anti-aircraft /ˌænti ˈerkræft/ *adj.* designed to destroy enemy aircraft 防空的

anti-aircraft fire/guns/missiles 防空火力；高射炮；防空导弹

Anti-aircraft guns opened up.

5. anti-submarine /ˈænti ˌsʌbməˈriːn/ *adj.* (of weapons, missiles, etc.) designed to combat or destroy submarines（兵器、导弹等）对抗潜艇的

anti-submarine warfare 反潜作战

The Navy is rebuilding its anti-submarine capability.

6. invincible /ɪnˈvɪnsəb(ə)l/ *adj.* too strong to be defeated or changed 不可战胜的

You couldn't help feeling the military's fire power was invincible.

7. specification /ˌspesɪfɪˈkeɪʃn/ *n.* a detailed description of how sth. is, or should be, designed or made 规格

the technical specifications of the new model 新型号的技术规格

The house has been built exactly to our specifications.

8. displacement /dɪsˈpleɪsmənt/ *n.* the amount of a liquid moved out of place by sth. floating or put in it, especially a ship floating in water 排水量

a ship with a displacement of 10,000 tonnes 排水量为1万吨的轮船

9. tonne /tʌn/ *n.* (also ˌmetric ˈton) a unit for measuring weight, equal to 1,000 kilograms 公吨（等于1000千克）

The study showed that a single new desktop machine created emissions of almost half a tonne during its manufacture.

10. cruise /kruːz/ *v.* (of a car, plane, etc.) to travel at a steady speed 汽车、飞机等以平稳的速度行驶

a light aircraft cruising at 4,000 feet 一架在4000英尺高度巡航的轻型飞机

a cruising speed of 50 miles an hour 每小时50英里平稳行驶的速度

11. mount /maʊnt/

① *v.* to fix something to a wall, in a frame, etc., so that it can be looked at or used 安装；挂载

② *n.* something, such as a piece of card, that you put something on to show it 底板；托板

The surveillance camera is mounted above the main door.

A black mount for this picture would look good.

12. amphibious /æmˈfɪbiəs/ *adj.* (of military operations) involving soldiers landing at a place from the sea（军事行动）两栖作战的

A third brigade is at sea, ready for an amphibious assault.

13. assault /əˈsɔːlt/ *n.* ~ (on/upon/against sb./sth.)(by an army, etc.) the act of attacking a building, an area, etc. in order to take control of it 攻击；突击；袭击

The army renewed its assault on the capital.

【词汇点拨】Proper Names

1. **Orville Wright** 奥维尔·莱特
2. **World War I** 第一次世界大战（1914—1918，帝国主义国家为瓜分世界、争夺殖民地和霸权而进行的首次世界规模大战）
3. **World War II** 第二次世界大战（1939—1945，又称世界反法西斯战争）
4. **the Royal Navy** 皇家海军
5. **Sea King helicopter** 海王直升机
6. **Search and Rescue** 搜救
7. **Merlin** /ˈmɜːrlɪn/ 梅林直升机
8. **Chinook support helicopter** 奇努克支援直升机
9. **Sea Harrier FA2** 海鹞 FA2 攻击机
10. **RAF Ground Harrier GR7** 鹞式 GR7 攻击机

11. Goalkeeper Weapon System 守门员武器系统

12. GPMG 通用机枪（General Purpose Machine Gun）

【长难句解读】

1. HMS Invincible is one of a family of aircraft carriers in service with the Royal Navy. (12-5)

> **句子分析**：HMS Invincible 这里指"无敌"号，是英国皇家海军无敌级航母的首舰；a family of 在这里指英国航母大家庭；in service 意思为服役。

> **翻译**："无敌"号是皇家海军现役的航母之一。

2. The aircraft are kept in hangars down below deck and lifted to the surface when they are ready to be deployed, which may be at any time, since sorties are flown at any time of the day or night. (12-6)

> **句子分析**：该句中 hangars 是指 hangar bay；which 引导定语从句，指代前面整句"飞机存放于飞行甲板下的机库里，准备使用时提升到甲板上"。

> **翻译**：飞机存放于飞行甲板下的机库里，准备使用时提升到甲板上，任务随时会有，不论白天或晚上，飞机随时出动。

3. For defence against attack by aircraft or antisurface missiles, HMS Invincible is fitted with the 30mm Goalkeeper Weapon System—which is a 7-barrelled cannon that can fire at a rate of 4,200 rounds per minute. (12-6)

> **句子分析**：前半句中的 by aircraft or antisurface missiles 为后置定语，修饰前面的 attack，译为"飞机或反舰导弹的攻击"；后半句中的 at a rate of 表示"以……速度"。

> **翻译**：为了防御飞机或反舰导弹的攻击，"无敌"号装备了30毫米的守门员武器系统，一门7管速射炮，每分钟射速4200发。

4. The ship is also equipped with two 20mm close range guns, which can be aimed and operated manually, and there are mounts for the Minigun and GPMG, which have been added to help guard against terrorist attack. (12-6)

> **句子分析**：该句中的短语 be equipped with 表示"配备"；manual 表示"可以手动操作的"；mount 此处是指"底座"。

> **翻译**：航母还配备了两挺20毫米近程炮，可手动瞄准操作，还有几个底座，加装微型炮和通用机枪，用来对付恐怖袭击。

5. Finally, our last slide, HMS Invincible must be ready to deploy anywhere in the world at very short notice, so the ability to play a number of different roles is very important. (12-6)

> 句子分析：该句中的 HMS 全称为 Her/His Majesty's Ship，皇家海军舰艇；at very short notice 意思为"在很短的时间内"。

> 翻译：这是最后一张幻灯片，"无敌"号必须保持待命状态，随时在很短的时间内部署到世界任何地方，所以一舰多能非常重要。

【口语输出】

1. What missions can a carrier battle group carry out?

【参考词汇】overseas deployment, training, humanitarian assistance, diplomatic exchanges, combat

2. What are the missions of HMS Invincible aircraft carrier?

【参考词汇】anti-submarine, anti-surface, anti-aircraft, amphibious assault ship

3. Can you make a comparison between the specifications of China's Fu Jian aircraft carrier and USS Ford?

【参考词汇】has a maximum beam, the same length as, is manned by a crew of

4. What are the members of the flight deck responsible for?

【参考词汇】aircraft handling, catapult and arresting gear, ground support equipment, crash and salvage

5. Why were mounts for Minigun and GPMG added to the ship's armament?

【参考词汇】guard against terrorist attack

Charlie Life onboard

【军事知识】

1. Underway

Underway, or Under Way is a nautical term describing the state of a vessel. "Way" arises when there is sufficient water flow past the rudder of a vessel that it can be steered. A vessel is said to be underway if it meets the following criteria:

It is not aground.

It is not at anchor.

It is not drifting.

It has not been made fast to a dock, the shore, or other stationary object.

If a vessel is adrift and not being propelled by any instrument or device, it is said to be underway, not making way. The concept of whether a vessel is or is not underway has important legal ramifications. For example, in many jurisdictions a child must be wearing a personal flotation device at the time the vessel is underway.

2. Flight deck shirt colors

The U.S. Navy super-carriers, for example, support more than 5,500 military personnel.

While there are many different job duties aboard a Navy aircraft carrier the primary purpose remains the same — to put aircraft into the air and safely recover the aircraft. Navy aircraft carrier personnel are distinguished by the color of their uniform.

Everyone that serves on the flight deck has a specific job:

Jersey Color	Crew Responsibilities
Yellow	- Aircraft Handling Officer - Catapult and Arresting Gear Officer - Plane Director
Green	- Catapult and Arresting Gear Crew - Visual Landing Aid Electrician - Air Wing Maintenance Personnel - Air Wing Quality Control Personnel - Cargo-Handling Personnel - Ground Support Equipment (GSE) Troubleshooter - Hook Runner - Photographer's Mate - Helicopter Landing Signal (LSE)
Red	- Ordnance Handler - Crash and Salvage Crew - Explosive Ordnance Disposal (EOD) - Firefighter and Damage Control Party
Purple	- Aviation Fuel Handler
Blue	- Trainee Plane Handler - Chocks and Chains (Entry Level) - Aircraft Elevator Operator - Tractor Driver Messengers and Phone Talker
Brown	- Air Wing Plane Captain - Air Wing Line Leading Petty Officer
White	- Quality Assurance (QA) - Squadron Plane Inspector - Landing Signal Officer (LSO) - Air Transfer Officer (ATO) - Liquid Oxygen (LOX) Crew - Safety Observer - Medical Personnel

【词汇点拨】Words and Expressions

1. petty officer a rank of non-commissioned officer in the navy, above leading seaman or seaman and below chief petty officer 海军军士

He was promoted to the equivalent of a petty officer after the rules were changed.

2. berthing /ˈbɜːrθɪŋ/ *n.* sleeping quarters on board a ship 铺位

3. rack /ræk/ *n.* a navy word for bed（海军人员）床

4. sea duty 出海执勤

After completing an internship and residency at the old South Baltimore General Hospital, now MedStar Harbor Hospital, he was assigned to active sea duty serving first on the battleship USS Texas.

【词汇点拨】Proper Names

1. Joe /dʒoʊ/ 乔（人名）

2. Paul /pɔːl/ 保罗（人名）

3. John /dʒɑːn/ 约翰（人名）

4. Evans /ˈevənz/ 埃文斯（人名）

5. Thailand /ˈtaɪlænd/ 泰国

6. Spain /speɪn/ 西班牙

7. Australia /ɔːˈstreɪliə/ 澳大利亚

8. Argentina /ˌɑːrdʒənˈtiːnə/ 阿根廷

9. Tom Cruise 汤姆·克鲁斯（美国男演员）

10. *Top Gun*《壮志凌云》（电影名）

【长句解读】

1. And in bad weather, being on a ship is like going to an amusement park, drinking five cups of coffee and riding a roller coaster non-stop. (12-7)

➤ 句子分析：该句使用了明喻（simile）的修辞手法，意在形象地说明在坏天气下，待在舰上的感受；non-stop 意思是永不停歇的。

➤ 翻译：遇上坏天气，待在舰上就像去游乐园，先喝上五杯咖啡，再不停地坐过山车。

2. At least on a carrier you get your own rack, but on submarines there aren't even enough beds for every member of the crew, and three sailors hot rack—that means one sailor is on duty and one is resting and then they change round. (12-7)

➤ 句子分析：该句对航母和潜艇进行对比，说明潜艇的空间小。其中，three sailors hot rack 是指三人轮流休息。

➢ 翻译：在航母上，你起码还有自己的铺位，要是在潜艇上，床位不够每个艇员一张，要三个艇员轮流休息，也就是说一个艇员值班，一个睡觉，然后轮流。

3. Onboard, you have to take what's called a navy shower. You get in the shower, turn on the water and then turn off the water. (12-7)

➢ 句子分析：第一句中的 navy shower 可以理解为海军所特有的洗澡方式，其特点是快；第二句是对 navy shower 的解释，说明舰上用水十分紧张。

➢ 翻译：在舰上，你不得不洗所谓的海军澡。你进去洗澡，打开水，马上就关掉。

【口语输出】

1. What do you know about watches at sea?

【参考词汇】shifts, stare at sea, rest, boring, responsibility

2. Why do you think privacy is an issue on board?

【参考词汇】floating city, crammed, limited space

3. What do you think life is like on a carrier?

【参考词汇】limited space and water, difficult, exhausting, opportunities, busy, exciting

Delta Navy commands

【军事知识】

1. Naval Language and Slang of the Royal Navy 皇家海军用语和俚语

The British Royal Navy has a language or slang all of its own which reflects both its long history and also the culture (both good and bad) of the seafarer.

Naval Expressions in Everyday Use

Many expressions we regularly use today, on dry land, originate from life on board ship in Nelson's day.

Take the expression "long shot" meaning attempting something with little chance of success. This originated from firing a cannon beyond its normal range.

What about "at loggerheads"? Loggerheads were hollow spheres of iron at each end of a shaft. They were heated and used to melt tar in a bucket. The expression arose because the two loggerheads can never come together.

"Swinging the lead" relates to a sailor dropping a lead weight on a line over the side of the ship in order to measure the sea depth. Sailors found this to be a handy method of avoiding real work.

On a more culinary note, "chew the fat" relates to the need for heavy mastication in order to break down the tough rind of beef that was stored in a barrel of brine for months on end.

The expression "all above board" refers to things on the top deck of the ship and therefore open to inspection.

2. Nautical metaphors in the English language 英语中的航海隐喻

Great Britain is an island nation; in the days before air travel, mastery of the sea was essential to the nation's defence and trade. In modern times ships play a less important role, and they tend to be powered by engines rather than sails. Yet many expressions derived from sailing remain embedded in the English language. Knowing this may shed light on some apparently obscure terms.

A flagship, for example, was the most important ship in a fleet, which carried the fleet's admiral and flew his flag. In modern English, however, the word is more likely to be used as a metaphor, so a company's flagship store is the one that has the most importance and prestige. A mainstay was originally a rope that supported the main mast of a ship, but now is a metaphor referring to any person or thing that provides crucial support, as in tourism is a mainstay of the economy.

The influence of sailing can also be seen in some idiomatic phrases. "To sail close to the wind" refers to the risky practice of attempting to fill a ship's sails with wind without losing control of it. This phrase is now used as an idiom: if you tell someone that they are sailing close to the wind you are warning them that they are doing something that is dangerous or possibly illegal. "To batten down the hatches" literally refers to closing the entrances to the lower part of a ship when a storm is expected, but metaphorically refers to any preparation to withstand a period of difficulty. If a ship has run aground and is unable to return to the water, it is said to be "high and dry", an expression we also use to refer to a person who is left in a difficult situation without any assistance.

Some similar phrases have now lost all their original associations with sailing. It may come as a surprise to learn that under way, meaning "in progress", was originally a nautical phrase meaning "in motion". Another example is "by and large": to the old sailors, this meant "in all conditions", whether sailing into the wind (sailing by) or with the wind (sailing large), but it is doubtful whether many current English speakers are aware of this when they use the phrase to mean "in general".

【词汇点拨】Words and Expressions

helmsman /ˈhelmzmən/ *n.* a person who steers a boat or ship 舵手
The old helmsman brought us about and we avoided a dangerous dash against the rocks.

【词汇点拨】Proper Names

1. Jones /dʒəʊnz/ 琼斯（人名）

2. Munro /mənˈroʊ/ 芒罗（人名）

3. Kosovo /ˈkɒsəvəʊ/ 科索沃（塞尔维亚自治省）

4. Gibraltar /dʒɪˈbrɔːltər/ 直布罗陀（英国海外领地，英国海军1704年在此击败西班牙和法国联军，直布罗陀落入英国之手）

5. Trafalgar /trəˈfælgər/ 特拉法加（位于西班牙，1805年英法两国海军在此交战，故称特拉法加海战）

6. Gallipoli /gəˈlɪpəli/ 加里波利半岛（位于土耳其，1915—1916年，英法盟军在此打败奥斯曼帝国，占领其首都伊斯坦布尔）

7. Bosnia /ˈbɑːzniə/ 波斯尼亚（位于原南斯拉夫中西部，现为波斯尼亚和黑塞哥维那北部地区）

8. Middle East 中东（从地中海东部、南部到波斯湾沿岸的部分地区，又称中东地区）

9. Northern Europe 北欧

【长难句解读】

1. A: Gangway, hot food.

B: Hey, Chief. What's for dinner?

A: Tuna and pasta.

B: Again! (12-8)

➢ 句子分析：该对话中 gangway 意思为"让路"；again 表示不想再吃同样的食物了。

➢ 翻译：A：让道让道，来热饭了。

B：你好大厨，是什么好吃的？

A：金枪鱼和意大利面。

B：又是金枪鱼和意大利面！

2. A: Officer on deck, ten-hut.

B: Good evening, gentlemen. As you were. Is Seaman Jones here?

A: Yes, sir. He's over there.

B: Seaman Jones?

C: Yes, sir.

B: Report to Chief Munro in the galley.

C: Aye, aye sir.

➢ 句子分析：该对话中 ten-hut 为"立正"之意；as you were 在这里表示军官要求士兵保持原来的姿势；aye 表示"是"。

➢ 翻译：A：长官到，立正。

B：先生们晚上好，该忙啥忙啥。琼斯在这儿吗？

A：在，长官，他在那儿。

B：琼斯？

C：是，长官。

B：去厨房找门罗班长报到。

C：是，长官。

【口语输出】

1. Do you know other naval language? Search the Internet and share with other students.

【参考词汇】all hands—the entire ship's company, toe the line—meaning to conform to rules and authority, copper bottomed—something worthwhile

2. What are the two speakers talking about in the third conversation?

【参考词汇】course, order, change

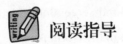 阅读指导

【文章导读】

This passage mainly introduces some naval terminology and vessels. After explaining the basic naval terminology, it elaborates the features and functions of the main naval vessels, including surface vessels and submarines. Finally, this passage makes a brief introduction of the the Royal Marines, including its history and missions.

【军事知识】

1. Fujian aircraft carrier "福建"号航母

Upon its completion, the gigantic ship will displace more than 80,000 metric tons of water, making it the largest and mightiest warship any Asian nation has ever built and also one of the world's biggest naval vessels of all time.

According to the People's Liberation Army Navy, the ship will use electromagnetic launch system, or electromagnetic catapult, to launch fixed-wing aircraft, which will give the carrier a much greater combat capability than its two predecessors that use a ramp to launch jets.

2. Nimitz class aircraft carriers "尼米兹"级航母

The Nimitz class aircraft carriers were the largest warships ever built until the commissioning of USS Gerald R Ford in 2017. With over 6,000 personnel (crew and aircrew), the carrier has a displacement of 102,000t and a flight deck length of 332.9m.

All ten nuclear-powered Nimitz-class carriers were built by Huntington Ingalls Industries Newport News Shipbuilding (now Northrop Grumman Ship Systems) based in Virginia.

Tasked with a multi-mission attack/ASW role, the first of the class, USS Nimitz, was commissioned in 1975. The last of the class, USS George H.W. Bush (CVN 77), was commissioned in January 2009.

3. Type 23 frigate "23型"护卫舰

The Type 23 frigate or Duke class is a class of frigate built for the United Kingdom's Royal Navy. The ships are named after British Dukes, thus leading to the class being commonly known as the Duke class. The first Type 23 was commissioned in 1989, and the sixteenth, HMS St. Albans was commissioned in June 2002. They form the core of the Royal Navy's destroyer and frigate fleet and serve alongside the Type 45 destroyers. Originally designed for anti-submarine warfare in the North Atlantic, the Royal Navy's Type 23 frigates have proven their versatility in warfighting, peace-keeping and maritime security operations across the globe. Thirteen Type 23 frigates remain in service with the Royal Navy, with three vessels having been sold to Chile and handed over to the Chilean Navy.

4. Spruance class destroyer "斯普鲁恩斯"级驱逐舰

The Spruance class destroyer was developed by the United States to replace a large number of World War II-built Allen M. Sumner and Gearing class destroyers, and was the primary destroyer built for the U.S. Navy during the 1970s.

First commissioned in 1975, the class was designed with gas-turbine propulsion, all-digital weapons systems, automated 5-inch guns. Serving for three decades, the Spruance class was designed to escort a carrier group with a primary ASW mission, though in the 1990s 24 members of the class were upgraded with the Mark 41 Vertical Launching System (VLS) for the Tomahawk surface-to-surface missile. Rather than extending the life of the class, the navy accelerated its retirement. The last ship of the class was decommissioned in 2005, with most examples broken up or destroyed as targets.

5. Vanguard class submarine "前卫"级核潜艇

The Vanguard class are a class of nuclear-powered ballistic missile submarines (SSBN) in service with the Royal Navy. Each submarine is armed with up to 16 Trident II missiles. The class was introduced in 1994 as part of the UK government's Trident nuclear weapons programme. The class includes four boats: Vanguard, Victorious, Vigilant and Vengeance. They were built at Barrow-in-Furness by Vickers Shipbuilding and Engineering between 1986 and 1999. All four boats are based at HM Naval Base Clyde (HMS Neptune), 40 km (25 mi) west of Glasgow, Scotland. Since the decommissioning of the Royal Air Force WE.177 free-fall nuclear bombs in 1998, the four Vanguard submarines are the sole platforms for the United Kingdom's nuclear weapons.

【词汇点拨】Words and Expressions

1. terminology /ˌtɜːrmɪˈnɑːlədʒi/ *n.* the set of technical words or expressions used in a particular subject（某学科的）术语

There is considerable imprecision in the terminology used.

2. vessel /ˈvesl/ *n.* a large ship or boat 大船；轮船

ocean-going vessels 远洋轮船

A plane with Danish markings was over-flying his vessel.

3. initial /ɪˈnɪʃl/ *adj.* happening at the beginning; first 最初的；开始的；第一的

My initial reaction was to decline the offer.

4. anchor /ˈæŋkər/ *n.* a heavy metal object that is attached to a rope or chain and dropped over the side of a ship or boat to keep it in one place 锚

at anchor 抛锚；锚泊

The ship lay at anchor two miles off the rocky coast.

5. knot /nɒt/ *n.* a unit for measuring the speed of boats and aircraft; one nautical mile per hour 节（船和飞行器的速度计量单位，每小时 1 海里）

Aircraft carriers can travel at speeds of more than 30 knots.

6. versatile /ˈvɜːrsət(ə)l/

① *adj.* (of a person) able to do many different things 多才多艺的；有多种技能的；多面手的

He's a versatile actor who has played a wide variety of parts.

② *adj.* (of food, a building, etc.) having many different uses 多用途的；多功能的

Eggs are easy to cook and are an extremely versatile food.

7. deploy /dɪˈplɔɪ/ *v.* to move soldiers or weapons into a position where they are ready for military action 部署；调度（军队或武器）

2000 troops were deployed in the area.

8. embargo /ɪmˈbɑːrɡoʊ/ *n.* an official order that bans trade with another country 禁止贸易令；禁运

an arms embargo 武器禁运

The United Nations imposed an arms embargo against the country.

9. surveillance /sɜːrˈveɪləns/ *n.* the act of carefully watching a person suspected of a crime or a place where a crime may be committed（对犯罪嫌疑人或可能发生犯罪的地方的）监视

The police are keeping the suspects under constant surveillance

10. disaster relief refers to the process of responding to a catastrophic situation, providing humanitarian aid to persons and communities who have suffered from some form of disaster 灾难救援

We will be providing both civilian and military disaster relief and humanitarian assistance.

11. backbone /ˈbækboʊn/ *n.* the most important part of a system, an organization, etc. that gives it support and strength 支柱；骨干；基础

Agriculture forms the backbone of the rural economy.

12. agile /ˈædʒ(ə)l/ *adj.* able to move quickly and easily（动作）敏捷的；灵活的

At 20 years old he was not as strong, as fast, as agile as he is now.

13. extensive /ɪkˈstensɪv/ *adj.* covering a large area; great in amount 广阔的；广大的；大量的

Extensive repair work is being carried out.

14. vanguard /ˈvænɡɑːrd/

① *n.* the leaders of a movement in society, for example in politics, art, industry, etc.（政治、艺术、工业等社会活动的）领导者；先锋；先驱者

The company is proud to be in the vanguard of scientific progress.

② *n. the part of an army, etc. that is at the front when moving forward to attack the enemy* 先头部队；前卫；尖兵

15. trident /ˈtraɪdnt/ *n.* a weapon used in the past that looks like a long fork with three points 三叉戟（旧时武器）

Today, American Ohio lass and British Vanguard class submarines are equipped with a sixth-generation Trident weapons system.

16. prime minister the main minister and leader of the government in some countries 首相；总理

The new prime minister was sworn into office.

17. tomahawk /ˈtɑːməhɔːk/ *n.* a light axe used by Native Americans 印第安战斧（美洲土著的一种工具）

The Tomahawk Land Attack Missile (TLAM) is a long-range, all-weather, jet-powered, subsonic cruise missile that has been primarily used by the United States Navy and Royal Navy in ship and submarine-based land-attack operations.

18. seaman /ˈsiːmən/ *n.* a member of the navy or a sailor on a ship below the rank of an officer 水兵；水手；海员

His father, Fince, was a merchant seaman, and his mother, Rose, was a homemaker.

【词汇点拨】Proper Names

1. **Nimitz** /ˈnɪmɪts/ 尼米兹

2. **Secretary of State for Defence** 防务大臣

3. **Kosovo** /ˈkɒsəvəʊ/ 科索沃

4. **admiral** /ˈædmərəl/ 海军上将

5. **Gibraltar** /dʒɪˈbrɔːltər/ 直布罗陀

6. **Trafalgar** /trəˈfælgər/ 特拉法尔加（西班牙）

7. **Afghanistan** /æfˈɡænɪstæn/ 阿富汗

8. **Bosnia** /ˈbɑːzniə/ 波斯尼亚

9. **All Terrain Vehicles(ATVs)** 全地形车

【参考译文】

海军术语和舰艇

英国海军有一套专门术语表述舰船的运动。

海军术语里，way 一词常用来指航行；underway 即指正在航行，通常指的是先前抛锚或停靠码头一段时间的船又开始航行。含 way 的常用术语还有：

- gangway：让路 / 避让；让开使别人通过
- to make way for someone：让行 / 让道；为别人腾出空间
- leeway：余量 / 余地；偏差

在海军术语中，速度是按节计算的。一节即一个速度单位，相当于每小时航行 1 海里（1.852 千米）。速度为 10 节，即相当于每小时航行 18.52 千米。

舰船运动的方向称为航向。表示舰船运动的常用短语有：

- to chart a course：为既定方向规划航线
- to set a course：启航；开始向既定方向航行
- to maintain course：保持航向
- to change course：改变航向
- to be on/off course：正朝着正确的或错误的航向航行

海军舰艇

海军舰艇可分为水面舰艇和潜艇。

水面舰艇

水面舰艇外形和大小各不相同，大的有航母，小的有河汊巡逻艇。常见的舰船有：

- 航母

航母很大，有"海上城市"之称，最大的航母长 300 多米，飞行甲板宽 70 多米，比如美军的"尼米兹"级航母，能搭载喷气式战斗机、直升机 85 架，容纳 5000 多人（包括船员和空勤人员）。尽管体形巨大，航母的航速却能达到 30 多节。航母不仅能运送飞机和人员，还能用作任务旗舰，作为海军作战或联合作战的指挥部。

- 护卫舰

护卫舰是多功能、多用途舰船，它可以执行反舰、防空或反潜作战，还能部署人员，执行禁运及侦查任务，协助救灾减灾。英国皇家海军拥有 16 艘"23 型"护卫舰，是英国海军水面舰船的中坚力量。"23 型"护卫舰长 123 米，宽 16 米，能搭载 185 名船员。其最大航速 28 节，舰上装备有舰炮、导弹发射器、反导系统、鱼雷和数架攻击直升机。

- 直升机攻击舰

攻击舰，如皇家海军的"海洋"号攻击舰，主要用于运送两栖作战部队（如皇家海军陆战队），搭载运输直升机和攻击直升机及两栖登陆艇。"海洋"号长 203 米，宽 35 米，可按 18 节航速航行。船员和空勤人员共计约 500 名，另外还能运载近 800 名海军陆战队员。

- **驱逐舰**

驱逐舰是一种快速灵活的战舰，航速超过 30 节，常用于支援其他舰艇编组，比如航母战斗群。驱逐舰最初设计是为了保护战舰和航母免受鱼雷艇和潜艇攻击，后逐渐演变成为一种多用途舰船，能反潜，能防空，能执行舰对岸作战。美国海军的"斯普鲁恩斯"级驱逐舰长 171 米，宽 16 米，能搭载约 380 名船员。

潜艇

英国皇家海军把潜艇分为两类，一类是弹道导弹潜艇，另一类是舰队潜艇。两类潜艇均有核动力发动机，能够在不加燃料的情况下航行数年，最大水下速度约为 25 节。出航执行任务时可连续航行数周，为了不泄露潜艇的位置，艇员不允许与家人联系。不过，潜艇提供充足的休闲活动，艇上有电影室、游戏区、藏书丰富的图书馆。两类潜艇的一个关键区别是武器装备。

- **弹道导弹潜艇**

弹道导弹潜艇，如皇家海军的"先锋"级潜艇，装备有携带核弹头的导弹。以"先锋"级潜艇为例，它装备了"三叉戟"式 D5 型导弹，精度高，射程超过 4000 海里（7200 多千米）。每一枚导弹可以携带 12 个弹头，能够攻击不同目标，每一艘潜艇有 16 个导弹发射管。这类潜艇有一套复杂的指挥体系，最上层是国防部长和首相，只有他们下达命令授权才能使用核武器。"先锋"级潜艇长 150 米，宽 12 米，搭载艇员 125 名。

- **舰队潜艇**

舰队潜艇，如皇家海军的"迅捷"级潜艇，携带"战斧"式等常规导弹，精度高，在常规战斗中用于攻击陆基目标。在 1999 年科索沃冲突期间，"辉煌"号潜艇发射过这种导弹。舰队潜艇还可以使用鱼雷攻击敌方的潜艇和水面舰艇。"迅捷"级潜艇长 83 米，宽 10 米，搭载艇员 116 名，能执行侦查、作战等各种任务。

皇家海军陆战队

皇家海军陆战队的历史可以追溯到 1664 年，当年招募了 1200 名士兵进入"海军上将"团。一开始，海军陆战队员的职责是既当战士，又当海员，既要参加陆岸战斗，还要操作舰船上的武器。1704 年，他们参加了直布罗陀战役，1805 年参加了特拉法加海战，1915 年参加加利波利战役，数战成名。现在，英国皇家海军陆战队满世界执行任务，一会儿在阿富汗，一会儿在波斯尼亚。

现代皇家海军陆战队是一支两栖部队，可以迅速部署到世界任何地方，中东的沙漠，北欧的群山，都有他们的足迹。除了乐手，皇家海军陆战队都是突击队员。他们在崎岖恶劣的环境里接受了长期艰苦的训练。训练结束后，海军陆战员一般都编入第 3 突击旅，该旅由驻扎在英国本土的 3 个不同的突击部队构成。

凭借其两栖作战能力，皇家海军陆战队可以在一个潜在的冲突地区附近隐蔽待命，一般是待在地平线附近的一艘舰船上，随时等待被召唤。他们可以乘直升机或登陆船快速部署到岸上。登陆船既有能搭载 8 名海军陆战队员、速度达 30 节的小型冲锋舟，也有长 27 米、能运载 120 名陆战士兵和 4 辆全地形车（ATV）、航速仅 9 节的大型运输舰。

词汇表

Alpha Boats for beginners
bow 舰首
bridge 舰桥
deck 甲板
elevator 升降机
flight deck 飞行甲板
hangar bay 机库
hull 船体
island 舰岛
port 左舷
starboard 右舷
stern 舰尾
waterline 吃水线
vessel 大船；轮船
hatch 舱口
bulkhead 舱壁
passageway 通道
galley 船上的厨房
overhead （舱）顶板
compartment 舱
embark 上船
disembark 下船上岸
warship 战舰
launch 发射

recover 返回（作战飞机返回机场或航母）
secure 保护，保卫；使安全
manoeuvre 机动；军事演习
catapult 弹射器
reactor （核）反应堆
accelerate 使加速
naught 零
cable 缆绳
arresting wire 阻拦绳

Bravo City at sea
aviator 飞行员
flotilla 舰队
knot 节
nautical mile 海里
sortie 架次
operation 行动
anti-aircraft 防空
anti-submarine 反潜
anti-surface 反舰
invincible 不可战胜的
specification 规格
beam 船宽
displacement 排水量

tonne 公吨
cruise 以平稳的速度行驶
mount 挂载；底座
amphibious 两栖的
assault 攻击

Charlie　Life onboard
berthing 铺位
petty officer 海军军士
rack （海军人员）床
shore station 海军陆上基地
underway 正在航行
watch 值更
sea duty 出海执勤

Delta　Navy commands
gangway 让路
ten-hut 立正
as you were 像往常一样
Aye, aye sir. 是，长官！
belay that last order. 停止执行上一命令。

Echo　Naval terminology and vessels
terminology 术语
vessel 船；舰
movement 行进（海军特定术语）
way 航行
initial 最初的
at anchor 抛锚；锚泊
dock 码头
leeway 偏航
knot 节
nautical mile 海里

course 航向
to chart a course 为既定方向规划航线
to set a course 启航；开始向既定方向航行
to maintain course 保持航向
to change course 改变航向
to be on/off course 正朝着正确的或错误的航向航行
river patrol boat 河汉巡逻艇
Nimitz 尼米兹
flight deck 飞行甲板
versatile 多才多艺的，多功能的
deploy 部署
embargo 禁运
surveillance 侦查
disaster relief 赈灾
backbone 支柱；骨干
amphibious combat troops 两栖作战部队
agile 机敏的
surface-to-land 舰对岸
Spruance "斯普鲁恩斯"级驱逐舰
ballistic submarine 弹道导弹潜艇
fleet submarine 舰队潜艇
extensive 广泛的，大量的
vanguard 先锋；前卫
trident 三叉戟
missile tube 导弹发射管
Secretary of State for Defence 防务大臣
Prime Minister 首相
tomahawk 战斧
Kosovo 科索沃
recruit 招募
Admiral's Regiment 海军上将团
seaman 海员

Gibraltar 直布罗陀
Trafalgar 特拉法尔加（西班牙）
Afghanistan 阿富汗
Bosnia 波斯尼亚
commando 冲锋队；突击队

rugged 崎岖的
commando brigade 突击旅
landing craft 登陆舰
raiding boat 冲锋舟
All Terrain Vehicles 全地形车

 拓展学习

In the Bohai Bay in the late autumn of 2012, my country's first aircraft carrier Liaoning was officially delivered to the Chinese Navy. At this moment, the whole country was jubilant. In the early winter of 2019, on Hainan Island, my country's self-developed and designed aircraft carrier Shandong was officially in service in the South China Sea Fleet. Since then, China has become more confident.

Ushering in the era of dual aircraft carriers is not only a milestone in the construction of naval equipment, but also an important sign that my country's military level has entered an international leading position.

On the steel giant ship, there is a group of "sea and air eagles"; they "dance on the tip of the knife" every day, and every time they land on the ship it is a battle of life and death. They are the carrier-based fighter pilots.

Wang Yong, the winner of the 25th "Chinese Youth May 4th Medal" and Navy special-class pilot, is one of them.

The road to the ship: the pursuit of the ultimate in technology

Every time the shipboard pilots land on the ship, there are dangers. It took only a few seconds from the commander's order to block the ship to the pilot's rigging. In these few seconds, it is hard to imagine how difficult it is to accurately land a fighter jet on the deck of an aircraft carrier that is only 200 meters under complicated sea conditions. "You must pay attention to safety when you get on the plane and take care of yourself..." Even though he has been flying for 10 years, before every takeoff, Wang Yong's mother still has to tell.

"Every time I hook the wire rope accurately and do one thing to the extreme, I particularly like this feeling." When Wang Yong drove the J-15 carrier aircraft nicknamed "Flying Shark" and landed the ship proficiently again and again, who would have thought that this seemingly easy landing action, he had practiced thousands of times before forming muscle memory.

Technically, pilots have to go through periods of ascent, flatness, repetition and collapse. "In fact, the most difficult thing is not how hard training is, but you have to overcome yourself and pass the psychological barrier", Wang Yong said. During the recurring period, he once doubted whether he was suitable for flying.

"But I firmly believe that I can do it, and I can do it well." It is this unwillingness to admit defeat that made Wang Yong and his comrades come up with various methods to train themselves. Some people "tied" themselves to the simulator repeatedly. In practice, someone flew for nearly a hundred hours just by landing on a subject.

There is no shortcut, only repeated practice. In the end, Wang Yong doesn't even need to look at the engine speed. As long as he puts his hand on the accelerator, he knows the speed, and pushes it to know how much speed has been changed.

Teaching Road: Cultivating Freshmen

In 2018, China's aircraft carrier construction entered a period of acceleration, but the large-scale training of carrier-based aircraft pilots faced a series of problems: shortage of teaching resources, weak backbone...Wang, who has won the first place on the ship for two consecutive years, is in urgent need of employment. Yong received a new order to join the carrier-based pilot training team.

"It is a soldier's bounden duty to obey orders." Although he was reluctant to give up on the blue sky, Wang Yong immediately rushed to a new position and devoted himself to the job of training a new generation of carrier-based pilots. From actual combat to behind-the-scenes teaching, the change of roles also brought him more pressure.

Faced with difficulties such as foreign technical blockades, lack of independent experience, shortage of teaching materials, and lack of equipment resources, he did not complain or flinch. With the drill energy of "conquering flying sharks", he began to lead the team to explore and cultivate a new generation of ships. Plan to carry pilots.

Because of his rigorous teaching, the students called him "black-faced instructor".

"Everything harsh on them now is the guarantee of flight safety in the future. They will definitely thank me in the future." Although Wang Yong is often "black-faced" in teaching, he is an enthusiastic "red-faced parent" in life.

A novice student encountered a bottleneck period in flight technology, and his psychological pressure was so great that he almost collapsed. Wang Yong saw that his mental state was not right, so he took the initiative to talk to him and use his flight experience to encourage him to break through.

Later, this student often consulted Wang Yong about flying skills and was diligent and able to make up for his weaknesses. In the final aircraft carrier qualification assessment, he succeeded in counterattack and won the first place.

"A lot of students are much better than I was back then, and they have participated in many special combat missions." Wang Yong said to his students as much as possible. "Seeing them flying well makes me even happier than I am flying." In just over a year, Wang Yong successively participated in the formulation of the J-15 flight training program and flight manual, and realized the rapid introduction of carrier-based pilot training.

The "Flying Shark" team is getting younger and younger, and this record is constantly being broken by newcomers brought out by Wang Yong.

Road to a strong army: growing up with the motherland

In 2000, Wang Yong was admitted to the Changchun Flight Academy of the Air Force. The first plane he touched was an old propeller plane, and the airstrip was grass; now Wang Yong can fly a J-15 fighter jet over the blue sea, guarding every inch of land in the motherland like an eagle.

I remember the first time I drove the J-15 fighter, Wang Yong's hand carefully stroked the skin of the "Flying Shark" and took a photo with the plane excitedly.

Now that he has successfully passed the aircraft carrier qualification assessment, his flight trajectory has also moved from yellow water to deep blue, flying to a wider sky.

He has participated in many major tasks such as the air refueling training of the J-15 fighter, the first live firing of the Liaoning ship, and the first crossing of the Miyako Strait by the Liaoning fleet to break through the first island chain. This "sea and air eagle" is constantly challenging itself. At the same time, they have also witnessed the wonderful moments of the Chinese aircraft carrier's rising combat effectiveness.

"The changes in equipment are all visible to everyone. What makes me more excited is that the concept is changing. Now our training is getting closer and closer to actual combat. We are training for war, and we must be able to fight and win wars!" Wang Yong said in peacetime. In the teaching, the students' combat awareness is also constantly cultivated. Once on the plane, they enter the battlefield. They can't relax at all, so that "the one who can fight, the one who fights will win" is truly engraved in the hearts of every soldier. (1195 words)

(Selected from https://www.tellerreport.com/news/2021-05-10-carrier-fighter-pilots--turning-extreme-technology-into-muscle-memory.HyWcW2fPuO.html)

【Notes】

1. **jubilant** 欢呼的
2. **self-developed** 自主研发的
3. **usher** 引导；引领
4. **milestone** 里程碑
5. **shortcut** 捷径
6. **rigorous** 严格的
7. **novice** 新手
8. **Chinese Youth May 4th Medal** 中国青年五四奖章
9. **J-15 fighter jet** 歼-15舰载战斗机

 单元练习

I. Complete the text with the following words

| bow | flotilla | knots | sorties | stern | aviators |

The Nimitz class carriers are nuclear-powered. They have the size of about three football pitches from __1__ to __2__. They can launch or land a plane every 25 seconds and generate between 140 to 160 __3__ per day. The __4__ are highly-trained pilots. These super carriers are fast and can travel at a speed of over 30 __5__ which allows forces to be deployed rapidly to almost anywhere in the world in less than two weeks. However, the carrier is vulnerable to enemy attack and so is always escorted by a __6__ of ships for its protection.

II. Multiple choice

1. A _____ is what sailors call a room on a ship.
   A. hull   B. passageway   C. compartment   D. hatch

2. When you leave a ship or submarine, you _____.
   A. embark   B. go topside   C. disembark   D. go below

3. Which word do sailors use for kitchen?
   A. Galley   B. Bulkhead   C. Overhead   D. Port

4. The _____ refers to the opening that goes from one deck to another.
   A. hatch   B. passageway   C. stern   D. starboard

355

5. The _____, located two decks below the flight deck, is where the aircraft are secured when not in use.

   A. port   B. island   C. hangar bay   D. elevator

6. The _____ is the body of a ship.

   A. waterline   B. hull   C. displacement   D. course

7. The _____ is a strong metal cable that is used for stopping aircraft.

   A. catapult   B. flight deck   C. arresting wire   D. tailhook

8. The _____ is a steam-powered/electromagnetic-powered machine that is used to launch aircraft.

   A. catapult   B. flight deck   C. arresting wire   D. tailhook

9. Which of the following terms is used when a ship is at sea?

   A. Gangway   B. Underway   C. Leeway   D. Knot

10. A senior officer tells a junior officer to ignore the last order he received and to change direction. He says, "_____"

   A. Aye, aye sir.   B. As you were.   C. Very well.   D. Belay that last order.

III. Translation: English to Chinese

1. The flight deck only has about 150 metres of runway space for landing which is not enough for high-speed jets, so aircraft land in a process known as an "arrested landing".

2. Carriers are open to attack from enemy ships, missiles and aircraft. For this reason, carriers are always escorted by a flotilla of other ships, called the carrier battle group (CBG).

3. The ship is manned by a crew of 685 sailors and officers, and there are also 386 Air Group personnel, who are responsible for all the aircraft on board.

4. The aircraft are kept in hangars down below deck and lifted to the surface when they are ready to be deployed, which may be at any time, since sorties are flown at any time of the day or night.

5. So, to sum up: the ship can act as an air defence platform or be used for anti-submarine, anti-surface and/or anti-aircraft warfare, or else as an amphibious assault ship and must be ready to conduct these operations anywhere around the world almost immediately.

IV. Translation: Chinese to English

航母是一种以舰载机为主要作战武器的大型水面舰艇。航母通常拥有巨大的甲板和位于右侧的舰岛，是一支航母战斗群的核心舰船。舰队中的其他船只为其提供保护和供给，而航母则提供空中掩护和远程打击能力。发展至今，航母已是现代海军不可或缺的武器，也是海战最重要的舰艇之一。依靠航母，一个国家可以在远离其国土的地方施加军事压力和开展行动，而不必依赖当地的机场条件。航母已是现代海军不可或缺的利器，也成为了一个国家综合国力的象征。

V. Oral practice

1. What are the basic components of a carrier? Can you describe them?

2. Please introduce the launch and recovery process of carrier-based aircraft.

附录：Unit 12 练习答案

I. Label the picture

1. bow 2. stern 3. sorties

4. aviators 5. knots 6. flotilla

II. Multiple choice

1-5 CCAAC 6-10 BCABD

III. Translation: English to Chinese

1. 飞行甲板只有约150米长的着陆跑道，对于高速喷气式飞机来说是不够的，所以舰载机采用"拦阻着陆"的方式着舰。

2. 航母易受敌舰、导弹和飞机的攻击，所以总是有一支由其他舰艇组成的舰队护航，这支舰队就称为航母战斗群（CBG）。

3. 该舰上有685名水手和军官，另有386名空勤人员负责舰上所有飞机。

4. 飞机存放于飞行甲板下的机库里，准备使用时提升到甲板上，任务随时会有，不论白天或晚上，飞机随时出动。

5. 综上所述，该舰既可作防空平台，也可以进行反潜、反舰和防空作战，还可以用作两栖攻击舰，必须随时准备在世界任何地方执行上述任务。

IV. Translation: Chinese to English

The aircraft carrier is a large surface ship which takes the carrier-based aircraft as the main combat weapon. Usually with a large deck and an island to the right, the carrier is the core ship of an aircraft carrier battle group. Other ships in the fleet provide protection and supply, while the carrier provides air cover and long-range strike capability. Up to now, the aircraft carrier has become an indispensable weapon of the modern navy, and is also one of the most important ships in naval warfare. With the help of aircraft carriers, a country can exert military pressure and conduct operations far from its own territory without relying on local airfield conditions. The aircraft carrier has become an indispensable weapon of modern navy and a symbol of a nation's comprehensive strength.

V. Oral practice

1. The flight deck is a flat surface on top of the carrier which is used to launch and recover planes. The bridge is a building on top of the flight deck where officers can direct flight and ship operations. The hanger bay is a zone located two decks below the flight deck where aircraft are stored when not in use. The elevator is a device surrounding the hanger bay which moves the aircraft to the flight deck.

2.

> **Taking off from a carrier**

Aircraft typically require long runways in order to gather enough speed so they can successfully take off. Since the runway length on an aircraft carrier is only about 300 feet, compared to the 2,300 feet needed for normal aircraft to take off from a runway, engineers have created steam-powered catapults on the decks of carriers that are capable of launching aircrafts from 0 to 150 knots (170 miles per hour) in just 2 seconds.

The crew hooks the aircraft's front wheel, or nose gear, to the catapult using a tow bar. The tow bar hangs off the front of the nose gear so the catapult can pull the aircraft. In order to prevent harmful jet discharge from going into unwanted places, a jet-blast deflector is placed directly behind the aircraft, pushing the discharge up into the air. The pilot then pushes the engine to full throttle, creating a forward thrust that would traditionally move a jet forward. A holdback bar is in place to prevent any motion at this time, despite the thrust of the jet. Once the force from the catapult is added to the thrust of the jet, the excess force will cause the hold-back bar to release and the jet will move. This is because the hold-back bar can only hold the force from the jet at full thrust, but not the additional force of the catapult.

> **Landing on a carrier**

Landing on an aircraft carrier is often described as the toughest task for a navy pilot. The pilot has to line up with the runway correctly, come in at the correct angle, and stop the plane in a short distance for a successful landing. For many this would be an unnerving task, but luckily engineers have devised two systems to help accomplish these tasks–the Fresnel lens and the arresting wires.

The optical landing system consists of a horizontal bar of green lights and a vertical bar of red lights on both sides of the "meatball". The "meatball" is the centerpiece that consists of five amber colored lenses. Certain lenses will light up one at a time depending on the angle the plane is in relation to the "meatball". This causes the center light to appear to be moving up and down in relation to the horizontal green bars on the sides. In order to safely land, the pilot tries to keep the center amber lens horizontal with the green bar throughout process. If the pilot gets too low, the amber light will turn red indicating that the aircraft is dangerously low and risks hitting the back end of the aircraft carrier. The red lights around the green horizontal bars will be flashing if the carrier is not able to receive the aircraft, and so the jet must keep circling or find another place to land.

The most dangerous part for the pilots is the touchdown and subsequent deceleration caused by the arresting wires. Not only does it take incredible skill to pull off this landing maneuver, but success also depends greatly on the ground crew avoiding any errors throughout the operation. Before touchdown, the pilot lowers the tail hook. The tail is a long metallic bar that hangs just inches above the surface of the carrier. When the aircraft lands, the hooked end of the tail snags one of the four arresting cables, stopping the aircraft. Although the cables are simple in structure, there is a great risk of something going wrong. Good pilots hit the second or third cables rather than the first or fourth, because these wires will keep the pilot from running into the back of the carrier while still allowing room for takeoff if they miss their target. Once the wheels hit the deck, the pilot immediately pushes the aircraft to full throttle. This is to ensure that if the tail hook misses the arresting wires, the aircraft can still have enough speed to quickly take off again at the end of the runway.

参考文献

［1］柯林斯英汉双解大词典［M］.北京：外语教学与研究出版社，2017.

［2］李杰.航母世界［M］.北京：中国和平出版社，2019.

［3］深度军事编委会.现代战机鉴赏指南［M］.北京：清华大学出版社，2020.

［4］深度军事编委会.航空母舰［M］.北京：清华大学出版社，2018.

［5］深度军事编委会.无人装备［M］.北京：清华大学出版社，2017.

［6］深度军事编委会.主战舰艇［M］.北京：清华大学出版社，2018.

［7］王传经，等.军事英语听说教程（新版）［M］.北京：外语教学与研究出版社，2023.

［8］王岚.军事英语：第三册［M］.洛阳：解放军外语音像出版社，2004.

［9］吴承义.美国军事概况［M］.北京：国防大学出版社，2010.

［10］张锦涛，杨丽娟.基础英语教程［M］.南京：南京大学出版社，2014.

［11］https：//www.thefreedictionary.com/